Library of Congress Cataloging-in-Publication Data

Demir, Ali, 1959—
Synthetic filament yarn : texturing technology / Ali Demir, Hassan Mohamed Behery.
p. cm.
Includes bibliographical references and index.
ISBN 0-13-440025-9 (hardcover)
1. Textile fibers, Synthetic. 2. Textured yarn. I. Behery, H. M. (Hassan M.) II. Title
TS1548.5.D45 1997
677 ' .4—dc20 95–23400
CIP

Acquisitions Editor: *Elizabeth Sugg*
Director of Manufacturing & Production: *Bruce Johnson*
Managing Editor: *Mary Carnis*
Production Editor: *Adele Kupchik*
Manufacturing Buyer: *Ed O'Dougherty*
Editorial Assistant: *Kahdijah Bell*
Printer/Binder: *Courier-Westford*
Cover Design: *Wendy Alling Judy*

To my beloved Mother, Father, and dear Wife
A.D.

To my Wife, Daughter, and Son
H.M.B.

Printed in the United States of America

10 9 8 7 6 5 4 3 2 1

ISBN 0-13-440025-9

Prentice-Hall International (UK) Limited, *London*
Prentice-Hall of Australia Pty. Limited, *Sydney*
Prentice-Hall Canada Inc., *Toronto*
Prentice-Hall Hispanoamericana, S.A., *Mexico*
Prentice-Hall of India Private Limited, *New Delhi*
Prentice-Hall of Japan, Inc., *Tokyo*
Simon & Schuster Asia Pte. Ltd., *Singapore*
Editora Prentice-Hall do Brasil, Ltda., *Rio de Janeiro*

Synthetic Filament Yarn Texturing Technology

Ali Demir
Istanbul Technical University
Department of Textile Engineering

Hassan Mohamed Behery
Clemson University
School of Textiles, Fibers, and Polymer Science

Prentice Hall
Upper Saddle River, NJ 07458

PREFACE

Texturing technology of synthetic filament yarns is significant for the textile industry and particularly to its experts who are already familiar with previous technical publications by Monsanto, such as Stretch Woven fabrics (1962), Textured Yarns (1963), Textured Yarn Technology (1968), and Draw Textured Yarn Technology (1974). These books, during those years, had become known as the Bibles of the world of textured yarns and fabrics. Nonetheless, texturing technology has advanced rapidly since these books were published. It has been realised in the industry that there are no similar books published or in preparation, at least since 1988. According to one of the reviewers of the present book, he stated that: "The book is needed. Texturing of synthetic yarns is a significant textile industry. Textbooks and general reference books on this topic will be helpful. Only a few recent books have been published. Most workers in this area have to use journals to keep up-to-date on advances in technology." Another reviewer had stated that: "The technology is quite complex and a comprehensive book of this type should be well received by industry. The book should also find acceptance in the many colleges and universities that have programs in fiber and textile technology." Also, the main readership will be members of the texturing and fiber-making industries as well as members of the Textured Yarn Association of America.

In this book, the outline was intended to cover the field adequately and thoroughly with as much details as possible.

In the authors' view, this book could be adopted as a textbook as well as a reference book. It is based on our experience in both teaching and research. The book was inspired by the authors' awareness that creativity is the lifeblood of texturing industry; that textured yarns are continuing to penetrate new markets through newly developed yarn and fabric types; and that the expertise of those industrialists, under whose able direction texturing has boomed, needs to be continually quickened by the infusion of new ideas and technology. The impetus that texturing has provided to the textile industry's economic development has, of course, been the mainspring of its present existence and further progress in the future.

The core of the book consists of an expanded portion on Air-jet Texturing with detail results obtained from research and study of one of the author's continuing research work. This subject can only be referred to in articles and publications in different journals and technical magazines.

Chapter 1 provides introductory and basic fundamentals into the understanding of synthetic yarns, especially for readers whose experience and work responsibilities are mainly in the texturing industry, and they are the prime users of these yarns. The other parts of the book cover indepth and in details all the different technologies and methods of textured yarn production.

The authors hope that this book will provide the knowledge and understanding of the subject to the readers -industrial or academic researchers, engineers, or managers -and giving an up-to-date account of the present state of understanding of the phenomena of yarn texturing as well as provide a source of reference for further and more detailed studies.

Finally, the authors would like to express their warmest thanks to the individuals who provided great help and encouragement during the writing of this book.

Ali Demir

Hassan M. Behery

ACKNOWLEDGMENTS

This work had been finalised after five years of work. During this time, I have received tremendous help, understanding, support and encouragement from my wife. Therefore, I do not hesitate to dedicate this work to her. Together with her, my daughters, Azize and Havva have also endured for many years. I thank them dearly. My special thanks to Chris for everything that she taught me.

I also greatly appreciate the help of the following, without whom this work may not have been completed as it is today; Dr. Gordon R. Wray, Dr. Memib Acar, Peter LeGood, and Janet Goodman.

A.D.

As co-author of this work, I also would like to express my appreciation to Dr. Ali Demir to include me in this endeavour and made me part of this valuable textbook. Through the multiple editing procedure, it was sure enlightening and I found it very interesting and of technological enhancement. It was of special interest and challenge to put the book in this final form. For this, I would like to express my deepest thanks and appreciation to my wife, Mervet, son, Mohamed, and daughter, Hala, whose continuous support and encouragement were the main reason for accepting the invitation as a co-author.

I should not overlook thanking Fulbright for providing me with the grant to work at Istanbul Technical University and gave me the opportunity to work with Dr. Ali Demir in bringing this valuable work and scientific contribution to light.

H.M.B.

ABOUT THE AUTHORS

Ali Demir

Ali Demir was born in Konya, Turkey; graduated as a mechanical engineer from the Mechanical Faculty of Ystanbul Technical University; and obtained a Ph.D. degree from Loughborough University of Technology for his work on air-jet texturing process. Dr. Demir worked in Switzerland for Heberlein Maschinenfabrik AG as development manager.

Dr. Demir is currently professor of textile engineering at the Department of Textile Engineering of Ystanbul Technical University, Turkiye.

Hassan Mohamed Behery

Professor Hassan Mohamed Behery is a graduate of Alexandria University, Egypt, with a B.Sc. (1950) and M.Sc. (1955) in Mechanical Engineering and a Ph.D. in textiles from Manchester University (1959). Professor Behery was the Head of the Textile Engineering Department at Alexandria University. He was a Visiting Professor at the American University in Cairo, teaching Polymer Science and Technology.

Currently, Professor Behery is (and has been) on the faculty of Clemson University for 23 years, conducting research and teaching in fiber science and fiber processing in the field of fiber/machine interaction.

Professor Behery's contributions to the texturing technology goes back to the 1971 when the first conference was organized at Clemson University with Dr. Behery as the Chairman. Dr. Behery is also a "Charter Member" of the Textured Yarn Association of America -- TYAA.

CHAPTER 1

SYNTHETIC FILAMENT YARNS AND THEIR PRODUCTION

1.1 INTRODUCTION

Synthetic fibers are now accepted as everyday textile material. The use of many synthetic fiber in modern life is as casual as the use of the wool, cotton, flax, and silk that have provided human's fabrics for centuries.

Until the 1930s, synthetic fibers existed as a few experimental filaments that showed little sign of any use in the textile industry. However, the growth of synthetics has been so spectacular during the last six decades that the world production of synthetic fibers is nearly equal to the world's cotton production (13,765,000 tons of synthetic fibers (filament and staple) against 15,196,000 tons of raw cotton in 1987. The man-made fiber production by the end of 1992 has been 20,240,000 tons, 85% of which being synthetic fibers.[1] Table 1.1 and Figure 1.1 illustrates this spectacular growth. Yet the analysts[2] comments that the meteoric rise of synthetic fibers seen during first two postwar decades may come to an end. However, there is little doubt that the synthetics will hold their share of about 41% of world fiber production in the foreseeable future. The forecasts[3] also states that by the year 2000, there will be, in average, a further 10 million tons of man-made fiber production.

The variation of regional production distribution has occured as follows; in the Western Europe, the production of synthetics has shown a 3% increase in parallel with the world trend. This, in turn, corresponds to 18% of the total world production in 1992. The U.S. production has maintained its share of 19% of the total production. This indicates that both the continental Europe and the United States has equal shares in total world production of synthetic fibers.

Table 1.1: World Man-Made Fiber Production						
Years	Cellulosic Fibers		Synthetic Fibers		Total	
	1,000 Tons	%	1,000 Tons	%	1,000 Tons	%
1970	3,585	43	4,809	57	8,394	100
1980	3,522	25	10,779	75	14,301	100
1985	3,234	20	13,025	80	16,259	100
1990	3,216	17	15,885	83	19,101	100
1991	3,057	16	16,499	84	19,556	100
1992	3,022	15	17,217	85	20,240	100

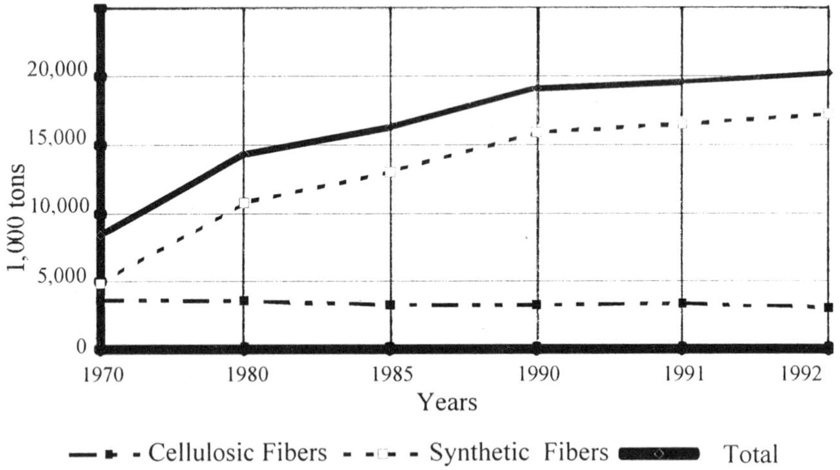

Figure 1.1 World fiber production trends

Japan has recorded a 1% increase and thus obtained a 9% share of the total World production.

Increases that may be described as above average have once again in 1992 been observed from the rest of the world (Figure 1.2). Fiber-producing countries in this category have increased their production by 4% and, hence, obtained a share of 54% of the total World fiber production. In Southeast Asia, there have been many capacity increases, especially in polyester. This naturally pressurizes the man-made fibers market. Taiwan, South Korea, and China together produce more than 28% of total fiber production.

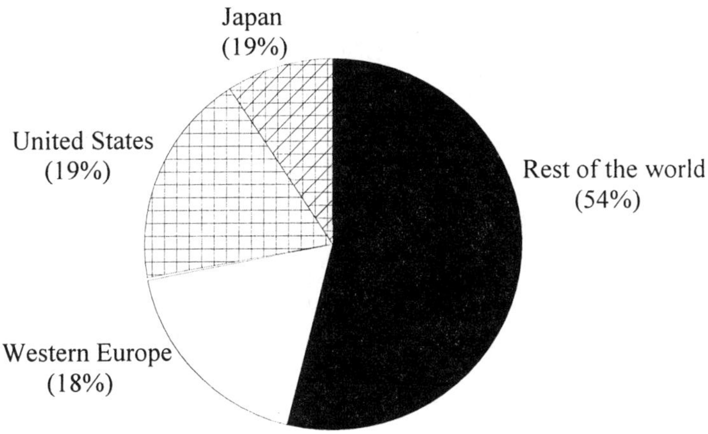

Figure 1.2 Regional distribution of man-made fibers in 1992

According to the mostly accepted fiber classification chart of the Textile Institute[4] (Figure 1.3), synthetic fibers are a subgroup of man-made fibers, which accounted more than one-half of the world production of fibers in 1992. In the developed countries, this number approaches to two-thirds (Table 1.2). The other subgroups are the natural polymers, which are of natural origin, such as cotton lints or wood pulp. The synthetic fibers, conversly, are made from substances by way of synthesis (building up) as the name suggests.[5] Chemical constitution of synthetic fibers also differs for making various fibers. Polyamides, polyesters, and polyolefins (especially polypropylene) are the synthetic fibers that are of commercial importance and mostly used in the texturing industry. Over the years, the share of each synthetic fiber of the total synthetic fiber production has been varying. In 1969, polyamides had a share of 50.5% compared with 28.9% share of polyester; in 1986, however, the share situation almost reversed as polyamides had 27% (3.5 million tons) compared with polyester's 53.9% share (6.99 million tons) of 13.76 million tons of synthetic fibers. The widespread use of polyester is due to its ease-of-care properties and low shrinkage, which result in good laundering characteristics. Polypropylene fibers, howeve, commercially came into the synthetic fiber scene as late as 1961, acquired a share of 5.8% (0.8 million tons in 1986) and seems to have a considerable potential for growth in the future (it has shown a 230% increase during the 1965 to 1984 period in world production).[6-8]

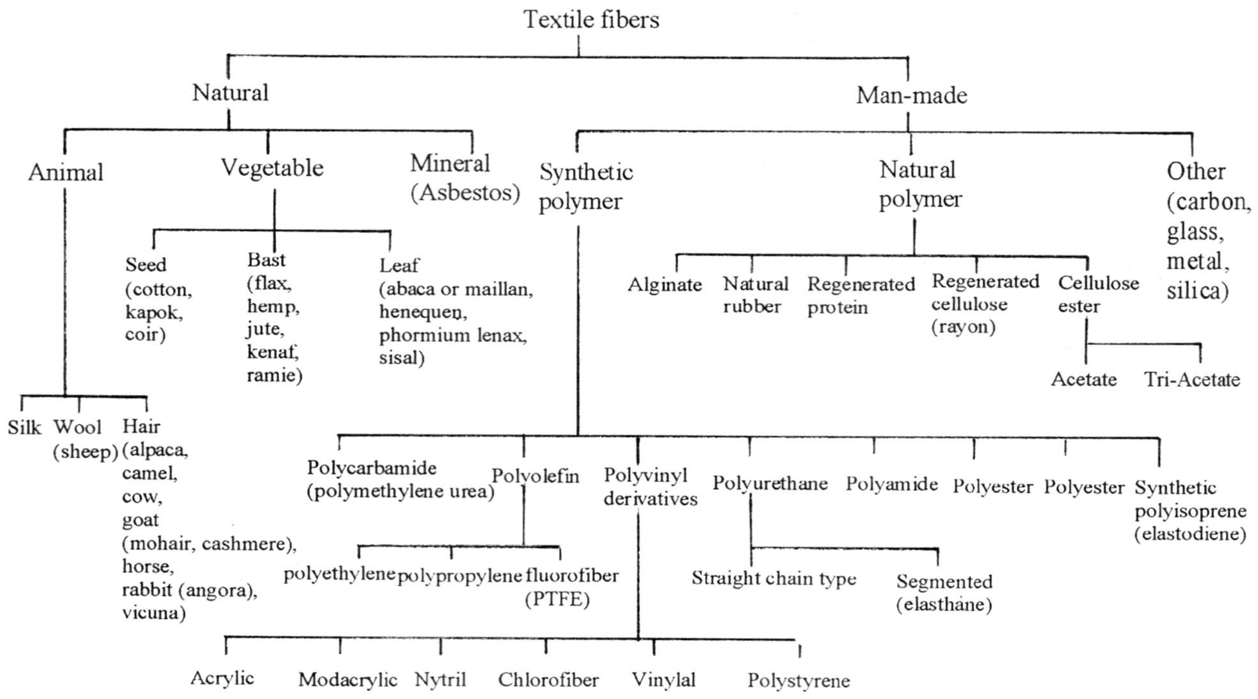

Figure 1.3 Classification of textile fibers

Table 1.2 World Production of Fibers, 1984 (1,000 Tons)			
Man-made Fibers		**Natural Fibers**	
Viscose	2,500	Cotton	18,500
Acetate	550	Flax	500
Nylon	3,100	Hemp	1,100
Polyester	6,150	Jute	1,650
Polypropylene	1,500	Sisal	300
Acrylic	2,300	Wool	1,650
Others	250	Silk	50
Total	**16,350**	**Total**	**23,750**

Although the individual fiber production is not specified in 1988, the world production of man-made fibers is reported to increase to 8.8 million tons.[9] Of this total, production of synthetic fibers accounted for 13.5-m tons.[9] All synthetic fibers are manufactured in continuous filament form by spinning of molten polymers through fine holes of spinnerets. However, almost one-half (57% in 1987[10]) of the world production of synthetic fibers is used as staple fiber and converted into textile yarns using conventional staple fiber-processing methods. The remaining 43% of synthetic fibers are used as flat or textured continuous filament yarn with nearly equal shares.[11] Synthetic continuous multifilament yarns are characterised by nearly complete filament alignment and parallelisation with respect to the yarn axis. Either a very low twist or a slight intermingling (see Section 4.3) is usually introduced into a multifilament yarn just adequate to produce interfilament cohesion. Such yarns are compact and smooth in appearance. Variety of processes have been developed to introduce texture in multifilament yarns in the form of bulk or stretch. These processes are designed to disrupt the high degree of filament alignment and parallelization and to introduce some form of structural feature associated with spun staple yarns.[12]

The uses of synthetic filament yarns are highly varied and cover the whole field of clothing, furnishing fabrics and technical fabrics (Figure 1.4). However, each synthetic fiber has its own pattern of end use associated strongly with technical merit. The rather different type utilizations of polyester and polyamide fibers are shown in Table 1.3.[13] As the table indicates, polyester and polyamide are the dominant synthetic material. The distribution of these materials according to the end uses is also another striking event observed from this table. The governing factor for this distribution is naturally the properties and characteristics of these materials. It is observed that, polyester is readily suitable for textile end uses, whilst polyamide is more suitable for technical and industial applications due its high tenacity and resilience.

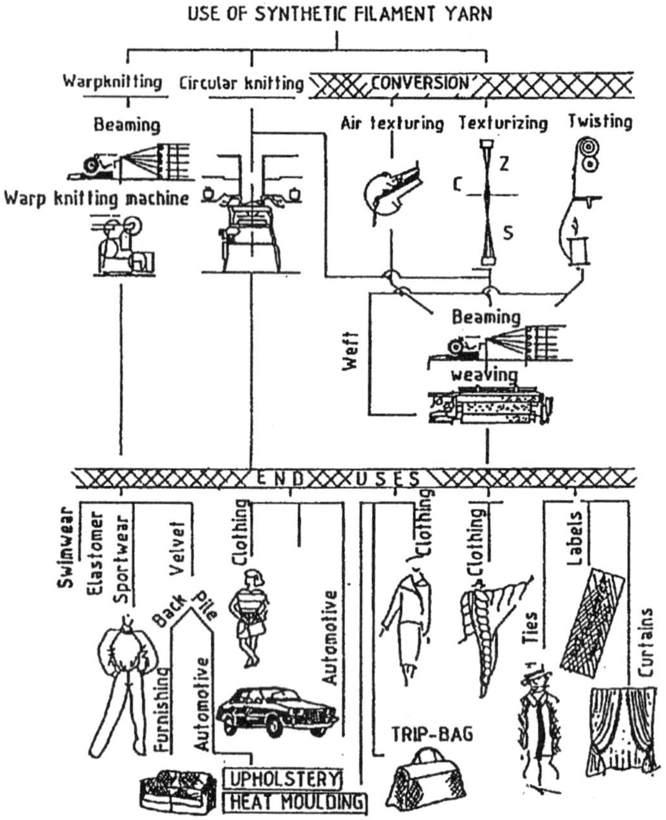

Figure 1.4 Uses of synthetic fibers

Table 1.3 Capacities for Polyester and Polyamide Fibers in 1984[13]							
Proportion % of World Capacities for Man-made Fibers, 1984							
	Textile Continuous Filament Yarn	Industrial Continuous Filament Yarn	Bulked Continuous Filament Yarn	Cotton-Type Staple	Wool-Type Staple	Carpet - Type Staple	Fiber Fill
Polyester	51	8	0	56	23	5	99
Polyamide	39	35	94	1	4	49	0

The principal end uses of synthetic fibers in the European Ecomomic Community (EEC) are given in Table 1.4.[14]

Table 1.4 Principal End Uses of Synthetic Fibers in EEC[14]

Fiber Type &and Major End Uses(*)	Technical/Aesthetic Basis of End Use
Polyester	
Trousers, skirts (s) dresses (y)	Improved wear, handle, ease of care
Shirts, bedlinen (s), net curtains (y)	Light resistance, ease of care
Nylon	
Carpet pile (y,s)	Durability, bulk, printability
Women's hosiery (y)	Elasticity, strength
Tyre yarn (y)	Tenacity, durability, adhesion
Weatherproof outerwear (y,s)	Durability, elasticity, luster
Polypropylene	
Carpet backing and pile (y,s)	Strength, durability
Ropes, cords, sack clothes (y)	Strength, low weight
Geotextiles (s)	Resistance
Upholstery (s)	Strength, light resistance
(*) y = continuous filament yarn, s = staple fiber	

1.2 MATERIALS

Synthetic fibers (and all fibers in general) consist of polymers _ compounds consisting of large molecules that contain recurring structural units repeated many times within each molecule. A single molecule might contain hundreds of repeating units, making its molecular weight up to 10,000. Such large molecules are called macromolecules, or chain molecules.[15]

Simply having a polymer does not mean having a fiber. Indeed, there are many molded plastics that are made of polymers but clearly are not fibers. The important characteristic that distinguishes polymers in fiber form is their orientation in a particular direction, which makes fiber anisotropic. Therefore, a fiber is much stronger lengthwise than transversely.[15] The physical properties of fibers are strongly influenced by the organization of polymeric chains, i.e., macromolecules, into crystalline and noncrystalline domains, and the disposition of these domains with respect to each other. Some of the structural characteristics of polymers that allow them to crystallize under appropriate conditions are the following:

Regularity. The polymer chains must be uniform in chemical composition and stereochemical form.

Shape and interaction. The shape of the polymer chains must allow close contact or fit to permit effective and strong intermolecular interaction.

This is generally achieved by linear macromolecules with no bulky side groups or with side groups that are regularly spaced along the backbone chain.
Repeat length. The ease of crystallization decreases with increasing length of the polymer repeating unit.
Chain stiffness. An optimal stiffness of the polymer is necessary.

Both chemical structure and physical processing play a part in determining mechanical properties of fibers, such as breaking strength, extension at break, stiffness, resistance to deformations, and recovery from imposed strains. Strength, for instance, is a function of molecular weight. Below a certain molecular weight, about 5,000 for most polyamides and polyesters, a polymer is unable to form fibers; just above that molecular weight a weak fiber is formed. As the molecular weight increases, so the strength of fiber increases. At values about 10,000 and above, commercial fibers can be made. High molecular weight alone will not ensure high breaking load. The degree of orientation is also crucial.

The most widely used synthetic fibers are based on polyamides, polyesters, polyolefins, acrylics, polyurethanes and certain vinyl derivatives.[12] The fiber-manufacturing processes from polymer to synthetic fiber can be outlined as given in Figure 1.5.

Identification of the synthetic materials is done by microscopic, chemical, and physical means as given in Table 1.5.[16]

Table 1.5 General Methods for Fiber Identification

Microscopic	Chemical	Physical
• Longitudinal shape • Cross-sectional shape • Refractive index • Birefringence	• Burning • Solubility • Chemical reaction • Differential thermal analysis	• Specific gravity • Melting point • Moisture regain • Infrared spectrum • Ultraviolet spectrum • Mechanical properties 1. Breaking strength 2. Modulus of elasticity 3. Others

1.2.1 Polyamides

These synthetic polymers have repeating units connected by amide linkages that are formed by the condensation of a dibasic acid with a diamine or by the self-condensation of a lactam. Polyamide fibers based on aliphatic monomers are generically called nylons; those with monomers that are at least 85% aromatic are referred as aramids.[12]

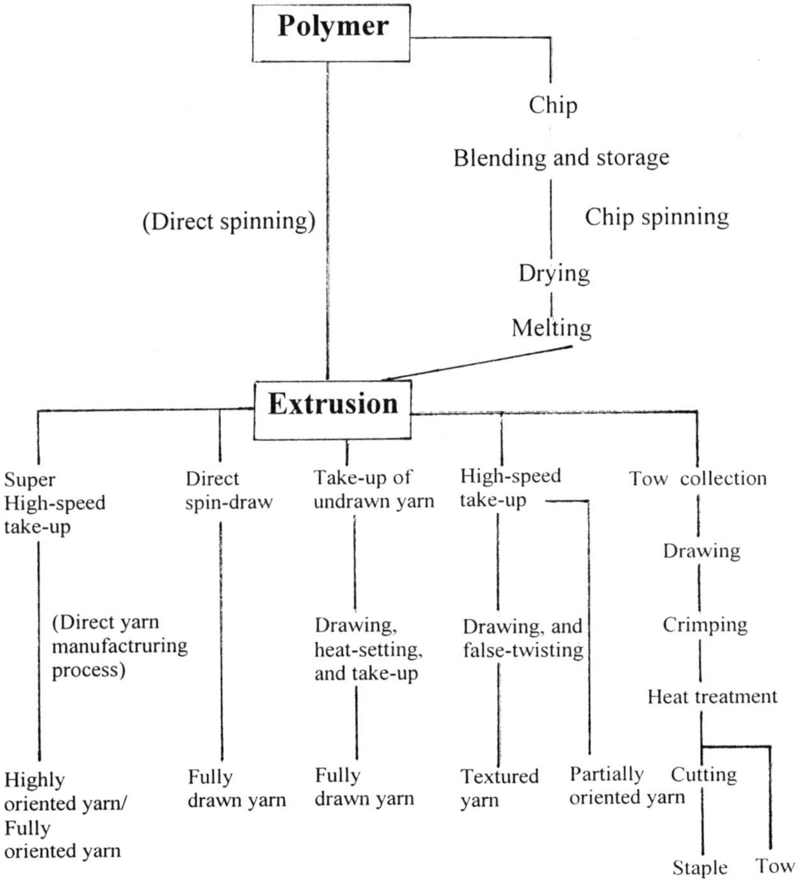

Figure 1.5 Outline of fiber-manufacturing processes

The word "nylon", although chosen by DuPont as a name for their new polyamide fiber, was not registered as a trademark. It has come to be a generic name for polyamides of different technical structures made by different manufacturers.[15]

The nomenclature commonly designates the number of carbon atoms in the repeating monomer unit. Thus, the polyamide from hexamethylenediamine and adipic acid is known as nylon 6.6, and that from caprolactam is known as nylon

6.[17] In 1980 when the world total polyamide production was 3.1 million tons, the shares of nylon 6.6 and nylon 6 were 38% and 62%, respectively. The total polyamide production has reached to 3.7 million tons in 1987.

Nylon 6.6

The first synthetic fiber, nylon 6.6 was launched in the United States in 1938[15] by DuPont. Hexamethylenediamine and adipic acid react together to form nylon 6.6.

$$H_2N(CH_2)_6NH_2 \;+\; HOOC(CH_2)_4COOH \;\Rightarrow\; [NH(CH_2)_6NHCO(CH_2)_4CO]_n$$

hexamethylene diamine — adipic acid — repeat unit of nylon 6.6

This reaction occurs under pressure and heat of about 275°C in an autoclave if a batch process is involved, or in a polymerization vessel if continuous process is involved. When the nylon polymer reaches required molecular weight, it is extruded from the autoclave in the form of a ribbon by falling the molten polymer on to a rotating water-cooled drum. This ribbon is later broken into small chips that are easier to store and transport. The chips can later be fed to the melting system of a spinning head. The continuous process may not require ribbon or chip formation, but feeding the molten polymer directly without solidifying to the spinning head.

Nylon 6.6makes strong fibers with excellent recovery from low and moderate extensions, stiffer than wool but less stiff than cotton and polyester, and with excellent abrasion resistance. These properties make it particularly valuable for ropes, tyre cords, carpets, stocking, and filament apparel.

Nylon 6

Self-condensation of caprolactam results in nylon 6.

$$n\ O{=}{=}C{-}{-}NH\ (\ )\ (CH_2)_5 \rightleftharpoons [{-}{-}(CH_2)_5CONH{-}{-}]_n$$

ε caprolactam — repeat unit of nylon 6

Pure caprolactam as synthesized from benzen is stable when molten; it polymerizes rapidly if an initiator is present. For fiber production, water is a common initiator. The process is carried out at 220°C to 240°C under pressure to prevent loss of water and lactam. The product is, however, of a lower molecular weight (usually 8,000 - 14,000) than ultimately required. Now the temperature is raised, and the pressure is reduced to below atmospheric to encourage removal of water and increase of molecular weight by a condensation mechanism. By further heating at about 270°C under nitrogen, a molecular weight of above 20,000 is reached, and polymer ribbon and chips are formed.

To remove remaining caprolactam, the chips are washed and dried. The nylon 6 polymer chips are now ready for melt spinning.[15]

These two aliphatic polyamides are by far the most extensively used nylon fibers. Only a limited number of other aliphatic polyamides, e.g., nylon 6.10, 9 and 11 are used in fiber form.[12]

Nylon fibers are manufactured by the melt spinning system. Because the T_g of nylon is below room temperature, nylon can be cold drawn.

The mechanical properties of nylon fibers depend on processing, drawing, and the nature of heat setting. Typical drawn nylon filaments are strong, highly resilient, and relatively insensitive to moisture. Characteristic properties of nylon are a low-elastic modulus at small strains, and a high-elastic recovery from large deformations. Typical properties of nylon are given in Table 1.6.

Table 1.6 Typical Properties of Nylon Fibers

Tenacity, N/tex	
65% relative humidity, 21°C	0.40 - 0.70
Wet	0.35 - 0.62
Extension at break, %	
65% relative humidity, 21°C	15 - 30
Wet	20 - 40
Elastic modulus, N/tex (65% relative humidity, 21°C)	3.5
Moisture regain, % (65% relative humidity)	4.0 - 4.5
Specific gravity	1.14
Approx. volumetric swelling in water, %	2 - 10
Glass transition temperature, °C	
0% relative humidity	107
100% relative humidity	- 8
Melting temperature, °C	215 - 250

Nylon 6.6 melt at 250°C, whereas nylon 6 melts about 215°C, indicating that nylon 6.6 possesses better intermolecular packing. Polyamide fibers are resistant to chemical and microbial degradation, and are electrical insulators. Nylon is readily dyed with disperse, acid, and reactive dyes to give a full range of fast colors.

Nylon fibers are used in many apparel, household, and industrial applications, especially where high strength is required along with high resilience; therefore, carpets constitute one of the principal markets for nylon yarns. Because fatigue resistance of nylon fibers is high, they are also excellent as tyre reinforcements. In the apparel area nylon has lost market share to polyester, because of "easy care" and better wrinkle resistance. Nylon has virtually a monopoly in hosiery, women's underwear, and stretch fabrics.

1.2.2 Polyesters

Polyester is polymerized by a reaction between an alcohol and an acid. The most usual way is the polymerization of terephthalic acid (this is made from para-xylene, which is distilled from petroleum and then highly purified) with ethylene glycol (the oil is cracked to give ethylene gas. This is oxidized in air with a catalyst to form ethylene oxide, which is then hydrated to produce ethylene glycol) in vacuum by a condensation mechanism at elevated temperatures (about 270°C) to form polyethylene terephthalate (PET) as illustrated below.

1. Esterification

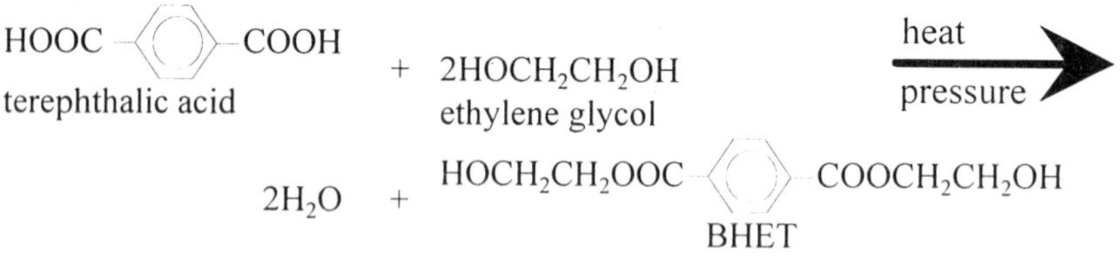

2. Polycondensation

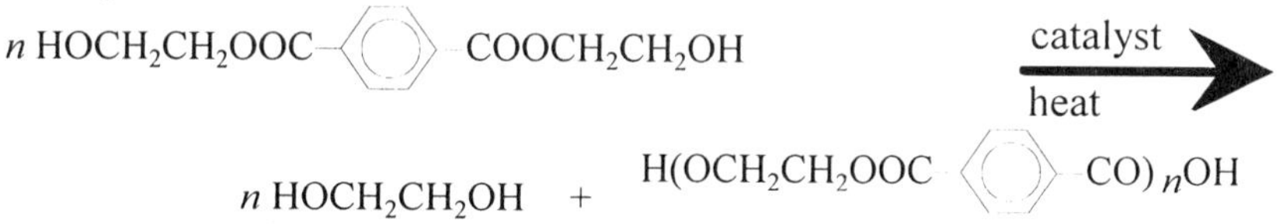

Fiber forming properties of polyester were first realized in 1941 by J. R. Whinfield and J. T. Dickson, and commercial production began in 1951 in the United Kingdom.[18, 19] According to ISO 1043/1978, the ISO letter symbol for conventional polyester fiber is PET, but the symbol PES is in widespread use in continental Europe, whereas the plastics industry uses PET.

Polyester fibers are also melt spun, normally through circular spinneret holes, although a variety of profiled fibers can be produced by specially designed spinneret orifices. Because the T_g of polyesters is 80°C, the filaments are hot drawn to develop orientation and crystallinity by a factor of about four. Properties of polyester fibers are strongly affected by fiber structure. The fiber structure, which has a strong influence on the applicability of polyester fibers, depends heavily on the process parameters in the fiber formation steps. The structure is determined by the spinning speed, hot drawing, stress relaxation, and heat setting processes employed.[19]

As the stress in the spinning threadline is increased by higher winding speed, the polyester molecules are extended, resulting in fiber with better uniformity, lower elongation, and higher strength, orientation, and crystallinity, as seen in Figure 1.6.

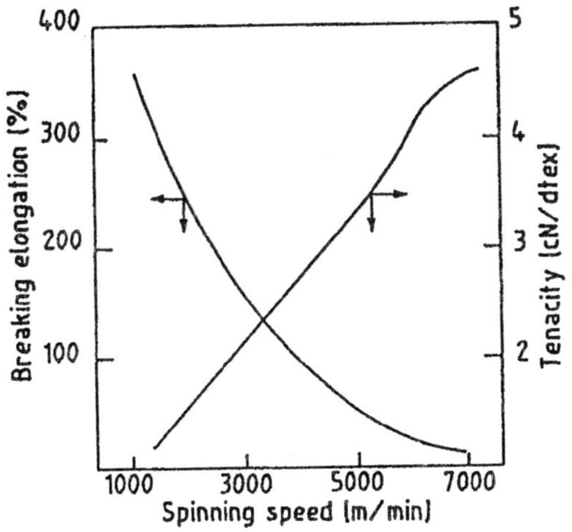

Figure 1.6 Effects of spinning speed on breaking elongation and tenacity properties of polyester

The physical properties of polyester fibers are given in Table 1.7, and stress-strain curves in Figure 1.7. The tensile stiffness or elastic modulus at low strains is much higher than corresponding polyamides. Polyesters have high elastic recovery, particularly from small deformations. The mechanical properties of wet polyester are virtually the same as those of dry polyester. Polyester fibers have excellent resistance to acids, alkalis, microbial attacks, and light. Low moisture regain (0.4% under standard conditions) makes polyester fibers subject to static electrification.

Dyeing of polyester fibers is difficult because of the lack of hydrophilic sites and the inherent stiffness of polyester molecules. Certain disperse dyes have been used by conventional methods, but new techniques of dyeing had been developed for full-shade ranges with adequate fastness. These methods are high-temperature dyeing, carrier dyeing, and the Thermosol method of dye fixation.

Table 1.7 Physical Properties of Polyester Fibers	
Tenacity, N/tex	
65% relative humidity, 21°C	0.35 - 0.53
wet	0.35 - 0.53
Extension at break, %	
65% relative humidity, 21°C	15 - 30
wet	15 - 30
Elastic modulus, N/tex (65% relative humidity, 21°C)	7.9
Moisture regain, % (65% relative humidity)	0.4
Specific gravity	1.38
Approx. volumetric swelling in water, %	None
Glass transition temperature, °C	80
Melting temperature, °C	260

High stiffness makes polyester fibers suitable for blending in staple form with cotton. Textured polyester yarns for dresses provide outstanding crease retention, and curtain nets that use the high ultraviolet stability when exposed to light behind glass are other important uses.

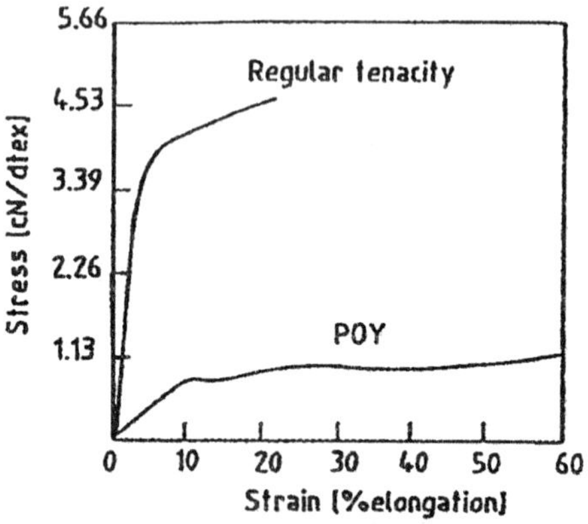

Figure 1.7 Typical stress-strain curves of polyester fbers

1.2.3 Polyolefins

Several crystalline polyolefins are capable of forming fibers, but only polypropylene is of practical importance. Propylene, CH_2-CH-CH_3, is a petroleum cracking product, plentiful and readily separated from other products and therefore cheap. But it cannot be easily turned into a high polymer. However, in 1954, the Italian G. Natta found that propylene could be polymerized by a new type of catalyst to a crystalline high polymer,[15] and the first commercial polypropylene fibers were introduced in 1961.

The principal method of manufacturing polypropylene filaments is melt spinning. A substantial volume of polypropylene fiber is also produced by slit-film technology that is favored for high linear density, i.e., 0.6 tex, filament production. The filaments may be produced by cutting or slitting a film, or mechanical or chemicomechanical fibrillation.

Polypropylene fibers are the lightest of the commercial textile fibers, with a specific gravity of 0.90 to 0.92. The fibers absorb virtually no moisture from the atmosphere because of the high crystallinity and lack of polar-sorption sites. As a result, the mechanical properties are almost completely insensitive to variations in relative humidity. Strength, extensibility, and stiffness in the wet state are identical to those under standard conditions. Typical properties of polypropylene fibers under standard conditions are given in Table 1.8.

Table 1.8 Physical Properties of Polypropylene Fibers

Tenacity, N/tex	
65% relative humidity, 21°C	0.44 - 0.79
wet	0.44 - 0.79
Extension at break, %	
65% relative humidity, 21°C	15 - 30
wet	15 - 30
Elastic modulus, N/tex	
65% relative humidity, 21°C	2.6 - 4.0
wet	2.6 - 4.0
Moisture regain, % (65% relative humidity)	0
Specific gravity	0.90
Approx. volumetric swelling in water, %	None
Glass transition temperature, °C	-10
Melting temperature, °C	165

The lack of water-sorption sites, the high crystallinity, and the general inertness of polypropylene poses a major dyeing problem. None of the existing classes of dyes provide an adequate range and depth of color with good fastness. The incorporation of pigment in the melt before extrusion provides a few basic colors at relatively high cost. Polypropylene fibers are used for home furnishings and industrial applications. The former include carpets and upholstery, and the latter rope, geotextiles and disposable and nondisposable nonwovens. Traditional, polypropylene fiber has limited use in apparel because of lower melting temperature. However, this market is increasing as manufacturers learn to take advantage of the wicking properties (moisture transport), as in lightweight sportswear and underwear.

1.2.4 Others

In this category, the polyvinyl derivatives are the most prominent fiber type. Although acrylonitrile as a chemical substance was known in 1893, the polymer made from it dates from 1925 because of the difficulties faced in dissolving and subsequently spinning into filament yarns. In 1942, a solvent-spinning system using dimethyl formamide produced satisfactory fibers of acrylic type and marketed as "orlon" in the United Sttes. Pure acrylic fibers, as first spun, were found to be similar to nylon in strength but low in dye affinity. To yield a textile fiber, further chemical modifications were introduced resulting in a dramatic (as much as about one-third of original) decrease in strength. This was a factor determining staple fiber as the main outlet for acrylic fibers. Therefore, acrylic fibers are generally available in staple form and are more often met in 100% than in blends. Acrylics more nearly resemble wool in texture, warmth, and handle than do other man-made fibers, and so are found in many wool-replacement uses.

Synthetic elastic “elastomeric” fibers and yarns were established as substitute for natural rubber, and are based on various polyurethanes.

Glass is a synthetic material based on silicon. It does not, like other fibers, consist of long chains of molecules: Its atoms are spatially connected in all directions; hence, it is nonfibrous and brittle. However, extruded glass fibers of sufficient thinness are flexible enough for handling as yarns and specialized fabrics.

1.3 SPINNING TECHNIQUES

In the context of synthetic fiber manufacture, spinning refers to the overall process of polymer extrusion and fiber formation. Because the natural polymers and their derivatives undergo decomposition at or below their melting points, they can be converted into a form suitable for spinning by only dissolution. Consequently, the two processes for extruding polymer solutions, known as wet and dry spinning, were developed well before the third process, melt spinning. Therefore, the three principal methods are melt spinning, dry spinning, and wet spinning. Although melt spinning is the most recent of the three, it is simplest in principal and most widely used in today's industry. Table 1.9 shows the spinning techniques and gives the list of materials that are manufactured by these spinning techniques. Melt spinning is the most economical, but can only be applied to polymers that are stable at temperatures sufficiently above their melting point or softening point to be extruded in the molten state without substantial degradation. In dry spinning, a polymer solution is extruded from a spinneret through a zone in which the solvent is rapidly evaporated. Wet spinning is characterized by the extrusion of a polymer solution through a spinneret into a liquid nonsolvent. The filament can be precipitated without a chemical reaction, or the fibrous polymer can be regenerated from a soluble derivative, such as salt.

Table 1.9 Spinning Techniques

Melt Spinning	Dry Spinning	Wet Spinning
Nylon	Acetate	Viscose
Polyester	Triacetate	Cupro
Polypropylene	Acrylic	Acrylic
Polyethylene	Spandex	Spandex

Other methods and variants of these basic processes include reaction spinning, dispersion spinning, emulsion spinning, dry-jet wet spinning, semimelt spinning, fusion-melt spinning, centrifugal spinning, electrostatic spinning, gel spinning, and tack spinning.[20]

1.3.1 Melt Spinning

Melt spinning was developed in the late 1930s for nylon 6 and nylon 6.6, and is now used for several organic fibers.[20]
An early commercial process, called grid spinning, was widely used for nylon and polyester fibers. In this process a grid consisting of metal tubes or bars heated electrically, or by other means of heating, is used to melt a polymer chip that hich falls on to it from a hopper or feed screw above. The melted polymer drops through the grid into a melt pool. To minimize degradation and prevent further polymerization, the melt pool is covered with oxygen-free nitrogen or, in the case of nylons, steam.
Most melt spinning plants based on solidified polymer supplies the melt polymer with the aid of screw extruders, which can achieve much higher melting rates than grids, can be fed directly with polymer in powdered or chip form (Figure 1.8). A single extruder may supply several spinning positions through a series of branching pipes or tubes. Figure 1.9 illustrates a conventional spinning plant.

Many nylon and polyester plants are based on continuous melt polymerization.

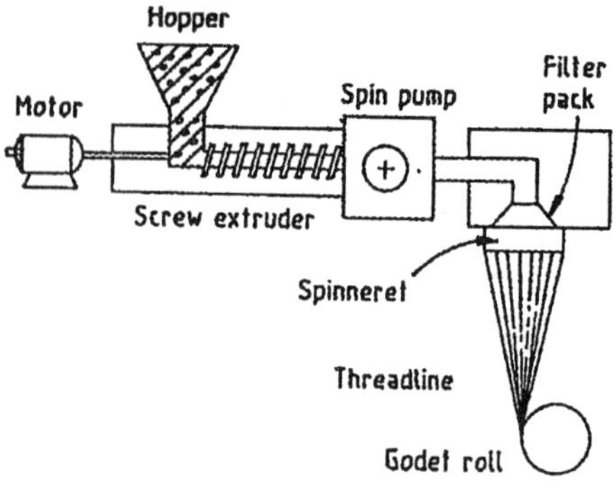

Figure 1.8 Polymer in powdered or chip form can be fed directly to melt-spinning equipment

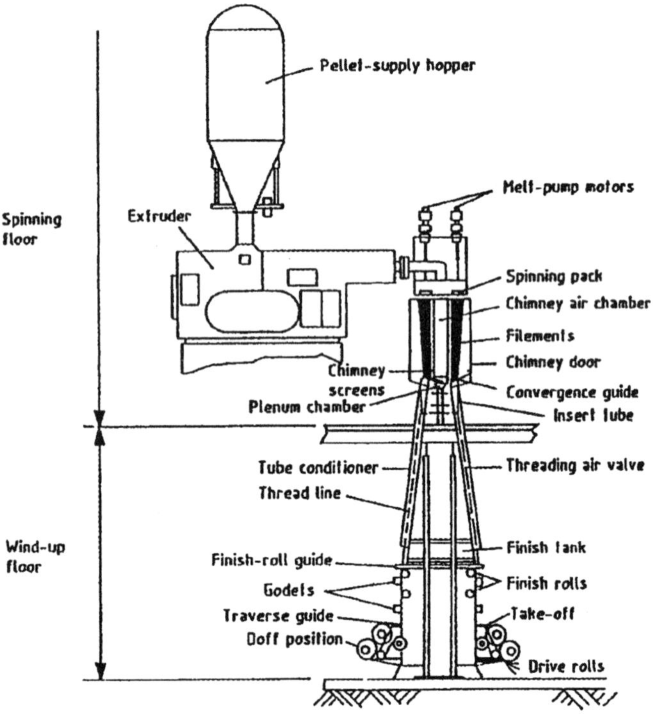

Figure 1.9 Conventional spinning plant

In these plants, the polymer is usually not solidified before spinning. Instead, the product is fed directly through a manifold from the polymerization plant to the spinning unit.

Whichever source of molten polymer is employed, the feed rate to individual spinning units is controlled by an accurately machined metering gear pump, which delivers molten polymer at a constant rate into a filter assembly.

Before reaching the spinneret, the molten polymer is filtered through a series of filters to remove large solid or gel particles that might block spinneret holes, or if passed through, occupy a sufficient cross-sectional area in the filament to affect its processing and tensile properties. Smaller particles, such as delustrants, are not retained by the filter.

After filtration, the molten polymer passes to the spinneret that is a plate containing orifices through which molten polymer is extruded under pressure as high as 20 MPa. Individual hole diameters are in the range of 175 to 750 μm at the exit. The capillary section is much shorter than the plate thickness which, in the case of melt spinning, is about 3 to 10 mm. The entrance to each hole is comparatively wide and gradually narrows down to give a gradual rather than abrupt reduction in diameter to avoid flow discontinuities (Figure 1.10).

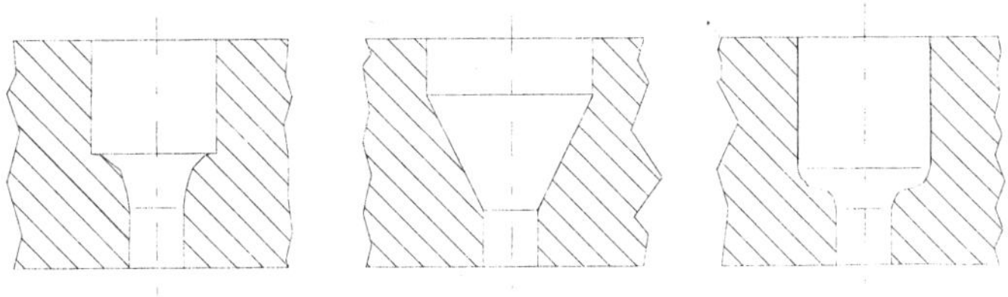

Figure 1.10 Longitudinal cross-section of spinneret holes showing streamlining of polymer flow

The number of holes in a spinneret may range from a few to several thousand. Spinneret holes for continuous filament yarn usually number in the tens or hundreds; spinnerets for tow production have a few thousand holes. The exit hole is usually circular, giving fibers with circular cross-sections. However, specially profiled fibers and also hollow fibers can be produced by specially designed spinneret orifices (Figure 1.11).

The spinneret hole does not directly control the linear density of the filament. Linear density is controlled by the throughput of polymer per hole, W (g/min) and the winding speed, V (m/min). The linear density (tex) of a melt-spun filament is then $10^3\ W/V$. The relationship between spinneret hole size and linear density involves the stretch ratio between the hole and the winding $V_O\ /V$,

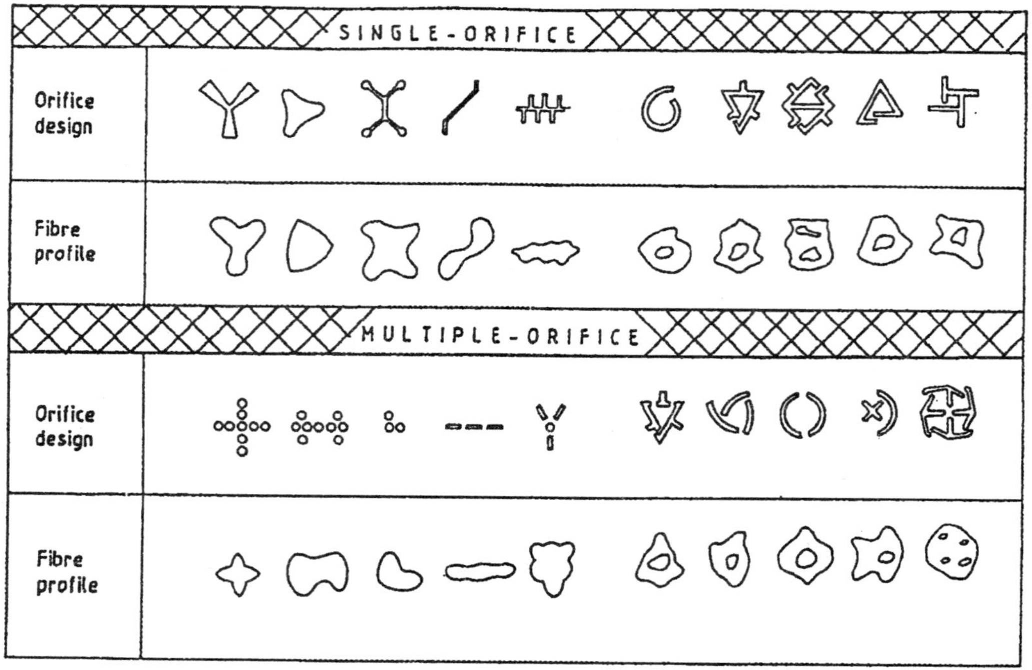

Figure 1.11 Spinneret hole (orifice) cross-section and its relation to the fiber cross-section

where V_o is the mean extrusion velocity:

$$\text{Linear Density} = \frac{A\rho W_p V_o}{V}$$

where A is the cross-sectional area (mm^2) of the orifice; ρ, the density (kg/m^3) of the extrudate at the extrusion temperature; and W_p, the weight fraction of polymer in the extrudate as it leaves the spinneret. For melt spinning $W_p = 1$.

As molten polymer passes through a spinneret hole, it is drawn away and attenuated by a drawing-down force applied by as forwarding or winding roll; simultaneously its temperature rapidly decreases. The diameter immediately below the hole and before attenuation begins is larger than the hole diameter.

This so-called die swell is due to relaxation of the viscoelastic stress induced in the hole. When spinning oxidation-sensitive polymers, it is useful to blanket a narrow zone immediately below the spinneret with inert gas to prevent degradation and deposition of polymers around the orifices.

Immediately below this region, cool filtered air is blown across the thread line at a controlled rate to encourage uniform cooling. Various arrangements have been described, where the air is directed from one side across the path of the filaments (cross-flow quench), radially inward (in-flow quench), or radially outward (out-flow quench). Thereafter, the thread line is protected from stray drafts by enclosing it in a protective chimney until close to the point where the filaments converge to form a yarn.

Soon after convergence, the filament bundle may come into contact with a device for applying spin finish, a mixture of lubricant and antistatic oils, often containing small amounts of biocides and corrosion inhibitors. It improves the cohesion of the yarn and reduces frictional drag against stationary guide surfaces. Yarn cohesion is essential for the yarn to move as a compact structure without filaments protruding from the bundle. Protruding filaments catch on processing equipment and cause breaks. They are trapped when yarn is wound on a bobbin, resulting in uneven tension and nonuniform texturing, knitting, or weaving. Static electricity, a common cause of poor yarn cohesion, can be alleviated by an antistatic finish. The spin finishes may be applied as neat oil or more often as an emulsion in water, using contact with a surface film or finish on a rotating roller or on a grooved ceramic guide fed from a finish-pumping system.[20] After evaporation of water, the concentration of finish on the yarn is usually 0.3 to 1 %. The proper choices of finish composition and concentration are critical for texturing and other processes.

The melt spinning processes are classified according to the degree of molecular orientation in the spun fiber. The spinning processes operating at speeds of 500 to 1,500 m/min give low-oriented spun yarn (LOY) and are referred to as LOY spinning process. Processes operating at speeds of 1,500 to 2,500 m/min give a medium-oriented spun yarn (MOY) and are termed MOY spinning process. Yarns referred to as partially oriented yarn (POY) are produced by spinning at speeds of 2,500 to 4,000 m/min. Between 4,000 to 6,000 m/min, highly oriented spun yarn (HOY) can be obtained. Above 6,000 m/min, fully oriented yarn (FOY) may be obtained. The demarcation line between LOY, MOY, POY, HOY, and FOY is not sharp, and is more clearly defined by orientation than spinning speed.

LOY spinning. In the 500 to 1,500 m/min spinning speed range typical of commercial LOY spinning process, the inertial contributions to the threadline tension are small, and the observed threadline take-up tension is virtually equivalent to the rheological force required to attenuate the filaments. As a result, the spinning tension is low and constant along the threadline. Under these conditions, the polymer responds to the applied tension in the melt portion of the threadline as a simple viscous elastic fluid, and little molecular orientation occurs.

Fiber obtained in this spinning speed range is practically amorphous. It has little strength, is highly deformable, and must be drawn to 3.5 to 5.0 times its original length to obtain useful properties.

MOY spinning. The demand for higher-productivity commercial processes led to MOY spinning. At speeds of 1,500 to 2,500 m/min, typical of commercial MOY spinning process, the inertial contributions to the threadline tension remain small, but air drag contributes to threadline tension. The resultant higher take-up tension attenuates the filaments more rapidly, and the polymer molecules begin to respond in a viscoelastic manner.

The spun fiber obtained in the MOY speed range is completely amorphous, but has higher strength and is less deformable than that in the LOY speed range. The maximum draw ratio is limited to 3.5 to 5.0.

POY spinning. Although the POY spinning was initially investigated in the early 1950s, it was not commercialized until the early 1970s, primarily because high-speed production winders were not available. In the late 1960s, simultaneous draw texturing provided the main impetus that led to the recognition of the desirability of POY properties. In the 2,500 to 4,000 m/min speed range typical of commercial POY spinning process, the air drag and inertial contributions to the threadline tension increases dramatically with distance from the spinneret. With the increased threadline tension, the filaments undergo rapid attenuation, and the polymer responds in a viscoelastic manner.

The fiber obtained in the POY speed range is almost completely amorphous, but exhibits substantially higher strength and lower extensibility than that obtained in MOY range.

HOY spinning. At speeds above 4,000 m/min, air drag and inertia continue to be the important contributors to the threadline tension, but structural effects also play a part: crystallization begins to take place and along with it the elongational viscosity changes, the exothermic heat of crystallization is released, and the solidification point is affected. Although the heat transfer coefficient and the cooling rates change only slowly with spinning speed, changes in these parameters are important because the sensitivity of the rheological properties and crystallization rate to temperature change is high.
In the 4,000 to 6,000 m/min spinning speed range, crystallinity develops rapidly, and amorphous orientation continues to increase.

FOY spinning. At speeds near to and above 6,000 m/min, air drag and inertia continue to dominate threadline tension. Heat transfer coefficient, cooling rates and radial temperature gradient continue to increase. However, at these speeds, the time available for crystallization is short, and crystallization can be inhibited. For instance, tenacity has been shown to decrease as the spinning speed exceeds 6,500 m/min and, the spinning process itself becomes more difficult because of the threadline breaks[21].

Figure 1.12 illustrates the stress-strain curves of polyester spun at various take-up velocities.

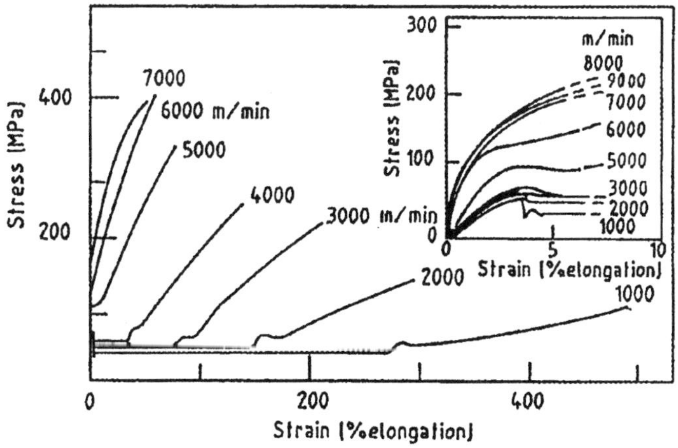

Figure 1.12 Stress-strain curves of polyester spun at various takeup velocities

1.3.2 Wet-Spinning

In the wet spinning process, the solutions are extruded into chemical baths that coagulate or regenerate the freshly extruded gel-like filaments to form solid filaments. Both aqueous and organic solvents are employed to produce solutions for wet spinning. Figure 1.13 schematically illustrates the wet spinning process.

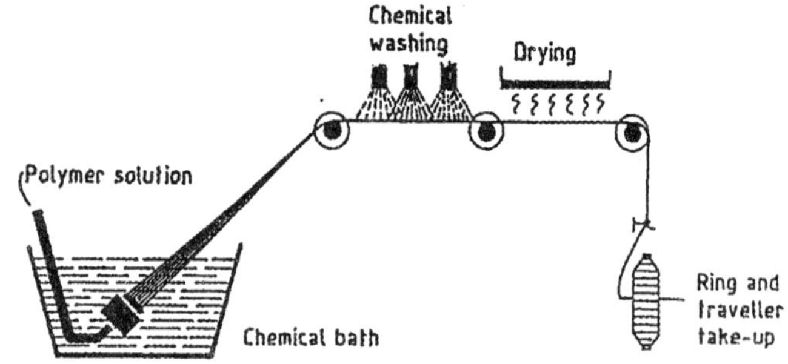

Figure 1.13 Schematic illustration of wet spinning process

The rate at which filaments can be drawn away from the spinneret holes without breaking is much lower in wet spinning than in melt or dry spinning[15].

Wet spinning processes fall into two broad classes, (1) those that employ a solvent/nonsolvent system, (2) those that employ a chemical reaction to dissolve and precipitate the polymer. Acrylics and modacrylics are usually spun by the former class of process, whereas viscose and cupramonium rayons require the second class[15].

1.3.3 Dry spinning

Dry spinning depends on extrusion of polymer solution through a spinneret into a heated gas or vapour so that the solvent is rapidly evaporated, leaving filaments (Figure 1.14).

Acetone and methylene chloride, the first important dry spinning solvents, are used in producing fibers from secondary cellulose acetate and cellulose triacetate. Acrylic and spandex fibers, conversely, are spun from much less volatile solvents, such as dimethyl formamide and dimethyl acetamide, which boil in the 150°C to 200°C range and require much higher temperatures to evaporate them[15].

1.3.4 Others

The production of fibers from film is a fast growing process, and it is of particular interest because of its great potential in terms of economy of production and variety of fiber product. In this process, polymer is extruded (usually melt extruded) through wide but narrow slits to form a sheet of film. Fibers are produced from the film by several techniques or a combination of techniques. Basically, the fibers are produced by stretching and orienting the film along its extrusion direction and then either breaking or cracking (fibrillating) the film into a mat of fibers by mechanical means. At present, mainly polypropylene fibers are produced in this method.

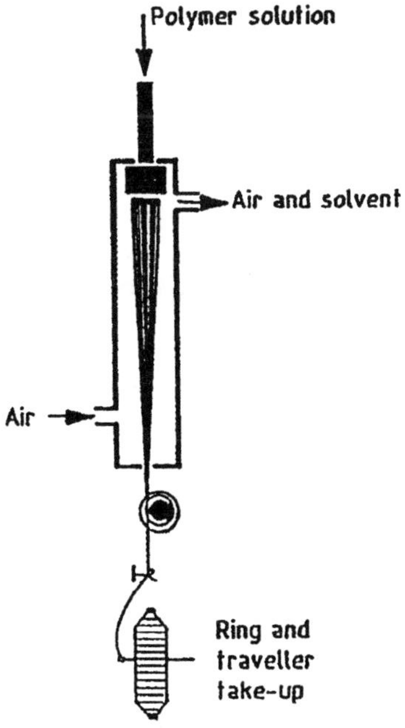

Figure 1.14 Dry spinning process

1.4 DRAWING

All textile fibers, natural or synthetic, have a common feature in their fundamental structure: the polymer molecules that make up the fiber are themselves fiber-like as far as their length to diameter ratios are concerned. They are made up by many molecular units joined end to end and repeated along the length of the polymer molecule as seen in Figure 1.15. A typical fiber may consist of thousands of such units[22].

To achieve fiber-like properties it is not sufficient that the molecules are long and chain-like; they must also be aligned more or less along the length of the fiber. When this alignment or "orientation" is achieved, with the molecules closely packed and substantially parallel, as in Figure 1.15, chemical bonds form between the chains and its neighbors with the result that the strength of fiber increases. Any extending force can thus be passed from one overlapping molecule to another along the length of the fiber[22].

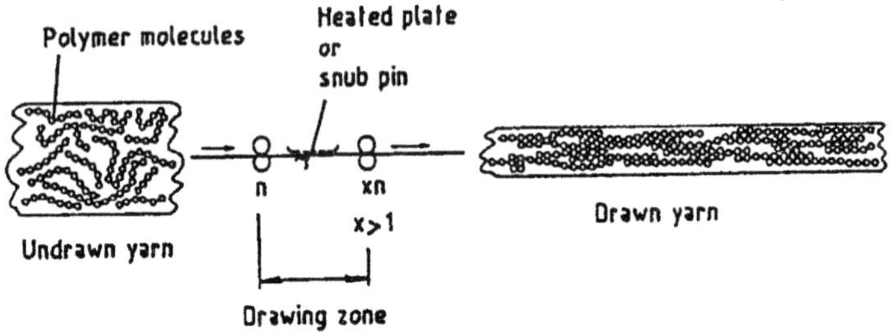

Figure 1.15 Molecular orientation before and after drawing process

In most instances the filaments formed by melt, dry, or wet spinning are easily deformable, have high elongation, and, in some cases, high shrinkage when exposed to heat, and are thus, unsuitable as textile fibers until they have been subjected to one or more successive drawing operations. Drawing is the hot or cold stretch and attenuation of synthetic filaments to achieve an irreversible extension, induce a molecular orientation with respect to the fiber axis, and develop a fiber-fine structure. The more closely the chains are parallel to the axis, the more highly oriented the fiber is said to be. This highly oriented structure is generally characterised by a high degree of crystallinity, and an orientation of both the crystallites and the polymer chain segments in the noncrystalline domains[12]. This results in an increase in tensile strength and a decrease in breaking elongation.

Whether the drawing is done cold or hot depends on a specific property of the material. This property is known as glass transition temperature, T_g, below which the material is rigid like glass, and above which the material is flexible. Below T_g filaments will not orient by drawing because the polymer chains are unable to slide over one another into a more extended configuration. However, if T_g is below room temperature, as in the case of nylon 6.6 and polypropylene, it is relatively unimportant, and such filaments are cold drawn. But if T_g is above room temperature, as in the case of polyester fibers, it becomes necessary to heat the fiber to orient.

For continuous filament yarn production, the orientation of the yarn depends on the speed at which it is spun. Usually, speeds below about 1,500 m/min are employed to produce LOYs and above about 2,500 m/min to make POYs. The orientation increases progressively up to about 6,000 m/min (Figure 1.16). In general, however, high-speed spinning produces less molecular orientation than mechanical drawing, because at spinline temperatures the relaxation of oriented molecules is significant, thereby reducing the net orientation.

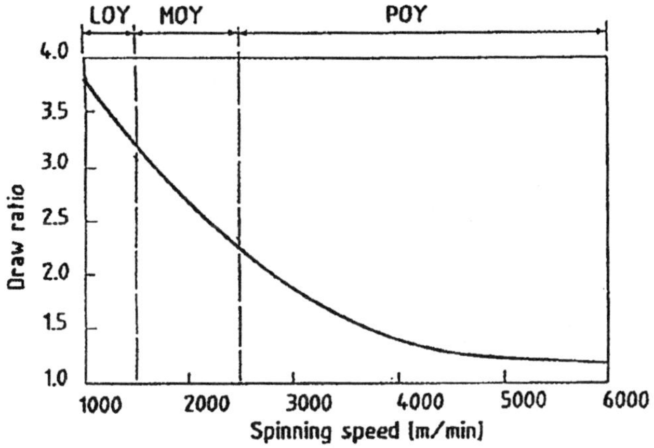

Figure 1.16 Orientation increases progressively up to about 6,000 m/min

Products of low orientation must be drawn to a relatively high draw ratio (4 - 7) to produce so-called FOY. Drawing may be carried out as a separate step in which the yarn is first wound up onto a bobbin, or it may be integrated with spinning as a spin-draw process in such a way that the yarn is removed from the spinning zone relatively slowly, forwarded to a drawing zone in which it is oriented, and finally wound up at a high speed (e.g., 6,000 m/min). POYs are usually wound up at speeds of 2,500 to 4,000 m/min and drawn at the following processes, such as draw texturing or draw warping.

A general structural model of a drawn fiber is given in Figure 1.17 as suggested by Dortmans[22]. In this two-phase model, the ordered (crystalline) region alternates with less ordered (amorphous) regions. It will be seen that a single molecule may pass through several crystalline and amorphous regions. Molecules that pass from one crystalline region to another are the so-called tie molecules. The crystallinity of such a fiber, seen in Figure 1.17, is in between 30% to 40%, the average molecular length is 1700 A°, and the orientation of the crystalline regions is approximately 0.98, and the average orientation of the molecules in the amorphous regions may vary between 0.70 to 0.85. The diameter of a filament is about 25 μm. From this diameter and the lateral size of a fibril, it can be calculated that within a cross-section of a filament millions of fibrils exist.

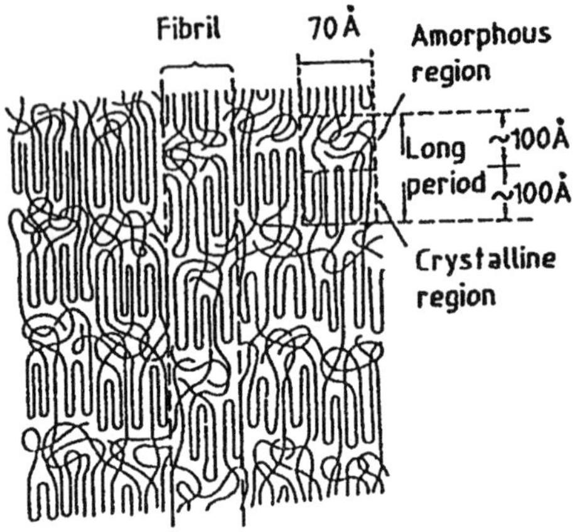

Figure 1.17 General structural model of a drawn fiber

Undrawn yarns may undergo a significant change in properties if stored for a long time. Morphological changes, such as further crystallisation, changes in moisture content with relative humidity, and, in some cases, absorption of spin finish component into the yarn may occur. It is, therefore, important to store the undrawn yarns under controlled conditions. In general, the more highly oriented yarns are less sensitive to storage conditions.

1.4.1 Cold Drawing

Some polymers of low-oriented filaments exhibit a phenomenon known as cold drawing, whereby each filament under a sufficiently great tension extends irreversibly to several times its original length at a localized draw point known as neck. This phenomenon occurs at temperatures close to the glass transition temperature (T_g) of the polymer. At lower temperatures brittle fracture may occur at high tension instead of drawing. At higher temperatures, drawing occurs at a lower tension and the process becomes uniform along the filament instead of occuring at a neck.

Nylon 6.6, having a glass transition temperature of -8°C (at 100% rh), draw in a stable fashion without applied heat. Passing the yarn around a stationary pin, the so-called snubbing pin, installed between the rolls, localizes the draw point and absorbs the heat evolved. The draw pins used are generally of titanium oxide with a diameter of 5 mm for low linear density and 8 mm for medium linear density filaments. A smooth-polished surface with a mean roughness of Ra 0.2 μ m is satisfactory for most cold-drawing purposes[24].

1.4.2 Hot Drawing

The essential differences between cold- and hot-drawing processes are in the way in which the draw point is stabilized. In cold drawing, the draw point is localized by means of a tension gradient created by wrapping the yarn around a peg situated in the draw zone. In the hot-drawing process, however, the draw point is located by means of a roll heated above the glass transition temperature of the fiber being drawn. After drawing, the yarn passes over a hot plate to stabilize the drawing and remove residual shrinkage by crystallization. Figure 1.18 illustrates the basic requirements of a hot-drawing equipment as used in draw twisting.

The technological advances in machinery production enabled to combine drawing operation with downstream textile operations. Draw warping and draw texturing are today's common operations that in the former POY is drawn, and warp reels are prepared in one operation; in the latter, POY is simultaneously drawn and textured by false-twist-texturing technique. As an even further development, spin-draw-texturing operation is being carried out on medium- to high-linear-density yarns with hot-fluid-jet-texturing technique.

1.5 SPIN FINISH AND ITS APPLICATION

Spin finish, spinning finish, dressing, treating agent, spinning lubricant, coating, fiber finish, processing agent, textile treating agent, textile treating composition, conditioning agent terms are all used to designate the same material: an antistatic lubricant applied to the surface of synthetic fibers as soon as practical after the fiber has been formed. The properties of the applied material are dictated by several factors, among which are the filament material and processing to which the filament will be subjected. In general, once the filament has been converted into yarn and then into fabric, the role of the finish is complete; it is usually removed before or during dyeing[25].

The potential for filament and yarn breakage is present in the fiber production as well as the downstream processes. Filament and yarn breakages are due to high-processing tensions, poor yarn coherency, or wear on yarn-contact surfaces. These problems are caused by friction, static electricity, surface wear, poor boundary lubrication, and chemical attack. Handling and processing are improved by the application of finishes, typically consisting of a mixture of components, including a lubricant to reduce friction, a long-chain aliphatic ester, an emulsifier so that the total system may be applied to the yarn from an aqueous emulsion, an antistatic agent to reduce the build-up of static electrical charges on the surfaces of the yarn as it passes over various machinery components, a bactericide, and an antioxidant or thermal stabilizing agent to reduce the build-up of resin deposits on the machinery and the yarn itself. Finish is generally applied during the spinning process before winding to reduce friction against guides, improve interfilament cohesion, and eliminate filament separation arising from electrostatic charge development. Although the exact details of finish formulations are usually proprietary, some of the desirable properties are as follows:

Lubricity. A lubricant is needed that reduces the coefficient of friction between the fiber-to-metal surfaces to prevent fiber abrasion and maintain low, uniform tension during processing.

Antistatic control. A lubricant must have an antistatic property to dissipate static electric charges built up during processing.

Cohesion. A balanced degree of cohesion is essential, because too much lubricity can cause fiber slippage resulting in package distortion in winding and other operations.

Oxidation resistance. After lubricants are applied, the fibers are often stored for prolonged periods, and lubricants must be resistant to discoloration, bacterial growth, and formation of insoluble resinous compounds in the presence of oxygen.

Controlled viscosity range. Too low a viscosity causes difficulties in slinging and low yarn frictional values, whereas too high a viscosity causes excessive add-on coupled with high frictional values.

Corrosion resistance. The yarn comes into contact with many metal surfaces during processing, and rusting tendencies would be detrimental to expensive machine parts; also, yarn pick up of rusts deposits would cause dyeing problems.

Color. The lubricant should be water-white and non-yellowing during processing or storage of yarns.

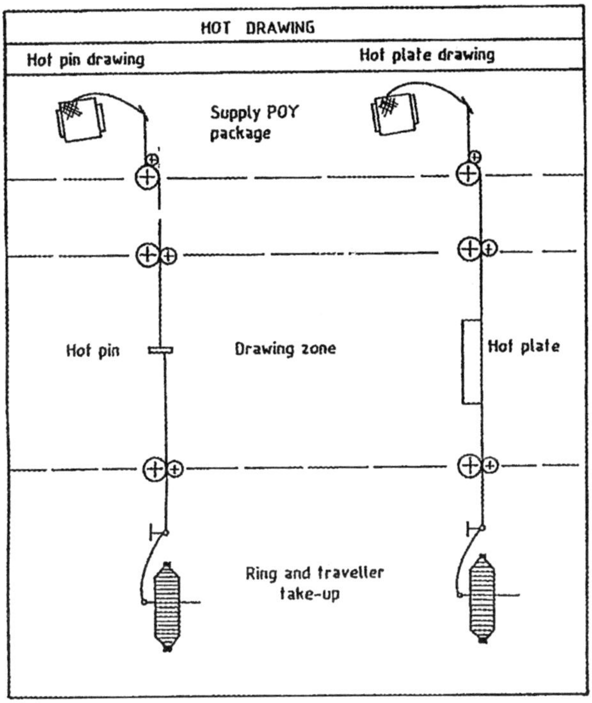

Figure 1.18 Basic requirements of hot drawing

The other desirable properties of the spin finishes are scourability, nonallergenic and nontoxic, odour resistance, product stability, nonvolatility[25].

Finishes are applied to continuous filament yarns undiluted or from an emulsion or solution, usually in water, with the help of a wick or rotating wheel. The most widely employed device for spin-finish application is one that comprises a ceramic roll, which, partially submerged, rotates in a pan containing the liquid to be applied. The filaments (or yarns) are caused to travel over and to contact the nonsubmerged surface of this roll, picking up absorbed liquid. The lack of uniformity in the application of spin finish by means of these devices is one of the drawbacks of these systems. As an alternative, spraying a liquid mist onto the moving threadline has provided somewhat improved finish application, but spraying involves added waste of finish and contamination of processing equipment. In some processes the finish is metered onto the yarn by means of a V-shaped ceramic guide fed from a small pump to maintain better control of the amount applied.

Spun and drawn synthetic filaments may then be subjected to further operations according to the final usage. Thermal setting and thermal relaxation provide dimensional stability; twisting and intermingling provide interfilament cohesion; texturing provides a voluminous yarn; crimping and cutting provide staple products similar in length and processing behavior to natural fibers[20].

REFERENCES

1. *Textile Horizon*, April 1992.
2. Anson R. and Simpson P., "World Textile Trade and Production Trends," *Textile Outlook International*, January 1989, 24-51.
3. Davies S., "Man-Made Fiber Capacity Growth - The Leading Suppliers of Plant and Technology," *Textile Outlook International*, 1988, 88-96.
4. The Textile Institute, "Textile Terms and Definitions," 7th Edition, 1975.
5. Ford J. E., "A History of Man-Made Fibers," *Textile,* 1983, Vol. 12, No. 2, 30-38.
6. Proceedings of *Polyester Textiles*, Shirley Institute S.51, Conference (17/18 May 1988) .
7. *Textile Organon*, 56, 4 (1985).
8. *Textile Organon*, 46, 3 (1975).
9. Davies S., "Escalating Raw Material Prices," *Textile Horizons*, 1989, Vol. 9, No. 9, 39-42.
10. Enka A.G., "Man-Made Fibers, 1987 - World Picture," *Textile Month*, April 1988, 58-60.
11. Ford J. E., "The Uses of Man-Made Fibers," *Textiles*, 1987, Vol. 6, No. 1, 2-10.
12. Rebenfeld L., "Fibers" in *Encyclopaedia of Polymer Science and Technology,* 2nd Edition, Vol. 6, 647-733.
13. Riehl L., "Development of the Melt-Spinning Process in the Past and in the Future," *CTI Man-Made Fiber Year Book*, 1987, 34.
14. Kilduff P. D. F., and Jackson K. C., "The Competitive Characteristics of the Man-Made-Fiber Industry in Western Europe," *Journal of Textile Institute*, 1989, Vol. 80, No. 2, 185-202.
15. McIntyre J. E., "*The Chemistry of Fibers,*" Edward Arnold, London, 1971.
16. Smith A. L., "Fibers, Identification," in *Encyclopaedia of Polymer Science and Technology,* 1st Edition, Vol. 6, 593-609.
17. Saunders J. H., "Fibers, Identification," in *Encyclopaedia of Polymer Science and Technology,* 2nd Edition, Vol. 11, 410-445.
18. Garbet T. W. J., "High Tenacity Polypropylene Yarns," in *Technology and Tomorrow's Fibers*, Textile Institute, Fiber Science Group Conference, December 1988, 7-8.
19. Davis E. W., and Talbot J. R., "Polyester, Fibers," in *Encyclopaedia of Polymer Science and Technology,* 2nd Edition, Vol. 12, 119-193.

20. McIntyre J. E., "Fibers, Manufacture," in *Encyclopaedia of Polymer Science and Technology,* 2nd Edition, 802-839.
21. Vasilatos G., Frankfort H. R. E., and Knox B. H., in "*High-speed Fiber Spinning,*" Edited by Ziabicki A., and Kawai H., Wiley-Interscience, New York, 1984, pp. 383-428.
22. Denton M. J., "Drawing and Texturing of Nylon and Polyester Yarns," *Textiles*, 1980, 50-55.
23. *Polyester Textiles*, Shirley Institute S.51, Conference (17/18 May 1988).
24. "Progress in Yarn Texturing," *Textile Month*, August 1975, 55-58.
25. Postman W, "Spin Finish Explained," *Textile Research Journal*, July 1980, 443-453.

CHAPTER 2

TEXTURE, TEXTURED YARNS AND TEXTURING TECHNIQUES

Appearance, feel, performance and cost are the important parameters in the design of yarns and, hence, apparel fabrics. According to the end use, the degree of importance of these parameters may change, but there is little doubt that visual and tactile properties that appeal to the human eye and skin, respectively, are always fundamental to yarn and fabric design. The appearance of a fabric is mainly governed by the interaction of color and the surface texture. The feel of a fabric, conversely, depends on not only the fabric itself but also several physiological-psychological complexities that are not fully understood. These two major parameters are functions of both basic yarn properties and the way fabric is manufactured.

The purpose of this part is to discuss the basic yarn characteristics that are relevant to human sight and touch. The synthetic filaments as spun through the holes of a spinneret have properties that attract little human attention. Therefore, the need for modification of synthetic filaments is introduced under the "texture" and "texturing" titles. The properties and production trends of textured yarns are discussed, and the texturing techniques are classified.

2.1 VISUAL CHARACTERISTICS OF YARNS

The appearance of a surface depends on the way light is reflected from it. A surface with uneven surface topography reflects light randomly as observed on an ordinary paper and gives a mat appearance, whereas a smooth surface reflects light in an organized manner like a mirror and, thus, creates a shiny appearance. Yarns and fabrics also behave in the same manner. A flat filament yarn as seen in Figure 2.1a reflects light in only one direction and creates a sparkle and sheen.

Fabrics made from filament assemblies are also highly shiny because of the well-organized light reflections from the surface. However, a spun staple yarn having quite an irregular surface characteristics reflects (Figure 2.1b) light in all directions and appears as a mat surface.

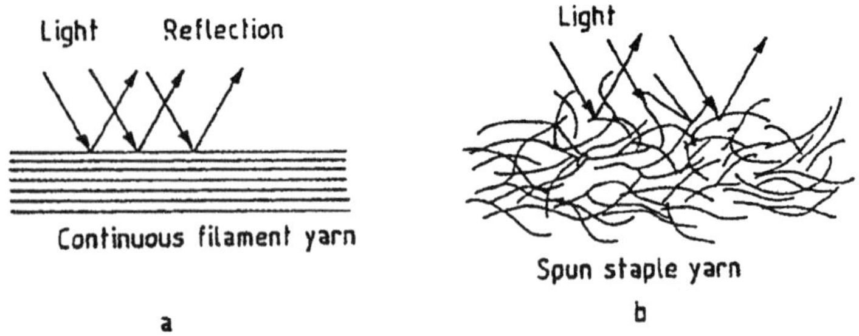

Figure 2.1 Light reflectance from surface of yarns

Not only the spatial organization of the filaments or fibers affects the appearance of the surface but also the variations in the cross-section of the filaments cause light reflected in random manner. For example, a seven or eight-lobed cross-section is known to eliminate the pin-point light reflection or sparkle associated with some textured filament yarns.

2.1.1 Color

Color is the sensation produced in the eye by rays of decomposed light from an object. This visual sensation is directly related to the molecular structure of the fibers rather than the surface properties or spatial arrangements of them. The colored compounds used as dyestuffs during synthetic filament production have chemical structures such that they absorb visible light of certain wave- lengths and so transmit only selected wave lengths to the eye.

2.1.2 Luster

Luster is an optical property that is largely determined by the presence of light-scattering centers on or within the fiber and by the geometrical arrangement of the fibers within the yarn. Although trained observers can place yarns and fabrics in the order of decreasing luster, luster is an extremely complex phenomenon. The extent of the light source, the directional properties of the light reflections, the general level of the reflected light intensity, the color and surface texture of the material, and visual perception of depth are some of the factors that influence luster.

If a fiber behaved as a perfectly reflecting circular cylinder, it would reflect light as shown in Figure 2.2. It is clear from this diagram that, if the light falls across the fiber, it is reflected at various angles, whereas if it falls along the fiber it is predominantly reflected at a constant angle. This is a basic feature of textile luster and shows the importance of causing the fibers to lie parallel to one another in a lustrous yarn or fabric[1].

Textile fibers depart from this ideal model in various ways. Finer fibers have a luster that differs from that shown by coarse fibers. Irregularities on the surface of the fiber and its cross-sectional shape will cause light to be reflected in various directions and will reduce the luster. To secure the type of reflection shown in Figure 2.2, it is essential that the fiber should be uniform along its length. For this reason, luster is greatest in regular filaments, such as those of silk and man-made fibers.[1]

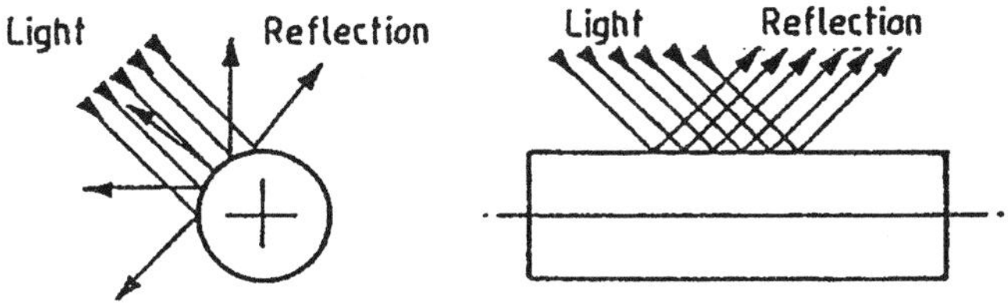

Figure 2.2 Light reflectance from surface of a single fiber

Fiber shape is itself an important factor. The particular types of luster associated with nylon, rayon, and silk must be due partly to the influence on the pattern of light-reflection of their respective circular, serrated, and triangular shapes.[1]

The early man-made fibers were bright and lustrous, because they transmitted light freely. At first this luster was unusual enough to help sell them, but eventually the time came when it proved a limitation and a matt appearance more akin to that of natural fibers was desirable. The most successful method of dulling viscose, discovered in 1929, was addition of colorless particles of refractive index different from that of the polymer. The most useful of these delustrants ever since has been titanium dioxide.[2]

Although filament draw or twist influences the luster, for a given delustrant content, the total light intensity scattered from the yarn surface at all angles in the plane of incidence is constant. Geometrical irregularities also diminish luster.

2.2 TACTILE CHARACTERISTICS OF YARNS

The human appeal of a fabric depends on the mutual interactions of the visual effect of fabric texture and its feel in the hand. It is not easy to separate the visual from the tactile effects, because most individuals are accustomed to seeing and feeling a fabric simultaneously. The tactile characteristics of a yarn and fabric are mainly their bulk and cover, warmth, and hand or handle.

2.2.1 Bulk and Cover

Bulk may be regarded as the specific volume of a yarn or fabric, i.e., the volume yarn and fabric occupies in a free space. By the very nature of all fibers and filaments, perhaps millions of them in a yarn or fabric occupy little space when they are organized in an orderly manner and, hence, create little bulk, leaving no room for still air to be trapped between. Fabrics manufactured from such yarns feel like a sheet of polythene, clammy and uncomfortable, and, thus, find little human attraction and use. To create bulk in a yarn and, hence, fabric, the basic elements, i.e., fibers and filaments, are so disorderly organized that although the fiber distribution in axial and radial directions is uniform, the yarn occupies maximum volume in space. Such a phenomenon is simply achieved by the geometrical irregularity, such as crimp, loop, and waviness of basic elements. Natural short staple fibers inherently possess such properties, cotton fiber has bean-shaped cross-section and twisted lengthways, and wool is curly and crimped. Synthetic filaments are also textured to create similar effects. When such fibers and filaments are assembled to make up a yarn, still air is entrapped between the fibers and improves the thermal insulation of the yarn; thus, the fabric manufactured from such yarn feels warm, bulky, and comfortable. These properties usually appeal to human nature; hence, such fabrics find many uses in everyday life.

The bulky yarn occupies considerably more volume than that needed for just the filaments themselves. This property of yarn is directly related to the cover of fabric made of yarns in that using less yarn, the fabric is still able to cover the area without obvious network of holes.

2.2.2 Stretch

The fabrics produced in Europe today have an elastic stretch of 8% to 12%. This stretch, also referred to as comfort stretch, confers optimal aesthetics and utility on fabrics made of textured yarns.[3]

2.2.3 Warmth

The imperfect fiber orientation within the staple yarn gives rise to voids in which air is trapped. Static air is a good insulator; thus, staple yarn feels "warm." By changing the straight cylindrical geometry of the filaments and their spatial disposition within the yarn, the warmth of continuous filament yarns can be improved. Textured, i.e., crimped and/or entangled, filaments also behave like the staple fibers of a staple yarn and still air is entrapped between the filaments.

Regardless of the fiber content, warmth or thermal insulation is dependent on the entrapment of air over a wide climatic range of wind, temperature, and humidity conditions. It is also true that, with textile structures, the rate of heat transfer depends much more on the real area of contact between skin and fabric surface than on the type of fiber used. Therefore, a textured yarn or fabric surface will feel warmer than that of an untextured yarn or fabric.

2.2.4 Hand (Handle)

Hand is described as "the subjective assessment of a textile material obtained from the sense of touch."[4] Hand is, thus, a psychological phenomenon. It implies the ability of fingers to make a sensitive and discriminating assessment, and of the mind to integrate and express the results in a single-valued judgment. Such a judgment cannot be obtained objectively, because the relevant mechanisms of the sensory organs, the nervous systems, and the brain are not known and will vary from individual to individual and with time for each individual.[5]

The sensation obtained when there is contact between the human skin, and a fabric is determined largely by the stiffness of the hairs or fiber loops projecting from the surface of the fabrics. The finer these outstanding hairs (or fibers), the softer will the fabric feel. However, the sensations from a light touch and from a heavy grasp will essentially be governed by different parameters The latter involve the bending and shear properties that mostly are functions of fabric construction, whereas the former is primarily related to the surface fibers and

loops. The deformation of these surface effects is, indeed, impossible to quantify by mechanical means; hence, purely subjective methods are the only true judgment of hand.

It is assumed that in the hand of a person there are four sensory centres corresponding to the four factors: (1) smoothness, (2) stiffness, (3) bulk or compactness (aspects of weight and thickness), and (4) thermal character (aspects of warmth, weight and thickness). These centres are (1) sensors of surface smoothness and roughness; (2) sensors of stiffness and flexibility; (3) sensors for evaluation of bulk properties, such as size, weight and form; and (4) sensors of heat and cold--the thermal centres.[5]

2.3 TEXTURE

The term *texture* defines and describes those attributes of an object that can be recognized by the human sight (visual characteristics) or touch (tactile characteristics).

The main texture properties desired in apparel and domestic textiles are warmth, opacity, flexibility, and good wearing properties, coupled with an attractive appearance and easy maintenance. All these desirable qualities are inherently provided with staple fiber yarn spun from the natural fibers, such as wool, cotton, and linen.

Attractiveness and customer appeal of the textile goods are closely related to their surface properties, which are in constant contact with human senses, such as sight and touch. These attributes of an object that can be recognized by the human sight and touch is described by the term texture, which covers bulk, cover, hand, and warmth as the tactile properties, and appearance and luster as the visual properties.

When a piece of fabric woven or knitted from a spun-staple yarn is handled by a light touch, a feeling of warmth and natural comfort will be observed, whereas a similar fabric made of a flat synthetic continuous filament yarn will give the feeling of clamminess and discomfort. Similar comparison is also valid for visual feeling in that a cotton or wool fabric will appeal to the human eye, whereas shiny, highly lustrous polyester or nylon plain fabric will distract human attention and disturb the vision. This superiority of the natural products is due to the random hairiness of the spun-staple yarn. Regarding to the tactile senses, these protruding hairs buckle under the effect of light touch and give the natural soft feeling; air pockets are also formed between these hairs of yarn, which create the warmth of the fabric. Furthermore, the hairs also give bulkiness to the yarn and, in turn, increases the cover properties.

The appearance of a textile surface, conversely, depends on the way the light reflected from it. When light is randomly reflected in every direction, the surface appears as matt and tends to be a natural part of the immediate environment, whereas a lustrous surface reflects light in an organized character and creates the shiny look that stands out but may be attractive to some (Figure 2.1). Stoastic structure, both surface, cross-sectional, and molecular, of natural fibers will, therefore, create a matt appearance. An array of parallel, and closely packed filaments, however, provides a shiny surface because of the highlight of each filament, which is parallel to the filament axis.

2.4 TEXTURING

Because of the growing world demand on textile goods and decreasing natural resources, synthetically produced yarns are inevitable supply of the textile industry today. These synthetic yarns, such as polyamide and polyester continuous filament yarns as spun from the spinnerets, however, lack the previously discussed human appeal resulting in a tendency toward natural products, though they offer some superior properties, such as high strength, good elasticity, stretch, and abrasion resistance.

Warmth, handle, natural texture, and appearance are considered to be desirable properties of most textile yarns. Flat, continuous synthetic filament yarns do not possess such qualities, but they are more often stronger and much more uniform than the natural fiber yarns. When producing textile yarns from synthetic filaments, the ideal objective would be to combine the desired properties of both natural and synthetic fibers, but this is an impossible task to achieve. Therefore, the primary objective of all synthetic filament yarn conversion processes is to imitate the features of the natural fiber yarns while maintaining the desirable properties of synthetic fibers.

One method of achieving this is to cut the continuous filaments into staple fibers and then process them into yarn form using conventional spinning methods as for cotton and wool; however, this is a time-consuming and, therefore, a wasteful process, although it is useful for blends of man-made fibers with the natural fibers. Alternatively, continuous filaments can be converted into yarns by various texturing methods at lower costs but often these inadequately simulate spun yarns.

To combine the desired qualities of natural and synthetic materials is an unobtainable goal. Certain aspects of natural products could be imitated by the synthetic filaments, and this is succeeded by the method of texturing. Therefore, the texturing in general is the modification process of regular structure of synthetic filaments into somewhat random structures. This is achieved in many ways, such as thermal, mechanical, and chemical deformation of the individual filaments, and their spatial arrangement in the yarn bundle. Another definition for texturing, as given by Denton,[6] could be that "texturing is the means whereby permanent fine distortions, crimps, loops, coils, or crinkles are introduced into the original straight filaments of synthetic yarns without destroying the continuity of the original filaments." The end product that is textured yarn is bulkier with bent, crimped, curled, or looped filaments. It, thus, attains soft and natural like feel, and increases the warmth and comfort of fabrics.

Textured yarn fabrics have better shape retention ability than those made from straight filament yarns. This is obviously due to the improved interfilament frictional properties. Flexibility is a important property desired in textile yarns and fabrics. It depends on the ability of the filament (in the outer layer) to deform easily. A straight filament is comparatively less flexible than a crimped, looped, and curly filament. This ease of deformation of filaments is a function of their bending and torsional rigidities determined by the linear density of filament and the geometry of the filament in the yarn. In a textured yarn, filaments lie in all sorts of geometrical configurations and are loosely packed so as not to inhibit their relative movement. This results in better flexibility and extensibility to yarns and ultimately to fabrics made from them. Texturing also results in softer yarns. Other factors being equal, the softer the yarn, the better is the yarn's covering power. Their high covering power is also a function of increased yarn bulk as discussed earlier.

Most of the thermoplastic synthetic fibers have low moisture content comparedwith natural fibers. However, the air entrapped between the interstices of the textured yarns retains moisture and, thus, makes the textured yarns have good comfort characteristics. Because little water is absorbed by the synthetic filaments, textured yarn fabrics are easily washed and dried.

2.5 TEXTURED YARNS

Textured yarn is a generic term for filament yarns that have been given notably greater apparent volume than conventional yarn of similar filament number and linear density. The apparent increased volume is achieved through physical, chemical, or heat treatments, or a combination of these. Hawthorne[7] also gives other definitions of texturing.

Textured yarns can be divided into the following two broad groups:

1. Stretch textured yarns
2. Spun-like textured yarns

Stretch textured yarns are characterized by their high extensibility and good recovery, but possess moderate bulk compared with spun-like textured yarns. The individual filaments in a stretch textured yarn may be in the form of a two-dimensional, zig-zag, crimp or three-dimensional helices and snaris. Therefore, these configurations can easily be flattened out by small loads. Because the zig-zag or helical forms are a function of molecular rearrangement, when the loads are removed the filament recovers and attain the original crimped shape. Stretch yarns, used extensively in stretch-to-fit garments in which extensibility and recovery from stretch are of primary importance, are produced by the thermo-mechanical texturing processes. When the stretch textured yarn, however, is heat treated in a relaxed state, the degree of stretch decreases whilst the bulk of the yarn is improved. Such yarns are termed as *set* or *modified stretch* textured yarns, and may be regarded as a subgroup of stretch textured yarns. Increased bulk and decreased stretch properties lead the set yarns to be extensively used in knitted fabrics because they impart excellent stitch clarity, greater smoothness and softness of hand, as well as better bulk and low extension under low loads.

Although the individual configuration of a straight filament remains still straight in a spun-like textured yarn, the spatial arrangement of the filaments are so arranged (or disarranged) that some of the filaments protrude from the yarn core, and form loops and snarls to implement the hairs of a spun-staple yarn. The filaments in the core of the yarn are also so entangled that the core is a compact filament assembly with spaces where still air is trapped. Spun-like textured yarns, therefore, show virtually no elongation under loads that are normally applied during processing and wear. They possess substantial bulk and degree of bulk virtually remain unchanged even under breaking loads. Only the air-jet texturing process produces such yarns; with spun-like characteristics, the air-jet textured yarns find applications in almost all textile spectrum in which mostly spun-staple yarns are used.

Figure 2.3 illustrates idealized models of textured yarns. Figure 2.3 also includes models of untextured (flat) synthetic yarn and spun-staple yarn.

There seems to be a significant lack of statistics concerning the production of textured yarns in the 1980s with the exception of certain individual countries. World production figures for the 1983to 1984 session, as estimated by Wilson and Kollu[8], is given in Table 2.1.

2.6 TEXTURING TECHNIQUES

To impart the desired qualities of natural yarns to the synthetic filament yarns, there are basically three routes to follow according to the supply material and the intended use of the product (Table 2.2).

Thermomechanical texturing method makes use of the thermoplastic property of synthetic yarns and basically consists of mechanical deformation, i.e., twist or torque deformation, stuffing and bending, of the filaments while heating them to semiplastic condition and setting this deformation by cooling to ambient conditions. Thus permanent crimps are imparted to the yarn, and these are due to the reforming of the molecular cross-links whereas the filaments are in their distorted shape. Yarns textured by thermomechanical texturing methods rather inadequately approximate those yarns spun from natural staple fibers. However,

Table 2.1 Estimated World Production of Textured Yarns, 1983-1984[8]	
Product	**1000 Tonnes**
False-twist textured polyester	1,540
False-twist textured Nylon	675
BCF nylon and polypropylene	540
Air-jet textured	110
Others	143
Total	3,008

they possess a new character (i.e., stretch) which offers many advantages over other types of yarns. False-twist, stuffer-box, knife-edge crimping, gear and knit-deknit texturing techniques are in this category of texturing, and false-twist technique accounts for about 85% of all texturing processes.

Table 2.2 Texturing Techniques		
I. Thermomechanical Texturing Techniques		
Technique	Linear Density Range	End use of yarns
1. False-twist	20-200 dtex	Hosiery, underwear
2. Stuffer-box	200-2000 dtex	Carpet, upholstery
3. Knife-edge crimping		
4. Gear-mesh		
5. Knit-deknit		
II. Mechanical Texturing Techniques		
Technique	Linear Density Range	End use of yarns
1. Air-jet	20-2000 dtex	apparel, industrial
2. Intermingling	20-2000 dtex	flat, knitted garment
III. Other Texturing Techniques		
Technique	Linear Density Range	End use of yarns
1. Bicomponent		
2. Differential shrinkage		
3. Chemical		

The parallel arrangement of the filaments can also be changed by purely mechanical means to create spun-like yarns with an entangled core and surface loops imitating the protruding hairs of spun staple yarns. This is achieved by a unique method known as the air-jet texturing. In this process, any filament yarn, not necessarily thermoplastic, is fed through a specially designed nozzle that creates supersonic, highly turbulent air jet at a higher speed than is taken away. Therefore, the extra length of the filaments provided by the overfeed is taken up as entanglement in the yarn core, and loops and arcs at the surface of the yarn. In contrast to the yarns textured by other texturing techniques, air-jet

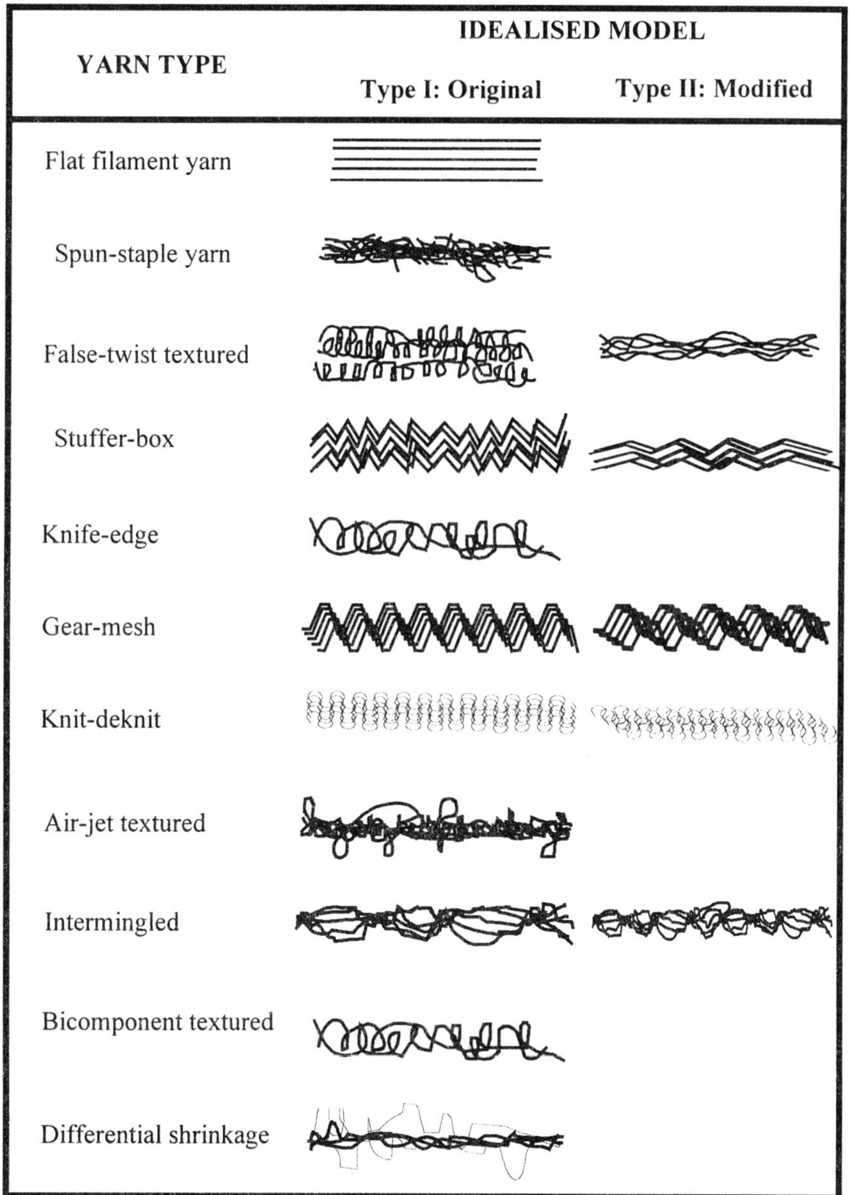

Figure 2.3 Idealized models of textured yarns
(models of flat and spun-staple yarn also included for comparison)

textured yarns are uniquely unstretchable just the same as spun-staple yarns. The surface loops are firmly fixed into the well-entangled core of the yarn and produce a voluminous, bulky structure that gives natural warmth because of the entrapped air between the loops and core.

Regular arrangement of the synthetic filaments may also be changed by various other techniques, such as chemical treatment, bicomponent forming and production of irregular-shaped individual filaments, so as to give certain amount of texture. As anticipated, these can only be achieved at the spinning stage of the yarn production; therefore, this type of textured yarn is generally called as producer-modified or textured yarns.

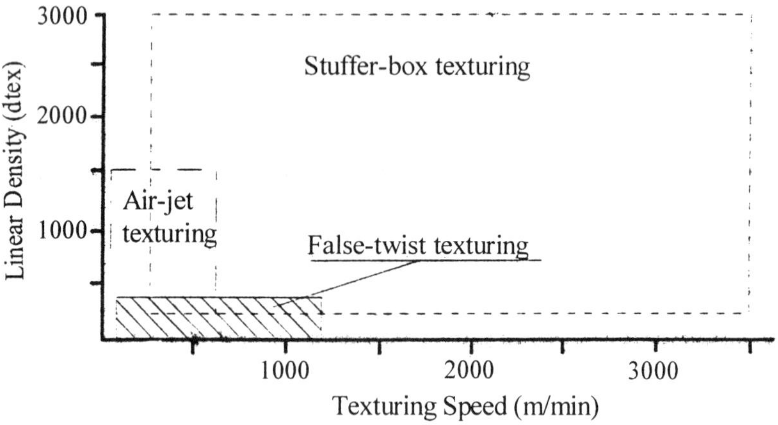

Figure 2.4 Relationship between texturing speed, yarn linear density, and suitable method of texturing

Texturing in a linear density range below 300 dtex, i.e., in the range dominated by the clothing sector, is indisputably the premise of the false-twist texturing method. In this respect, the false-twist texturing technique enjoys a popularity of more than 90% of the market, whereas the remainder is accounted for by air-jet texturing and various stuffer-box texturing methods. Especially in high linear density range as used in the production of textile floor covering and furnishing, the latter techniques gain great importance.[9] Figure 2.4 on page 46 illustrates the relationship between the texturing speed, yarn linear density involved, and the suitable method of texturing.

REFERENCES

1. Morton W. E. and Hearle J. W. S., "*Physical Properties of Textile Fibers*," Heinemann, London, 1993.
2. McIntyre J. E., "The Chemistry of Fibers," Edward Arnold, London, 1971.
3. Derder G, "Textured Polyester Filament Yarns in Spun Yarn Weaving," *Melliand Textilberichte* (English Edition), September 1985, 699-713.
4. Textile Terms and Definitions, The Textile Institute, 7th Edition, 1975.
5. Ellis B. C., "A Review of Techniques for the Assessment of Hand," *Textile Research Journal*, April 1980, 231-238.
6. Denton M. J., "Texturing of Continuous Filament Synthetic Yarns," *The Journal* (Bradford Textile Society), 1987, 51-56.
7. Hawthorne B. L., *Woven Stretch and Textured Fabrics*, John Wiley and Sons, New York, 1964.
8. Wilson D. K. and Kollu T., *Textile Progress*, The Textile Institute, Manchester, 1987.
9. Acar M., Alexander A. J., Turton R. K., and Wray G. R., "The Mechanism of the Air-Jet Texturing Process," in *Texturing Today* (conference papers), Shirley Institute Publications S.46, Manchester, 1983, 207-230.

CHAPTER 3

THERMOMECHANICAL TEXTURING TECHNIQUES

3.1 INTRODUCTION

Immediately after the spinning of polymers, the molecular structure in some cases is only partially drawn; therefore, the filaments are not suitable for textile use. Parallel arrangement of these macromolecules by drawing render the desired properties to the filaments. The binding force between these parallel and regularly arranged macromolecules is the cohesion force that can be released by heat. At this stage of minimum internal energy, if the filament is mechanically deformed the molecular structure will accept this new filament configuration. A subsequent cooling of the filament will cause the permanent setting of the new configuration. Therefore, many of the synthetic filament yarns are called thermoplastic. This is the characteristic on which thermomechanical texturing techniques depend.

The basic requirements of thermomechanical texturing technique are (Figure 3.1) as follows:

1. Heat the thermoplastic filament above glass transition temperature T_g (but below melting temperature T_m).
2. Deform the filaments to the desired shape, (e.g., twisting).
3. Cool the filaments below T_g while they are held in the desired shape.
4. Rearrange the cooled filaments so that they occupy a greater volume (texture).

Steps 1 and 2 can be reversed or may be coincident.

As a result of these steps, the individual filaments acquire permanent crimp, crinkle, waviness, or helices depending on the type of deformation applied.

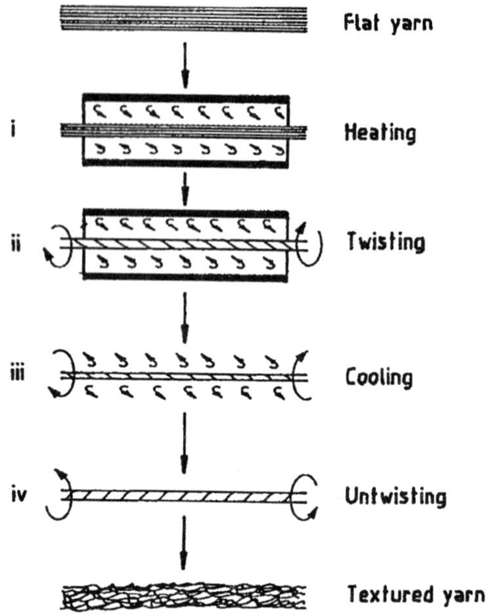

Figure 3.1 Basic requirements of thermomechanical texturing technique

Thermomechanical texturing processes had been applied to cellulosic filaments, such as viscose and acetate rayon; however they were not widely accepted because of the deterioration of such materials by the effect of heat. Several deformation techniques (i.e., thermomechanical texturing techniques) have been developed since the advent of synthetic filament starting with nylon in the early 1930s.

These techniques follow:
1. Twist-untwist (true-twist) texturing
2. False-twist texturing
3. Stuffer-box textring (recently known as hot fluid jet texturing)
4. Knife-edge crimping
5. crinkle yarn texturing
   a. Knit-deknit texturing
   b. Gear-mesh texturing

The following sections discuss these techniques in further details.

3.2 FALSE-TWIST TEXTURING

3.2.1 Historical Development of False-Twist Texturing

In the early days of the synthetic filament yarn production and texturing, a rather long and tedious route has been adopted for the texturing which produced stretch type yarns with crimped filaments. Although this route is no longer used by today's industry, it will be briefly mentioned for it forms the foundations of today's most common texturing technique, i.e., false-twist texturing.

According to Piller[1] who calls this process "the conventional process" in his book, *Bulked Yarns* published in 1973, seven steps are taken to impart the desired texturing. These are winding, twisting, heat setting, rewinding, untwisting, doubling, and winding. These many steps had to be taken because of the lack of appropriate machinery. With the development of the appropriate machinery, however, the number of steps was reduced to only four zones: supply, twisting with heating and cooling, untwisting, and winding (Figure 3.2).

The mechanical deformation of filaments are achieved by twisting of the filament yarn –the number of the twist per meter is highly dependent on the linear density of the yarn and intended use of the textured yarn– on perforated, heat-resistant packages. This highly twisted yarn is then heated in an autoclave by pressurized steam. To obtain a uniform heating of all parts of a package, the duration of the heat treatment is needed to be accurately set. As previously mentioned, during the heat setting operation the cohesion forces between the macromolecules will decrease, and the macromolecules will adopt the twisted configuration. Because a bundle of filament is twisted together, the individual filaments will be twisted around both their own axis and the yarn axis, and, in turn, will have a three-dimensional helical configuration. Once set, the filament –resulting from the configuration of macromolecules– will tend to keep this configuration. Steam-set packages are then dried and relaxed for a certain length of time, and the yarn is finally untwisted to zero twist. At this level, the crimped filaments will separate from each other and, in turn, will increase the apparent bulk of the yarn. Because of the open crimped structure, the textured yarn is stretched even under slight loads, and the yarn bulk decreases. To keep the filaments together, several processes, such as doubling of S and Z twisted yarns and insertion of extra twist, may be applied.

The rate of production with such a technique was slow, usually less than 250 g (of 70 dtex) per spindle per week, and large amounts of plant and floor space were occupied.[2] It is evident that such a process is too tedious and uneconomical for today's industry; therefore, it is an asset of history now.

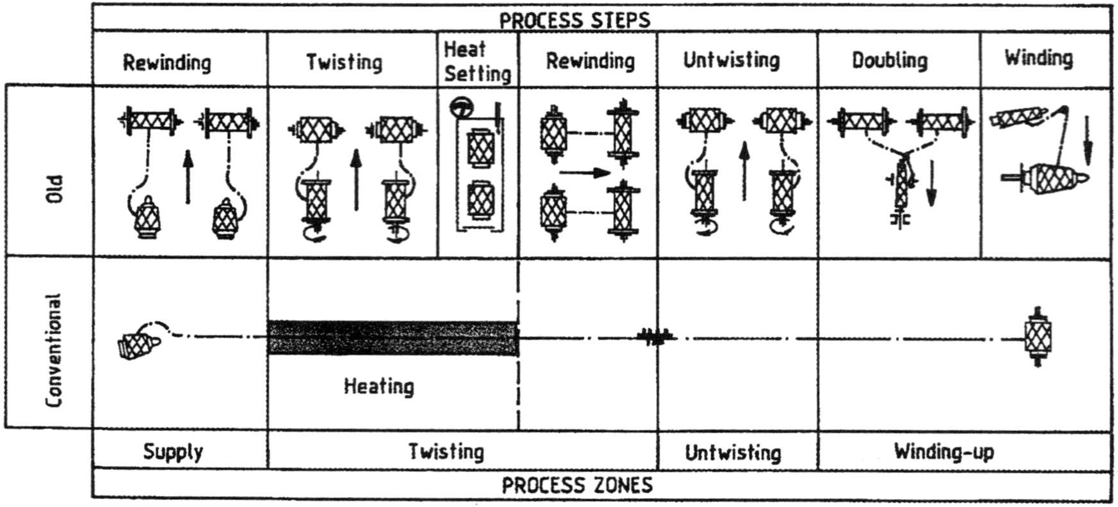

Figure 3.2 Multisteps conventional texturing process in comparison with zones of modern texturing process

As an alternative to this multistep time-consuming process, a continuous process has been developed in early 1930s. The models of pin false twisters appeared in the textile industry as early as 1933, and these, when developed on a large scale after 1945, were capable of running up to 100,000 revolution per minute (rpm)[2].

Although the pin spindles were small, they absorbed appreciable amount of energy and created unacceptable noise levels. The pressure from the noise regulations has tended to force these into early obsolescence.[3] In the early 1960s, alternatives to pin twisting were suggested in the form of internal type (bush) and band twisters, which work by friction[2] (Figure 3.3).

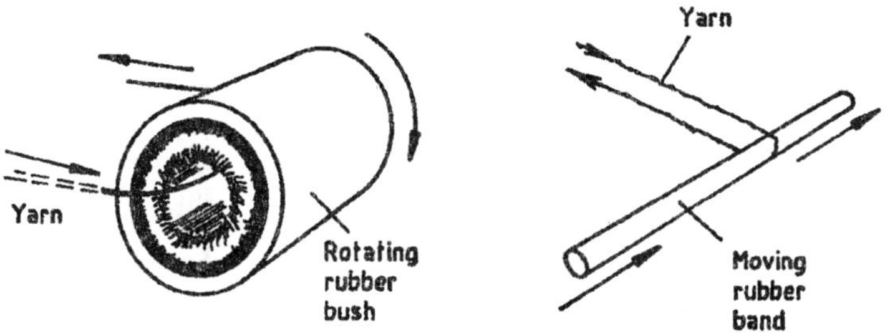

Figure 3.3 Internal type (bush) and band twisters[2]

The major breakthrough in false-twist texturing has been achieved in mid-1970s (Scragg introduced friction disk in 1973) with the introduction of friction false twisters.

The evolutionary stages observed by the continuous false-twist texturing process over the years can be graphically shown in Figure 3.4.[4]

The stage A is the first mode of operation and could be regarded as a separate texturing technique. If the drawing of spun yarn is carried out immediately after in-line with the spinning, the texturing of fully drawn yarn forms the stage B. Advances in processing technology brought the stages C to E, with E being the ideal case. Most of the modern-texturing machines operate within these forms and possess necessary equipment. The additional features imposed by this development are as follows:

1. Drawing of POY during the texturing process.
2. Heat treatment of textured yarn to give an increased bulk and decreased shrinkage.
3. Intermingling of textured yarn to increase the filament cohesion to eliminate the need of further twisting or sizing that is necessary for further processing.
4. Computer-controlled machines to provide the maximum flexibility and ease of setting. This feature also reduces the machine down time by warning the operator of the faulty or down positions.

The draw texturing that combines the drawing and texturing operations simultaneously is the current standard of the false-twist texturing industry. However, the basic process of false-twist texturing will be studied first and the particulars of the draw texturing will be given later.

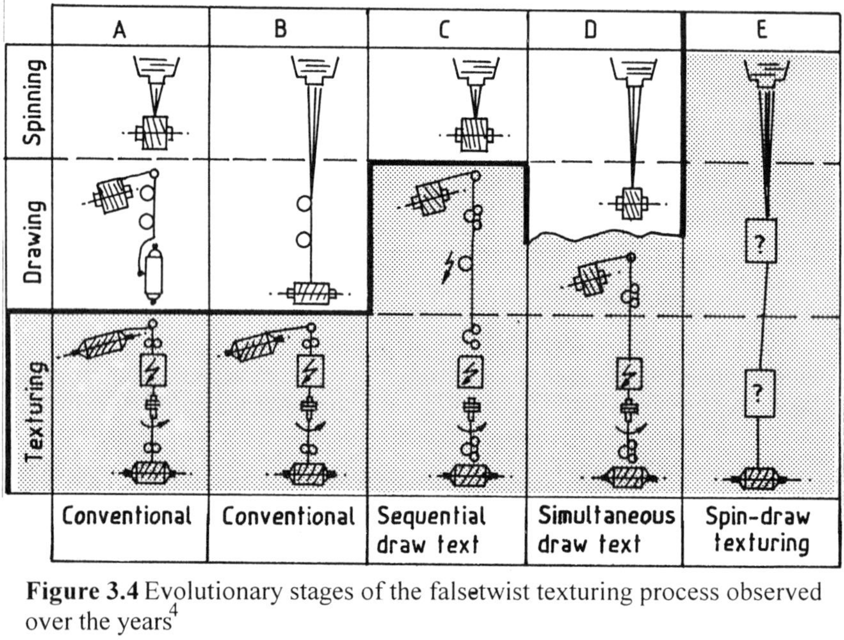

Figure 3.4 Evolutionary stages of the falsetwist texturing process observed over the years[4]

In the batchwise multistep false-twist process, production speeds were 1 to 2 m/min. In the mid-1950s, the continuous false-twist texturing process with pin spindles reached the speed of about 40 m/min. At just under 300 m/min in the early 1970s, this process was taken to the technological limits. Another major development has been successful with the introduction of friction texturing in the early/mid-1970s. The speed of 600 m/min was the commercial production speed by 1975. The current speeds are around 1,200 m/min depending on the type and linear density of the supply yarn.

In the linear density range of 20 and 300 dtex, predominantly for apparel end uses, the false-twist texturing process claims more than 90% of the market.

Improved and also fully or partly automated doffing systems are now becoming common features of texturing machines and a necessity with the high-production speeds.

3.2.2 The Principle of False-Twist Texturing

The false-twist texturing principle makes it possible to join all the operations of the previously described multistep true-twist texturing technique into a continuous process. These operation are twisting, heat setting, and untwisting, and even doubling.

Being the heart of the technique, the false-twist principle deserves explaining. It can be visualized that if a stationary multifilament yarn was to be held at both ends and twisted in the center by a suitable false-twisting device, then equal amounts of twist would be imparted on each side of this device (Figure 3.5a): S twist on one side, and Z on the other. However, if either half of the yarn were considered separately, it would be observed that the twist was real as long as the yarn remains stationary, and the algebraic sum of twist throughout the whole yarn is zero.

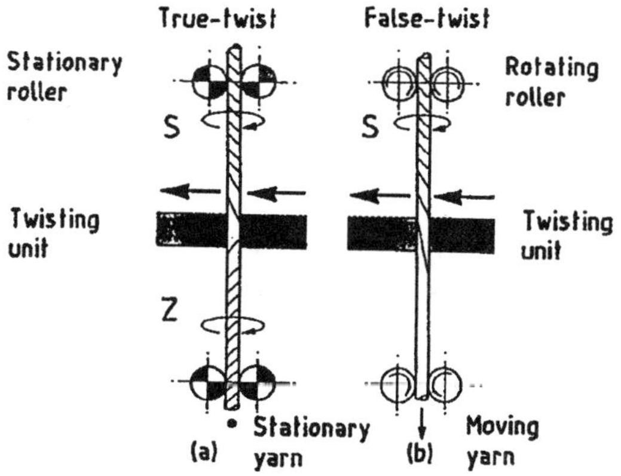

Figure 3.5 Principle of false-twist insertion

With the false-twisting device rotating continuously and with the yarn passing forward, a canceling out of twist on the delivery side occurs (Figure 3.5b). When the system reaches a state of equilibrium, no twist exists after the yarn has passed through the false-twisting device. Thus, the resultant effect is for the yarn to be twisted on the feed-in side of the false-twisting device and to be untwisted on the delivery side.

Thus, a heater placed between the feed roll and the false twister can cause the yarn to be heat set in the twisted state (Figure 3.6). If the yarn is cooled before reaching the false-twisting device and is suitably overfed, the emerging parallel filaments tend to coil or snarl separately and, thus, produce a bulk in the yarn that was not present in the feed yarn.

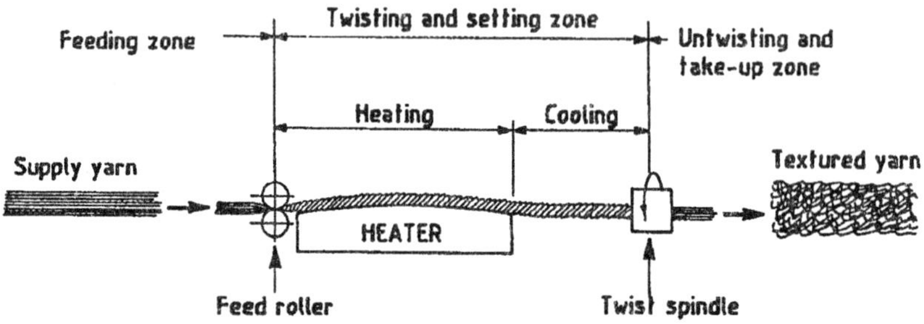

Figure 3.6 Heater placed between feed roller and falsetwister causes yarn to b heat set in twisted state

The stretch of the false-twist textured yarns is created by the action of both snarling and the formation of helices.[5] When a section of textured yarn is observed under suitable magnification, the effect and contribution of these snarls and helices can be appreciated. A twisted straight filament (or yarn) will snarl when allowed to contact. The snarl formed will be either the usual bi-component helical snarl, which looks like a doubled yarn and has its axis perpendicular to the unsnarled filament axis as seen Figure 3.7a, or a single-component helical snarl, which looks like a close-coiled helical spring and has, as its axis, a line parallel to, but possibly slightly displaced from, the unsnarled filament axis as seen in Figure 3.7b. These snarl types will be referred as normal and cylindrical snarls, respectively. It is observed in practice that cylindrical snarl does not occur until the filament is subjected to high twist, depending on the filament's thickness and mechanical properties. Under normal circumstances, the formation of cylindrical snarl is not desired because it causes irregularities in the yarn and may cause damage to the machinery.

Both of these snarl types may easily be observed on monofilaments under a twisting action. However, the case of filaments within a false-twist textured yarn are largely different from that of a monofilament because they are under both torque and bending deformation. Yet the study of monofilament behavior under twist partly simulates the filaments in the core of the false-twist textured yarns and also provides some information regarding the filament behavior and deformation leading to bulk and stretch of textured yarns.

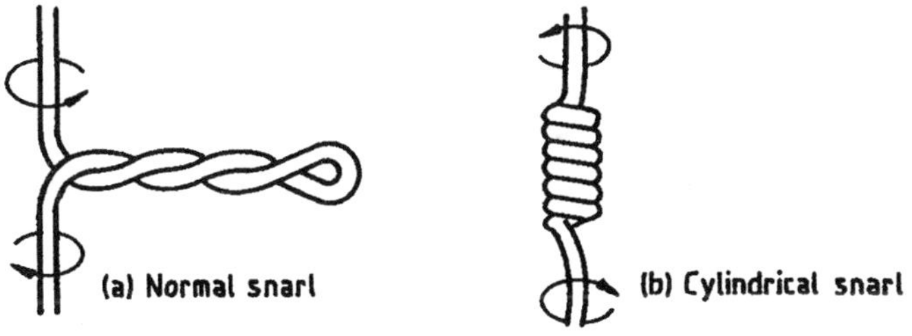

Figure 3.7 Typical snarl formation in a false-twist textured yarn

By calculating and comparing the strain energy remaining in the filament in the case of normal and cylindrical snarl, Hearle and Yegin[5] found that for a given contraction of the filament in cylindrical snarling, the torsional strain energy in the filament is much lower than that in normal snarling; however, the bending strain energy is much higher. At low-twist levels, the bending strain energy, put in the filament by snarling, is the dominant factor. However, at high-twist levels, the decrease in torsional strain energy caused by snarling becomes dominant. At low-twist levels, normal snarling is, therefore, preferred to cylindrical snarling.

3.2.2.1 Means of false twist formation

3.2.2.1.1 *False twisting with pin twisters*

The principle of pin twister depends on the fact that the rotational speed of the twister is equal to that of the yarn passing through. For the friction-type false twisters where the speed unity is no longer valid, however, the slip phenomenon should be considered. This analysis is given below.

The yarn path in a false-twist texturing operation should be considered in three sections, (Figure 3.8).

1. Heat-setting zone before the spindle in which the yarn is in contact with the heater for part of its path
2. Untwisting zone between the spindle and the take-up rollers
3. Take-up zone between the take-up rollers and the take-up package

In the heat-setting zone, twist is inserted into the yarn by rotating spindle at a constant rate (a constant spindle speed and zero slippage is assumed). At the same time, the yarn passes through the spindle. The twist loss from this zone is proportional to the twist insertion rate; therefore, after a short transition, Denton[6] makes the following approximations and develops the following theory;

1. The yarn and spindle speeds are assumed constant, and it is assumed that no twist slippage occurs at the spindle. (It is a known fact that no slip occurs in a pin spindle).
2. The effect of yarn heating is neglected.
3. The contraction of yarn resulting from twisting is neglected.
4. The effects of overfeed or underfeed are neglected.
5. The initial twist is zero throughout.

Symbols

s Spindle speed (rev/s)
v Yarn speed (m/s)
l Length of heat-setting zone (m)
h Length of untwisting zone (m)
k Length of take-up zone (m)
n Twist in the heat-setting zone at a time t (turns/m)
r Twist in the untwisting zone at a time t (turns/m)
q Twist in the take-up zone at a time t (turns/m)
N Final twist in the heat-setting zone (turns/m)

In a time dt, the yarn moves forward by a distance $dx = v \cdot dt$, and the spindle rotates by $s \cdot dt$ revolution.

Heat-Setting Zone

Turns gained by rotation of spindle $= s \cdot dt$
Turns lost through spindle $= n \cdot v \cdot dt$
Net twist gain,

$$dn = \frac{s - n \cdot v}{l} \cdot dt \quad \text{turns/m}$$

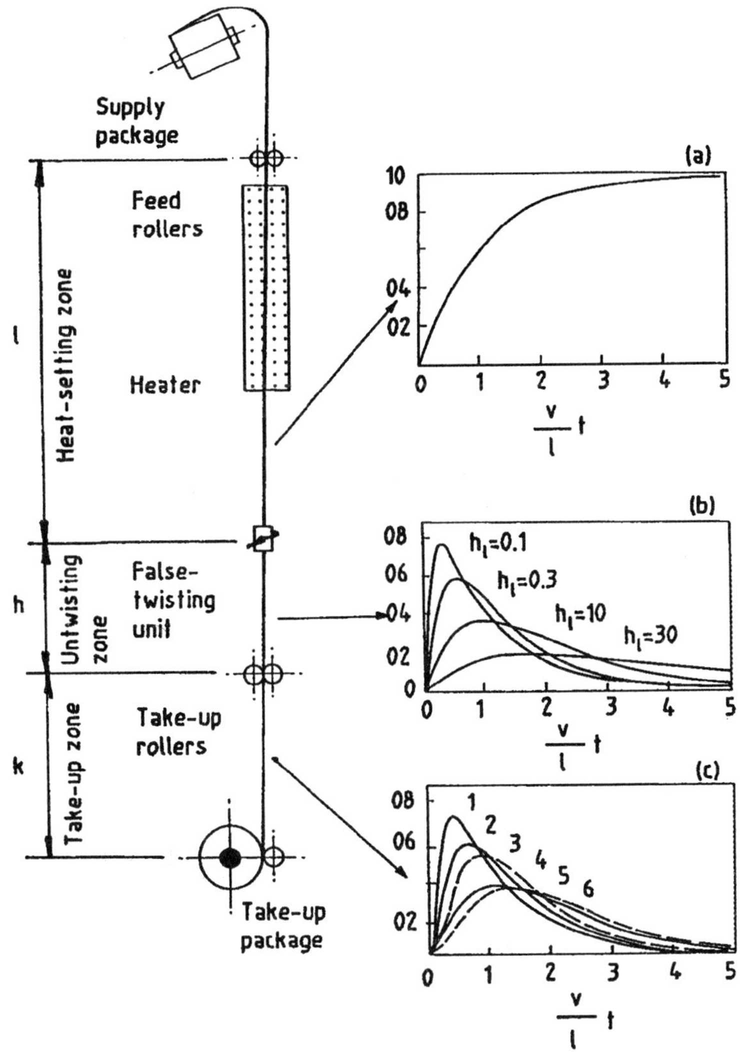

Figure 3.8 Yarn path in a false-twist texturing operation and twist distribution in these zones

Hence,

$$\frac{dn}{s - n \cdot v} = \frac{dt}{l}$$

Integration, together with the initial condition $t = 0$, $n = 0$, gives

$$\ln(1 - \frac{n \cdot v}{s}) = -\frac{v \cdot t}{l}$$

from which

$$n = \frac{s}{v} \cdot \left\{1 - e^{\left(-vt/l\right)}\right\}$$

Now,

$$N = s/v$$

Hence,

$$n = N\left\{1 - e^{\left(-vt/l\right)}\right\} \text{ turns/m}$$

Untwisting Zone

Negative turns inserted by rotation of spindle = $- s \cdot dt$.
Turns passing through spindle from the heat-setting zone

$$n \cdot dx = n \cdot v \cdot dt \text{ turns.}$$

Turns passing through take-up rollers into the take-up zone

$$C = r \cdot dx = r \cdot v \cdot dt \text{ turns.}$$

Net twist gain,

$$dr = \frac{n \cdot v - s - r}{h} \cdot dt \qquad \text{turns/m}$$

$$dr = \frac{1}{h}\left[Nv\left\{1 - e^{-\frac{vt}{l}}\right\} - s - rv\right]dt$$

But

$$N \cdot v = s$$

Hence,

$$dr = \frac{1}{h}\left\{-Nve^{-\frac{vt}{l}} - rv\right\}dt,$$

i.e.,

$$dr + \frac{v}{h} \cdot r \cdot dt = -\frac{v}{h} \cdot N \cdot e^{-\frac{v \cdot t}{l}} \cdot dt$$

Multiplying through by $e^{\frac{vt}{h}}$ and integrating, together with the initial condition $t=0$, $r=0$, give

$$r = -N \cdot \frac{1}{l-h} \cdot \left(e^{-\frac{v \cdot t}{l}} - e^{-\frac{v \cdot t}{h}}\right) \text{ turns/m}$$

Take-up Zone

Turns passing through feed rollers from the untwisting zone $= r \cdot dx = r \cdot v \cdot dt$ turns.

Turns wound onto take-up package $= q \cdot dx = q \cdot v \cdot dt$ turns.

Net twist gain,

$$dq = \frac{r-q}{k} \cdot v \cdot dt \text{ turns/m}$$

Hence,

$$dq = \frac{v}{k} \cdot \left(-N \cdot (l-(l-h)) \cdot \left(e^{-\frac{v \cdot t}{l}} - e^{-\frac{v \cdot t}{h}}\right) - q\right) \cdot dt$$

i.e.,

$$dq + \frac{v}{k} \cdot q \cdot dt = -\left(\frac{v}{k}\right) \cdot N \cdot \left(\frac{1}{l-h}\right) \cdot \left(e^{-\frac{v \cdot t}{l}} - e^{-\frac{v \cdot t}{h}}\right) \cdot dt$$

Multiplying through by $e^{\frac{vt}{k}}$ and integrating, together with the initial conditions $t=0$, $q=0$, give

$$q = -\frac{Nl}{(l-h)(l-k)(h-k)} \left\{ l(h-k)e^{-\frac{vt}{l}} - h(l-k)e^{-\frac{vt}{h}} + k(l-h)e^{-\frac{vt}{k}} \right\}$$

The ratios l/v, h/v, and k/v are the times taken for the yarn to pass through the heat setting, untwisting, and take-up zones, respectively.

Figure 3.8 also shows the twist distribution in the three zones plotted against time on a nondimensional basis. The effect on the twist in the untwisting and take-up zones of the relative lengths of the three zones is also shown. Although the real twist passing onto the take-up package reduces as h/l or k/l increases, the total number of turns wound up is independent of both. As might be expected, this total number of turns is equal to $-N \cdot l$ turns, $N \cdot l$ being the equilibrium number of false-twist turns held in the heat-setting zone[6].

3.2.2.1.2 *False twisting with friction twisters*

In contrast to the form-based twist insertion of the pin false twisters impart false twist to the running yarn by friction. This naturally involves great amounts of slippage and, therefore, makes a mathematical analysis rather difficult and uncertain.

However, because of enormous gear ratio between the friction disk and the yarn, the friction false twister possesses tremendous speed potential compared with the gear ratio of unity of pin false-twisters. Despite the huge amount of possible slippage, the friction twisters offer much higher twist insertion potential and, therefore, found extensive application.

It is now a fact that, except certain specific applications, the friction false-twisters are the dominant twist insertion means in the texturing industry. Therefore, the friction false twisting requires separate and extensive treatment. The mathematical analysis of the various forms of friction false twisting will be given in the sections entitled "false-twist insertion systems".

3.2.2.2 Process

When the yarn path of a typical false-twist texturing machine (Figure 3.8) is observed, the basic requirements of the process are the following.

Multifilament, fully drawn synthetic yarn is first fed into a heater where it is highly twisted. The twist configuration is imparted to the individual filaments by the effect of heat and immediate cooling after the heater. Then the yarn is untwisted to zero-twist level after passing through the twisting element and wound onto package as textured stretch yarn.

3.2.2.3 Heat setting

All solids are held together by a bond of some kind between the molecules. These bonds are loosened by heating; in some cases, as with ice and many metals, the bonds break when a certain temperature is reached (the melting point) at which point the substance changes suddenly from a hard solid to a liquid.

Because synthetic polymers used in textiles have their long-chain molecular structure, the various types of bonds holding the long, chain-like molecules together have a wide range of energies that cause them to break over a wide temperature range. It is for this reason that these materials become progressively softer as their temperature is increased. For example, polyester will begin to lose its almost rigid structure at about 90°C (the glass transition temperature), will become progressively softer and plastic as its temperature rises to about 200°C, and will eventually melt at about 250°C.[8]

The inter-molecular bonds in a thermoplastic fiber are deformed by the strains imposed on the fibers such as twisting, stretching or bending, and quite large restoring forces are generated. Take, for example, the simple process of bending a filament, as schmetically shown in Figure 3.9. Bond deformations produce a restoring resistance in the bent filament that would restore the shape if the filament were released. If the bent filament is held in that shape and heated, however, many of the strained bonds will break, reducing the stress, and will then reform in a new position when the yarn cools, thus stabilizing the yarn in the bent state. This process is known as heat setting. It makes the deformed state permanent by the action of heat.[8]

Heat setting is the most crucial stage of the texturing process. During this process, the yarn has to be heated to softening point, the stresses in the filaments must be allowed to decay, and the yarn permitted to cool before being untwisted at relatively high tension.[9]

3.2.2.4 Filament migration

Conversion of a loose assembly of parallel filaments into a tightly twisted bundle is not unique to the false-twist process. The textile industry has been twisting continuous filaments of man-made fibers since the turn of the century, and silk existed before that time.

Such twist conversion occurs at the entry of new material into a rotating torque-driven segment of yarn. The path of each filament in the twisted yarn is generally helical, but the radius of each helix is observed to vary from position to position along the length of the yarn.[10] This phenomenon is referred to as filament "migration." This migration is not accidental but arises from the requirement that every filament must have the same mean length per unit length along the yarn axis. The studies[11] of the pattern of migration in a variety of structures have shown that this pattern contains both random and periodic (or regular) features, the degree of regularity depending on the type of twisting process, and possibly also on the material[12] and the number of filaments in the yarn.

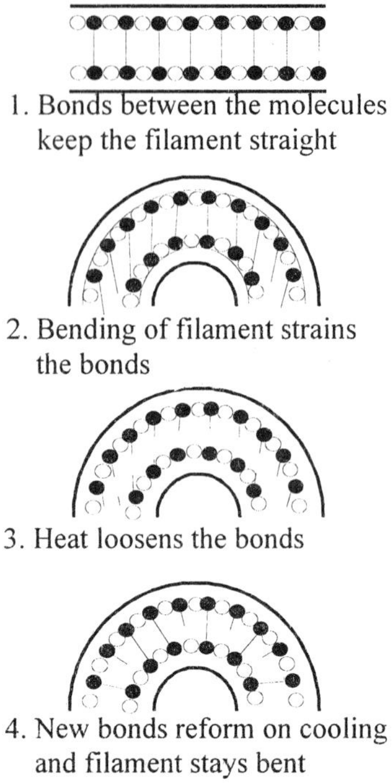

Figure 3.9 Principles of heat setting

Briefly stated, filament migration occurs in response to tension differences among the various filaments - the differences arising from the strain differences inherent in the various path lengths which the inner filaments assume in comparison with the outer filaments. It has been possible to calculate the tension build up on the outer filaments and the tensile drop on the inner filaments - even the development of axial compression and buckling. Under such conditions the outer filaments press inward while the inner filaments are pushed outwards, and periodically the filament shift radially to relieve the resultant unstable stress build ups[10].

A typical filament of a false-twist textured yarn migration is shown in Figure 3.10. The filament migration intensity affects the yarn bulk. It also affects the yarn strength, but there is usually more strength than needed; therefore, there is little need to consider filament migration in this regard as far as most filament yarns are concerned.[13]

By false twisting a 94/28 polyester yarn containing one dyed filament and exposing the yarn to a fluid of similar index of refraction before photographing,

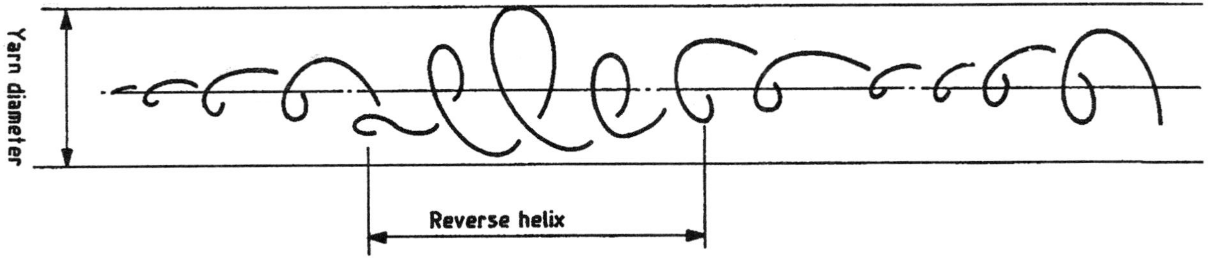

Figure 3.10 Typical migrated filament of a false-twist textured yarn

Backer and Yang[10] have been able to study the migration of the dyed filament on yarn entry into the cold zone. Typical migration traces obtained by these researchers are shown as line drawings in Figure 3.11, where the filament is seen to dwell at the outer radius in one case Figure 3.11a and at an intermediate radius in Figure 3.11c. In Figure 3.11b, the filament migrates from an intermediate position to a central one, and in Figure 3.11d from the outermost radius to the central position[10].

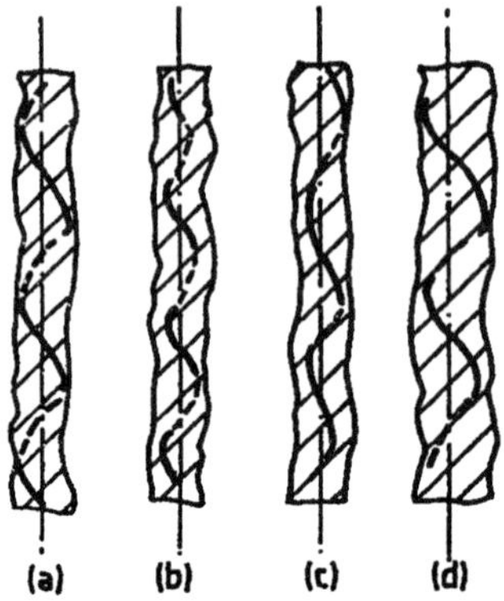

Figure 3.11 Typical migration traces as observed by Backer and Yang[10]

3.2.2.5 The supply yarn

Because the high-level twist insertion and setting of the twist is the essential of the false-twist texturing process, the yarn linear density that is to be textured by this method is limited to lower end of the linear density range that is up to 350 dtex but mostly less than 167 dtex. However, the high-production rates of 1,500 m/min makes the process cover nearly 85% of all textured yarn production.

3.2.2.6 Machinery

Being the heart of the false-twist texturing process, the twist insertion spindles deserve special attention. The progress and present situation of the spindles and other machine components together with overall texturing machinery will be studied in the following sections.

3.2.3 False-Twist Insertion Systems

In parallel with the development of the process in general, the twisting elements have also observed many changes varying from true-twisting untwisting operation to "one rotation per twist" type pin spindles and friction twisters, and then to belt- and air-driven false-twisting elements. The following sections will study different types of false-twist insertion systems in detail.

3.2.3.1 True twisting-untwisting operation

This process consists of inserting a high twist into a continuous filament yarn, setting this high twist by a heating and cooling process, followed by completely untwisting the yarn.

A simple calculation will show that 1 kg of 78 dtex nylon has a length of 128,000 m, and, at a twist level of about 3,200 turns/m that is used in texturing process, would need the insertion of 400 million turns to twist it and a further 400 million turns in the opposite direction to untwist it after the package of twisted yarn had been steamed to heat set the twist. At a spindle speed of 12,000 rev/min this total operation takes 46 days! Nevertheless, this three-step so-called conventional method was used for several years because of the novel and valued properties that it conferred on nylon yarns.[14]

3.2.3.2 Pin spindles

The proceeding slow process was soon replaced by a continuous "false-twist" process in which yarn is passed through a rotating spindle or a similar twisting device that imparts a temporary twist to the yarn traveling toward the spindle and removes as it passes beyond the spindle[14] (Figure 3.12).

The pin spindle causes the yarn to rotate with the same speed as the spindle itself, inserting one twist for each complete revolution; therefore, the yarn twist is determined by the ratio between the spindle and yarn speeds. As long as the ratio is constant, the twist remains constant, and the spindle automatically supplies the torque required to produce the twist.

In the early days of texturing, processing speeds were limited by the spindle speeds that were available. Many pin units are now in operation at about 1,000,000 rpm, and spindles capable of operating at speeds exceeding 1.2 million rpm are also available. These speeds have been achieved by improved design and manufacturing techniques of the units, particularly the driving wheels, by the use of finer spindle tubes and improvements in the design of the magnets that hold the spindle tube in position. Figure 3.13 shows a selection of pin-spindle driving systems, of which magnetic friction-type driving system is the most successful.

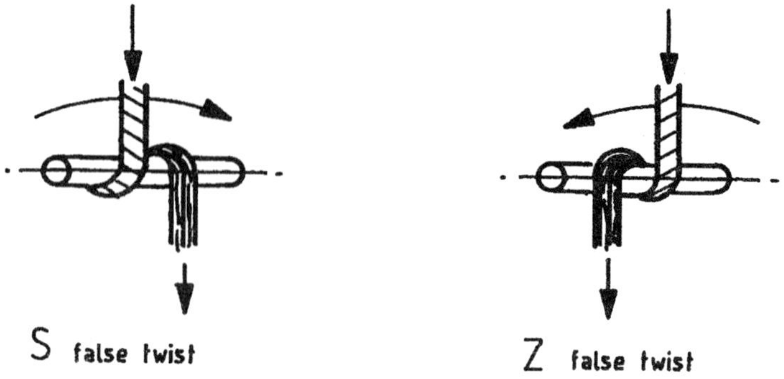

Figure 3.12 False-twist action over a pin spindle

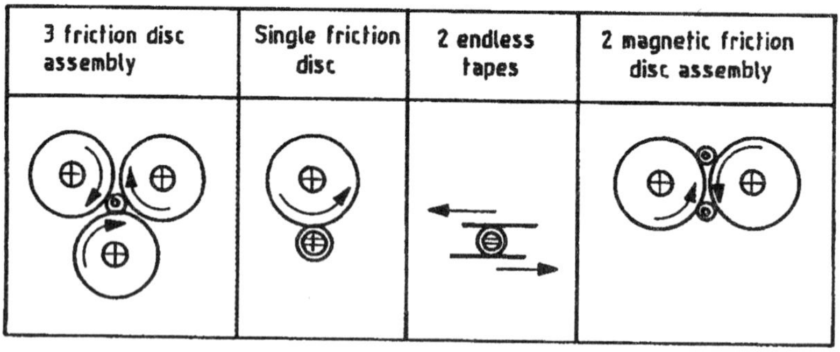

Figure 3.13 Selection of pin-spindle driving systems

In the simplest form, a pin spindle could be a tube with a cross bar over which the yarn is wrapped once and, hence, forced to rotate as it passes through the tube (Figure 3.14). The twist spindle is rotated continuously by an external drive so that both the yarn and the tube are revolved simultaneously.

The late 1960s and early 1970s were the years that pin spindles were then main false-twisting element. Both FAG Kugelfischer Georg Schafer and Co. and Heberlein and Co. AG manufactured similar spindle-drive mechanisms that were capable of driving the pins at approximately 800,000 rpm. In this so called biroll drive mechanism, the small pin spindle is placed in the intersection of two rotating rollers and held there by the magnetic attraction. Figure 3.15 shows the Heberlein's magnetic pin spindle.

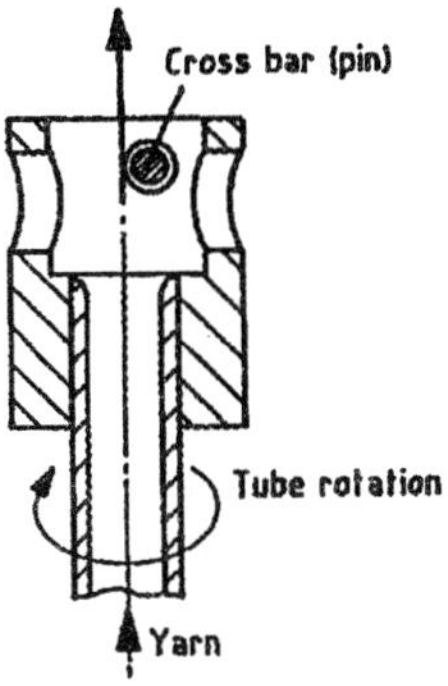

Figure 3.14 Simplest form of a pin spindle

Some of these units were also equipped with individual speed-monitoring probes. A central computer facilitated the speed check of all positions of a 160-head machine in as short as 15 minutes. Monoroll magnetic spindles were also developed and used by ARCT of France on their machines. In this design, the pin is held against the single drive roll again by the magnetic attraction force (Figure 3.16).

Figure 3.17 shows the development of the spindle speeds during a 25-year period. Through such combined efforts in sectors of magnetic bearings, material for the friction roll drives as well as the pin design and arrangement has it been possible to achieve this progress.

Depending on the material to be processed and with full utilisation of the pin spindle speeds, the maximum texturing speeds lie between 280 and 350 m/min, compared with the speeds of 800 to 1500 m/min with friction texturing[15].

From the viewpoint of texturing quality, the pin spindles are superior to any other twist insertion device because they guarantee better uniformity of the textured yarn even in the case of higher variation in tension, denser contraction, and so on. Nevertheless, they severely limit the texturing speeds and make the process uneconomical. These economical reasons ultimately causes the pin spindles to disappear from the most recent textile industry. Out of eight of the most important texturing machine manufacturers today (Barmag, Rieter-Scragg, Murata, Giudici, ICBT, RPR, Teijin-Seiki, Aiki), none uses pin spindles on their latest machines.

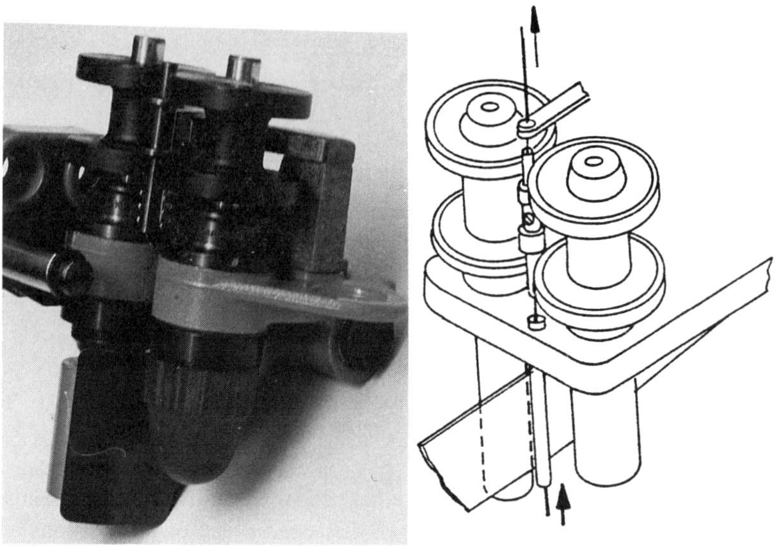

Figure 3.15 Photograph and drawing of Heberlein magnetic spindle

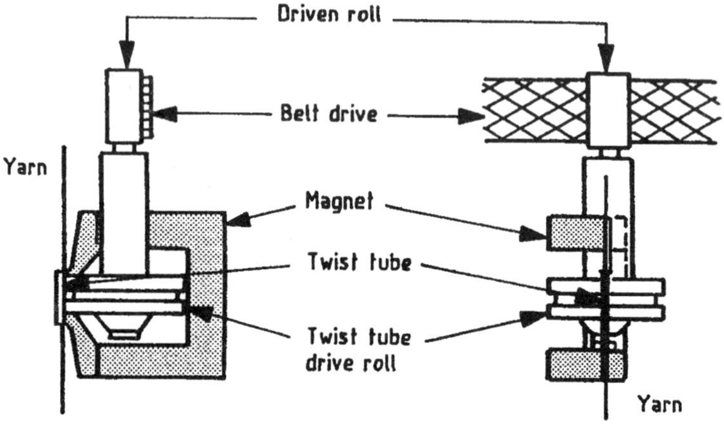

Figure 3.16 Monoroll magnetic false-twist spindle
(pin held against single drive roll by magnetic attraction force)

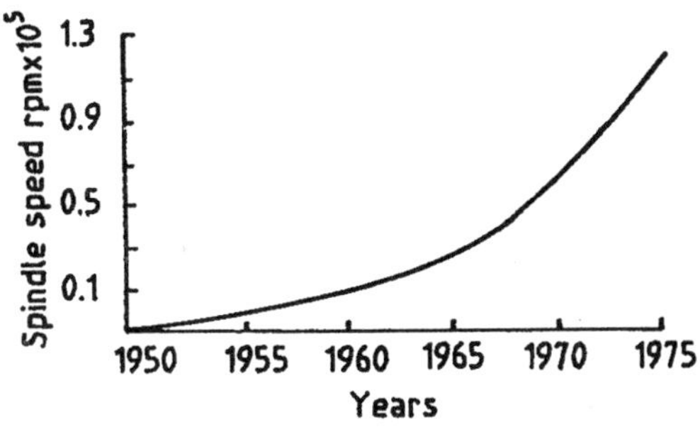

Figure 3.17 Development of spindle speeds during a 25-year period

The theory of pin twister is so simple that it can be summed up in the formula

$$T = \frac{n}{v}$$

where T is the twist, n the rotational speed of the twister, and v the surface speed of the feed rollers. Thus, with a pin twister, the twist can be controlled simply by controlling two parameters. Of these, the surface speed of the feed rollers, v, present no problem. The control of the rotational speed of the twister, n, at the high speeds that are used today is monitored by electronic means. Therefore, maintaining a certain yarn quality pose no problem with the pin twisters.[16]

It is perhaps impossible to list all the development and improvements that were necessary to achieve the above-mentioned high speeds. The three main ones, however, concerned the material used for the driving rollers, the vectorial arrangement of the magnetic field that ensures the stability of the twister, and the eccentric arrangement of the diabolo pin.[16]

Regarding the driving rollers, it must be remembered that, at 1,000,000 rpm, they are subjected to very centrifugal forces, corresponding to approximately 35,000 times the gravitational accelerating force. To withstand such forces, special materials had to be developed whose main characteristics were a low coefficient of thermal expansion and a high degree of uniformity regarding density and shore hardness.[16]

In the arrangement of the magnetic field that maintains the pressure between the twister and the driving rollers and counteracts the drag of the yarn tension, the main problem was to overcome the effect, because of Coulomb's law, whereby the attractive force diminishes with increasing distance from the magnet. The solution to this problem was found in the vectorial system illustrated in Figure 3.18. This diagram shows that the attractive forces between the rotor of the twister and the two magnetic poles diminish as the twister moves away from the driving rollers. Nevertheless, the resultant restoring force increases with increasing distance; thus, the whole system operates like a spring, in which the restoring force increases with increasing displacement. The system is, therefore, self-stabilizing[16].

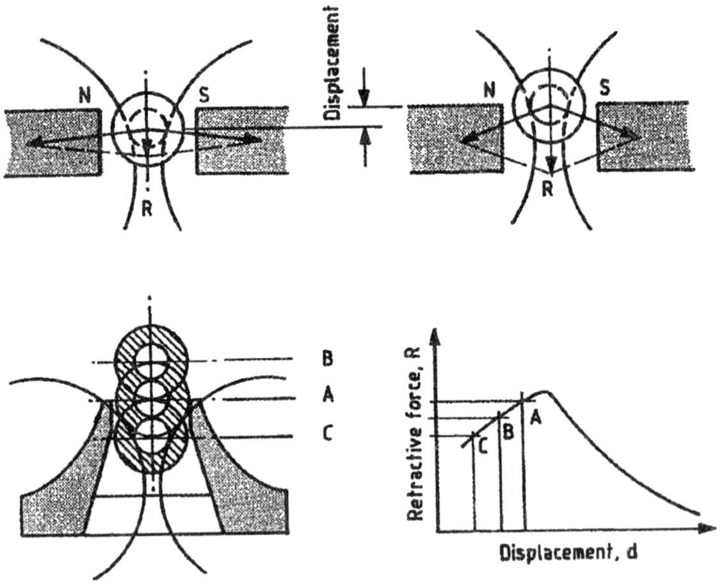

Figure 3.18 Attractive forces between rotor of twister and two magnetic poles diminish as twister moves away

Figure 3.19 shows a front view of the system; here, too, the system works as self-stabilizing using the vectorial system. As the yarn tends to drag the twister in its running direction, the restoring force is increased and ensures the stability of the twister in position.[16]

Regarding the position of the diabolo pin, in the conventional arrangement, the pin is arranged symmetrically in relation to the axis of the twister, and the yarn, therefore, enters the twister eccentrically. Because this arrangement leads to excessive ballooning of the yarn at high rotational speeds, the position of the diabolo pin is changed as shown in Figure 3.20, which enables the yarn to enter the twister along its axis and thus reduces the ballooning tendency.[16]

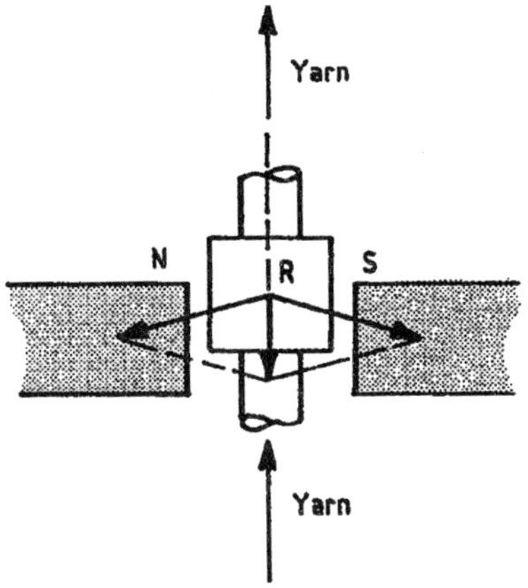

Figure 3.19 Front view of the magnet/pin system

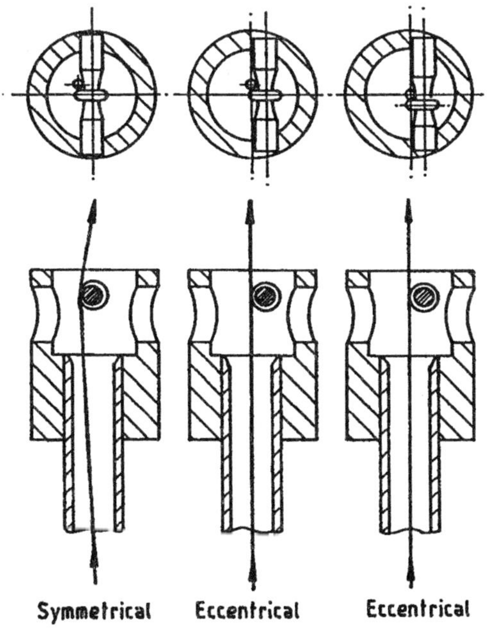

Figure 3.20 Eccentrically positioned pin enables yarn to enter twister along its axis and reduces ballooning tendency

At the peak of the technology, a speed of 1.25 million rpm was reached with pin spindles. To achieve this high speed Murata has departed from all previous technology and have designed a small air-turbine running in air bearings with the pin spindle also forming the shaft of the turbine. This step is in itself relatively easy, but the difficult stage is the control of the spindle speed at the desired value. This has been done by measuring the speed of each spindle continuously and applying closed-loop automatic control to each individual spindle. In this way it is claimed that spindle speed is controlled to within ±0.5%.[17]

3.2.3.3 Friction twisters

Friction twisting is not new to the texturing processes. It has been used at least as far back as the last century to provide temporary twist and, thus, temporary cohesion during delicate stages of staple yarn manufacture. However, it was first applied to false-twist texturing after the discovery of strong thermoplastic continuous filament yarns.

The principle of imparting twist by means of friction, i.e., the rolling of the yarn over a friction surface, eliminates rotating machine elements, such as, pin spindles, which would have to rotate at the same speed as the yarn thus the limiting speed of pin spindles are now far exceeded by modern friction twisters. The friction application is basically achieved by two means: being internal friction twisters and external friction twisters. Figure 3.21 systematically shows some typical examples of these friction twisters. In general, yarn is passed around the inside surface of a bush or the outside surface of a disk, or more usually a series of bushes or discs. In most cases, the tension in the yarn pulls it

against the high-friction surface of the bush or disk, and motion of the friction surface causes a torque to be produced in the yarn, which, thus, produces twist

In all cases, the yarn has to slide along the friction surface to move through the system, and the value of yarn-to-surface friction influences the level of twist developed in the yarn and the tension build-up in the system.

Before moving on to further details about different types of friction twisters, a comparison between the pin spindles and friction twister is of value. Broadly speaking, the pin spindle is a twist generator whereas a friction twister is a torque generator. With pin spindle, the amount of twist imparted can be mathematically determined and preset by machine adjustment, whereas the friction twisting involves certain slip; therefore, no simple mathematical relation could be established.

Because the friction twisters are capable of inserting more than one twist for one revolution of the twist head, friction twisting results in a high rate of twist

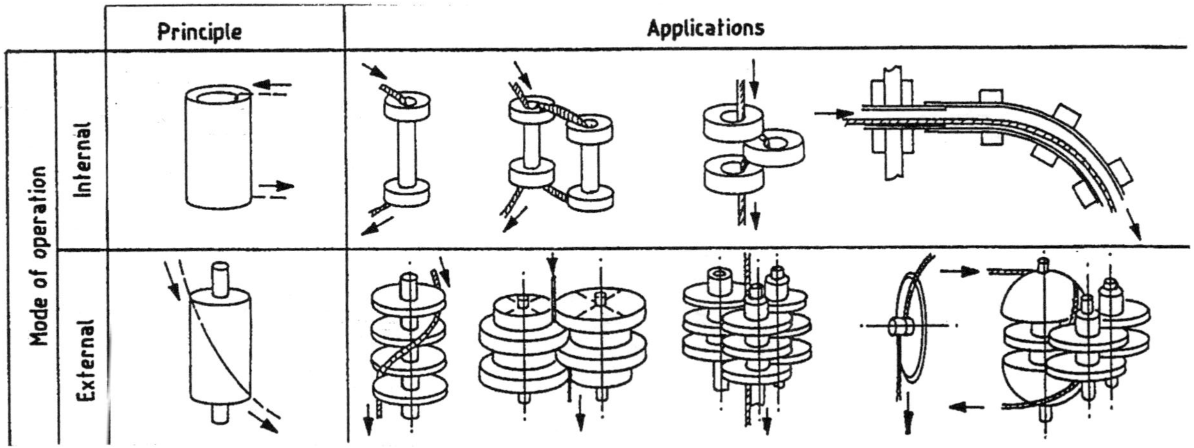

Figure 3.21 Some typical examples of friction twisters

insertion for a relatively slow speed of the friction twisting elements; with friction twisters, it is also possible to minimize the tension increase in the yarn path. By adjusting the contact angle between the yarn friction surface, the tension ratio across the friction twisting unit could be made equal to or less than unity. In contrast to pin spindle, in which the available torque is generally much greater than that needed to twist the yarn, the level of twist attained by the yarn in friction twisting is limited by the torque that is available and the torsional rigidity of the yarn.

To generate the friction forces to produce a high-twist level, a high value of the product of the total angle of wrap of the yarn over the friction surfaces and the coefficient of yarn-to-surface friction is necessary. This combination, however, may cause a build-up of tension in the yarn and, thus, high exit tension, which, in turn, may damage the yarn and reduce its bulk. It can further lead to instability (surging) of the false-twist threadline.

The following relationships are valid for all friction twisters:

1. Magnitude of force at the yarn surface depends on yarn-to-surface friction, and its direction depends on relative speed.
2. The values of the twist and of the yarn tension ratio depend on the yarn friction, input tension, and relative speed.
3. The yarn is pulled along its axis over the friction surface and must slip in this direction of motion of the friction surface to obtain any twisting forces.
4. To achieve stability, use of relatively high levels of slip seems to be advantageous.

3.2.3.3.1 *Internal friction twister (bush twister)*

This is the earliest form of friction twister to achieve any success. The first texturing machine (the Hobourn/BNS machine) that used internal or bush friction twisters was introduced in mid-1950s. An internal friction twister consists of a rotating metal tube fitted with a lining or bush of resilient material, usually rubber. As seen in Figure 3.22, the yarn passes through the tube in contact with the bushes and either leaves from the same side of the axis or leaves after crossing to the other side. Some fine yarn passes, in turn, through more than one such tube.

Because the contact between the yarn and rotating surface of the twister is a frictional contact, it is reported[2] that no adjustment need be made for a change of yarn linear density, because the method automatically provides a constant twist factor. The advantage of the internal-type friction twister over the pin twister is that it is only one-quarter as sensitive to yarn and twist head-speed changes as a pin false twister.[2]

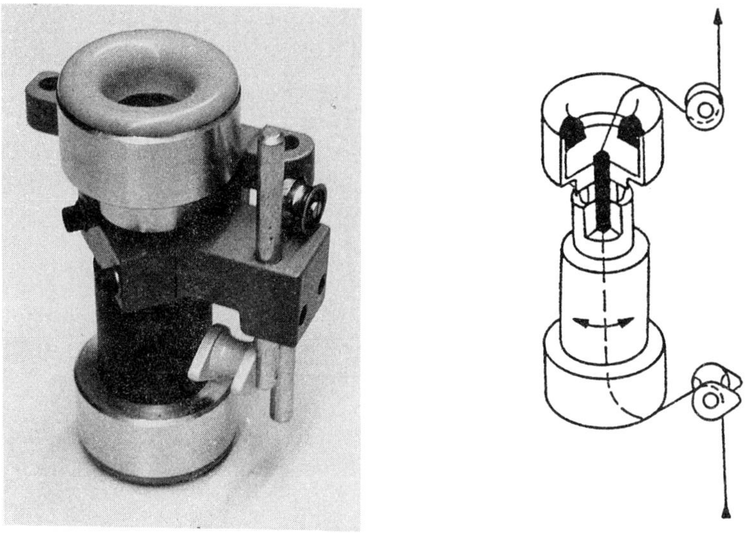

Figure 3.22 Internal friction twister (bush twister)

Quick wear of the friction surfaces and consequent problems in texturing resulted in the disappearance of such twisting elements. Although they were popular in the early 1970s, they are hardly used today.

3.2.3.3.2 *External friction twisters*

The advances in false-twist technology in the 1970s were undoubtedly due to general adoption of external friction systems. They also followed a gradual development and appeared to reach a final design with multistacked friction disk spindles. This development involved the following steps:

Two spindle friction disc twister. This type of twister imparts twist to the yarn by contact with opposite rotating rollers. In one of the most original designs, only one roller pair was used, but the number of pairs was later increased to three as seen on Figure 3.23. One member of each pair is a low-friction metal roller called an anvil running in contact with a high-friction polyurethane or nitrile rubber disk. Both driving shafts rotate in the same direction and turn the disks, which, in turn, drive freely rotating anvils.

The rotational drags provided by the two types of contact surface are in different directions, but because the coefficients of friction are different, the yarn is driven predominantly by the high-friction disks.

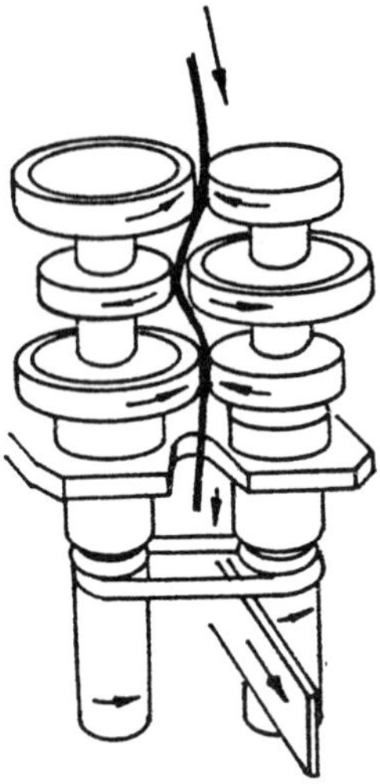

Figure 3.23 Two-spindle friction disk twister

Tension keeps the yarn close to the roller-pair intersection and is said to eliminate the rapid wear associated with bush twisters where the yarn can cut the surface by traveling round with it a little way before sliding back to the mean position. A further advantage is that the geometry supposedly results in a greater force between yarn and friction surface than would follow from yarn tension and wrap angle alone.

Despite many advantages over bush twisters, the two spindle friction disk-type false twister does not seem to enjoy much success in industry and is a part of history today.

False-twisting over a sphere. This device basically consists of a sphere having an equatorial groove into which a rotating disk is projected. As seen in Figure 3.24, the yarn path approximates to a meridian of the sphere. The yarn is "wedged" between the sphere and the disk so that a mechanical advantage is gained that makes it possible to generate a high torque (twist) with a comparatively low-yarn tension.

One of the distinguishing features of the twisting over a sphere is the yarn path, which is such that the yarn first makes contact with the sphere at its pole, where the circumferential speed is zero. The yarn then follows roughly the line of a meridian until it reaches the equator, i.e., the slot that contains the disck Between the pole and the equator, the yarn passes through a region of increasing circumferential speed of the friction surface, whereas its own circumferential speed virtually remains constant. The twister surface, therefore, initially lags behind the yarn surface (negative slip), and later it runs ahead of the yarn (positive slip).

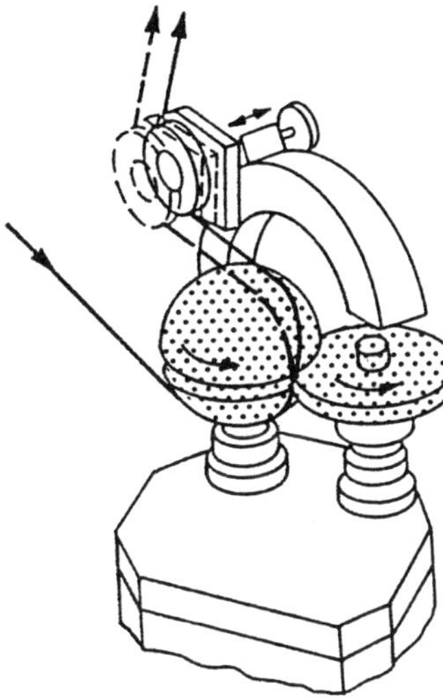

Figure 3.24 False twisting over a sphere

At the change-over from negative to positive slip, the yarn passes through a point of zero slip. Clearly, the twisting of the yarn can occur in only the region of positive slip, but the existence of a change-over point to negative slip has a self regulating influence on the system. If, for instance, the yarn twist diminishes, the point of zero slip moves toward the pole and, thus, increases the region of positive slip. This increases the twisting action and, therefore, compensates for the reduction in twist.[18]

A reasonable success has been achieved by this type of twisting device because of relatively easy yarn inspection and threading up of the twister. In practice, it has also been found possible to texture yarns with a wide range of linear densities without change in the twister setting. However, uncontrollable yarn path over the sphere results in variation in twist insertion and, hence, leads to poor texturing quality.

False-twisting over a rotating cylinder. For production of low-twist yarns (reduced number of turns with a high level of elasticity) this type of friction twisting elements have been developed by Rieter-Scragg[19]. As seen in Figure

3.25, it consists of one single rotating "cylinder" around which the yarn is wrapped. By adjusting the height of a yarn guide at the exit end, different wrapping angles are obtained between the cylinder axis and the yarn. The wrap angle determines the degree of twist insertion[19].

The fine denier polyamide torque yarns for the sheer look in ladies' hosiery are conventionally produced by the pin-twisting method at low speeds. The availability of this "Hitorq" unit gives the potential for yarn production at speeds up to 1,000 m/min.[20]

Triple-stack multidisk spindles. The final success came through with the introduction of triple-stack multidisk spindles in 1960s and 1970s. An Ernest Scragg and Sons Ltd. patent, British Patent BP1.085.052, filed on 30 April 1963 describes such a device, and Barmag exhibited first design in ITMA 71. Later, other machine manufacturers have also designed their version of similar twisters. The apparent success of this type is due to the contact angle between yarn and friction surface, which provides velocity components as required in both the forwarding and rotational directions.

Figure 3.26 shows a typical design of such twisting units. As seen from this figure, these units comprise three stacks of overlapping disks mounted in an equilateral triangular configuration. The yarn angle on the friction disk surface is carefully controlled so that the velocity vectors impart the desired twisting

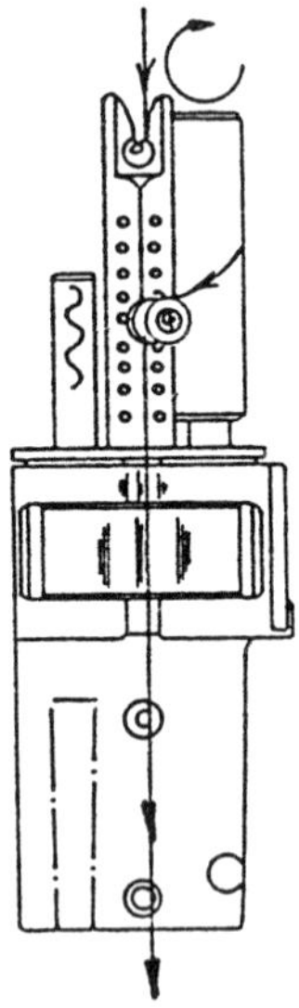

Figure 3.25 False twisting over a rotating cylinder

orientation of the and yarn-forwarding actions. Each disk acts as a guide in controlling the yarn on adjacent disks, and the lack of this control on the first and last disks are sometimes corrected by additional smooth-edged disks acting as guides but not as driving surfaces because the yarn may not have the required orientation over them.

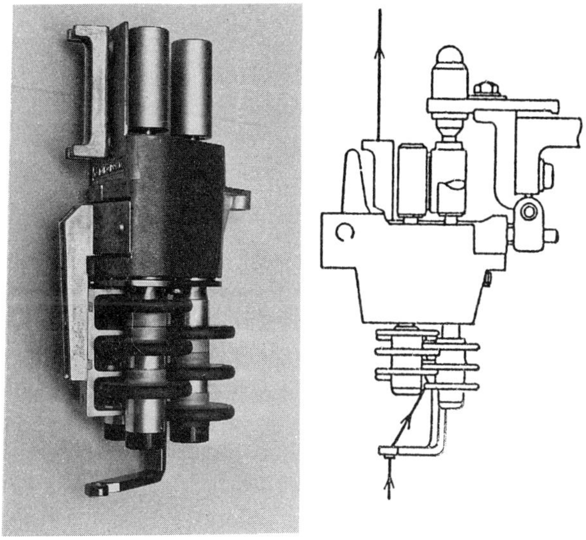

Figure 3.26 Triple-stack multidisk spindles

The multidisk unit has the further advantage that it is easier to provide a multiplicity of friction elements than it is with bush systems. Hence, the total angle of wrap through the system can be made large. At first sight, the "gearing ratio" looks extremely high, but it must be appreciated that there is considerable slippage between the disk and the surface of the yarn.[3]

If the yarn had to be pulled through the system, the tension build-up would be extremely high, but the forwarding action of the device eliminates such difficulties while still maintaining the high lateral forces on the yarn that enable high-twist values to be attained. For example 167 dtex polyester yarn could be twisted to a twist level of 3,200 turns/m, whereas the output tension is little more that that of the input tension.

Development work on disc profile, surface structure, finish and configuration, has resulted in units capable of operating at speeds up to 1500 m/min. Typical speeds used in practice are, however, 1,000 m/min for 78 dtex nylon and 750 m/min for 167 dtex polyester. And nearly all of the modern texturing machines use this type of twisting element of varying designs. Currently, the mostly accepted types of disks may be plasma coated, diamond coated, solid ceramic, or polyurethane. Changing the disk overlap is the primary means of changing the angle of attack, and changing the number of disks then changes the angle of wrap. These all provide twist and tension control to the twisting unit.

3.2.3.3.3 *Rotating curved tube*

A patent claimed in 1974 described a method of false twisting by passing yarn through a curved rotating tube constrained to be in a fixed plane. The rotating member resemble a flexible hose in that it consists of a helical spring with an elastomeric lining (Figure 3.27). This design can be regarded as the anticlastic analogue of the synclastic spherical twisting elements.

Although it may appear to have many advantages such as equal tension at input and output zones, and higher twist potential over other twisting units, it did not find any commercial exploitation and appears to remain in the patent literature.

Figure 3.27 False-twist generation over a rotating curved tube

3.2.3.4 Nip twisters

One apparent disadvantage of stacked-disk friction-twisting units is their rather difficult twist flexibility to change the twist level, i.e., the amount of twist to be inserted, the only way to achieve is to change the disk-stacking configuration. This difficulty is overcome by the nip- twisting systems (belt or ring), which can apparently provide a good twist-level flexibility. By changing the crossing angle of belt twisting or amount of overlapping of ring twisting, the twist level can easily be adjusted.

3.2.3.4.1 *Crossed-belt twisters*

The belt-twisting principle has been used in the textile industry for many years, and it is also applied to textured yarn production.[22] The most recent technological aspect of belt twisting is the use of two crossed belts to twist the yarn running at the center of the contact area of the two belts. This principle has received a great deal of attention because of claims that it is capable of producing uniformly textured yarns at production speeds higher than the stacked disk spindles.[22]

A belt-type nip twister consists of two endless belts that are placed against each other with a certain crossing angle and are driven, respectively, while yarn is nipped between the belts and false twisted (Figure 3.28). The twisting principle is simple. The belts move in such a way as not only to induce twist in the yarn by frictional contact with the surface of the yarn but also to "pump" the yarn through the system. This system improves the twist control compared with the stacked disk, but the belts tend to have a rather short life. With this type of twister, inner twist surface is particularly critical and is perhaps the major parameter that hinders the growth of this twister unit. Although the manufacturer claims that the surface rotational speed of the twisted yarn is the same as the speed of the twisting belt, i.e., no slip, the industrial experience through quick wear of the belts indicate that some slip occurs between the yarn and belt surfaces.

Figure 3.28 Crossed-belt false twisters

Conventional friction disk spindles are approaching their speed limitation of 1,000 m/min for high-inear density yarns. However, these high speeds adversely affect the quality of the textured yarn and its further processing suitability of dyeability. The belt twister is supposed to be more controllable than stacked disk spindles because of its increased frictional forces which is created by belt pressure and these forces are independent of yarn tension. In stacked disk spindles, however, friction is generated by yarn tension and curvature of the disks[22].

A further development in the belt twisting has been the possibility of twisting two yarns side by side by one unit (designated as Twintex by Murata), hence, doubling its capacity. To accommodate the yarns, Twintex uses one broad and one normal-width belt (Figure 3.29). It is pointed out that many texturizers often take up two ends of yarn from two separate texturing spindles onto one package via, for example, air entanglement. One half of the winders of the texturing machine are then not used. One possible advantage of the Twintex system is then by doubling the spindle usage, all the take-up winders can be used in such a situation.[23]

Figure 3.29 Twin-yarn crossed-belt false twisters

Despite all the good features of the belt twisting, it has little acceptance from the modern industry mainly because of the high-wear problems of belts and complicated driving mechanisms.

Theory of crossed belt false-twisting. Kang and El-Shiekh[22] developed a theoretical analysis regarding the relationships between the primary variables of the belt-twisting system. As the yarn leaves the belts, it changes from a twisted to an untwisted state. Therefore, the yarn speed increases because of the removal of its twist. The uncertainty of surface geometry and characteristics of the belts and the yarn make the theoretical investigation complicated. Consequently, the researchers[22] had to make the following assumptions to obtain a reasonably uncomplicated analysis:

1. The upper and the lower belts have the same physical properties.
2. The yarn path remains steady and straight, i.e., the yarn path between the two belts does not change with time.
3. The yarn maintains continuous contact with the two belt surfaces.
4. The changes in the yarn resulting from the frictional forces and changes in yarn diameter during untwisting are neglected.
5. The yarn bending moments, shear forces, and yarn inertia forces are negligible.
6. The yarn and belts have uniform properties, such as surface characteristics.
7. Belt pressure is constant along the yarn in the contact area, i.e., the normal forces are constant and uniform.
8. The frictional force is equal to the normal force time of a coefficient of friction. This coefficient of friction is independent of variables, such as belt pressure, velocity, and temperature.
9. The yarn is slipping over the belt surface everywhere or slipping nowhere. (Experiments show that the system is in the slipping configuration rather than the gripped mode.)

Under the proceeding assumptions and considering the practical applications, i.e., the yarn passes through the center of the two similar belts moving at the same speed, the analysis is simplified, and we have $\phi = \theta / 2$, $V_u = V_B$, and, hence, the slippage factors $\delta_1 = \delta_2 = \delta$. Using these simplifications, Kang and El-Shiekh[22] reached the following equations:

$$\frac{T_{out}}{T_{in}} = 1 + \frac{2\,\mu NS}{T_{in}} \cos\delta$$

$$\frac{Q_{out}}{aQ_{in}} = \left(1 - \frac{T_{in}}{T_{out}}\right) \tan\delta$$

$$\frac{Q_{out}}{aT_{in}} = \left(\frac{T_{out}}{T_{in}} - 1\right) \tan\delta$$

where

T_{out} = yarn tension after belt twister
T_{in} = yarn tension before belt twister
Q_{out} = torque created in the yarn
μ = coefficient of friction between the belts and yarn

N = normal force between belt and yarn
a = yarn radius
S = length of yarn path on belts
δ = slippage factor, or the relative speed between the surface of the yarn and the surface of the belt.

These equations show the relationships between the primary variables of the system. The overall performance of the belt twister is determined from the values of the tension ratio and torque. Figure 3.30 shows the change of tension ratio and dimensionless torque with belt-speed to yarn-speed ratios. The system is optimum when the tension ratio is equal to unity, i.e., the yarn is neither dragged nor pumped by the twister at that condition. In this condition, the forces act perpendicular to the yarn axis and exert only pure torque. As the surface of the belts reaches a certain velocity at a certain belt angle, the friction force may have a component in the direction of the yarn movement. When this occurs, the belt will pump the yarn. Because the whole dragging force can be directed perpendicularly to the yarn path, the belt twister can be operated with pure torque and no tension change as shown in Figure 3.30. The operating belt-speed / yarn-speed (B/Y) range for setting the same level of twist in the yarn is lower than that of stacked disk spindles because of the higher torque that can be generated in the belt-twisting system.[22]

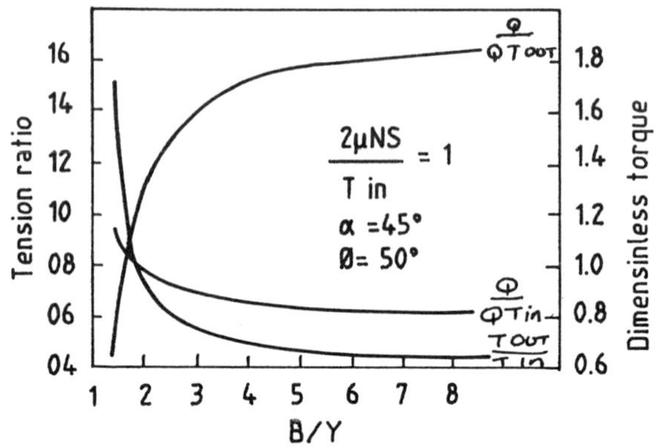

Figure 3.30 Change of tension ratio and dimensionless torque with belt-speed to yarn-speed ratios

The tension ratio changes greatly with belt angles as well as the ratio of the velocities, and it can vary from a drag to a pump mode. For the system to operate, the belt angle must be greater than twice the yarn helix angle. Thus, the belt angle primarily determines the level of twist in the yarn.[22]

The forwarding frictional force generated by the belt twister in the slipping mode can be calculated for the tension difference in the threadline as

$$F_{forward} = T_{in} - T_{out} = 2 \cdot \mu \cdot N \cdot Cos\delta$$

and the twisting torque in threadline can be calculated from the rotational force as follows:

$$F_{rotational} = 2 \cdot \mu \cdot N \cdot S \cdot Sin\delta,$$

$$Q = a \cdot F_{rotational} = 2 \cdot \mu \cdot N \cdot S \cdot a \cdot Sin\delta,$$

where δ is the slippage factor in the system. Thus, the total frictional force generated by the belt twister is given by

$$F = \sqrt{(F_{forward}{}^2 + F_{rotational}{}^2)}$$

When T_{in} equals T_{out}, the total slippage is only in the rotational direction, and δ approaches 90°. Thus, the frictional force acts in only the rotational direction of twisting; at this condition, the system applies pure torque. In this case, the system will run under no drag and no pumping conditions.[22]

The frictional coefficient between yarn and belt is an important factor that affects the twisting efficiency of this system. It depends on the surface characteristics of both yarn and belt. The spin finish of the yarn is important because it will change the surface characteristics of the yarn and its interaction with the belt.[22]

The desirable speed ratio (B/Y ratio), which determines the optimum condition of the system, i.e., $T_{out}/T_{in} = 1$, depends on the belt crossing angle (θ). The tension ratio becomes unity when the ratio of speeds equals $(\cos(\theta/2))^{-1}$, i.e., B/Y = $1/\cos(\theta/2)$. Thus, the optimum speed ratio increases with twister angle θ as shown in Figure 3.31. If we consider the system running at its optimum condition, i.e., no drag and no pumping conditions, then the changes in the belt-crossing angle θ cause changes in yarn-twist and tension ratios.[22]

The operating modes, therefore, play an important role in determining the performance of the belt-twisting system. Kang and El-Shiekh[22] found that certain conditions under which the system can be operated in a fully gripped state exist. Otherwise, there is always a certain amount of slippage because of the high level of tension and torque in the threadline. The total frictional force generated by the belt twister can be calculated from the tension difference in the threadline and slippage factor in the slipping mode.[22]

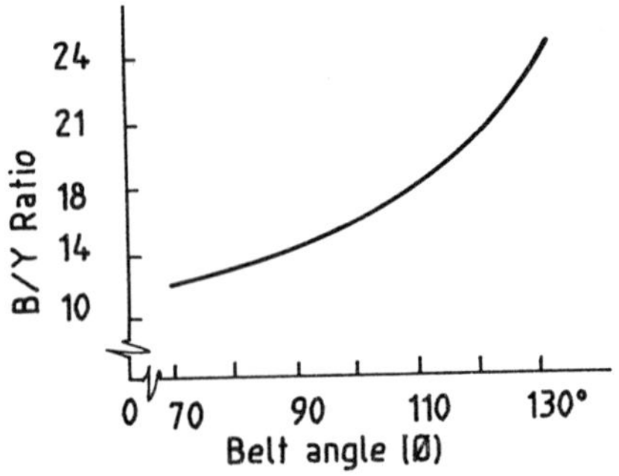

Figure 3.31 B/Y ratio against the belt-crossing angle

The primary variables in the belt-twisting mechanism are belt pressure, B/Y ratio, belt angle, and yarn and belt characteristics. These factors determine the overall performance of the belt twister. There is a unique value of B/Y ratio to give equal tensions across the belt at a certain belt angle. The belt angle is the predominant factor affecting the optimum B/Y ratio and the twist level of the yarn.[22] The effects of change in belt angle on B/Y ratio, twist inserted, and bulk level of 167f34 polyester yarn is given in Figure 3.32.

3.2.3.4.2 *Ring twisters*

In another type of nip twister, the yarn is nipped between two overlapping rings, one of which is flexible and the other is supported by a rigid steel ring (Figure 3.33). The yarn is driven by the circumferential speed of the rings in the nip point. By changing the distance between the center of the rings, the twist level and speed can be adjusted.

Because this type of twisting is comparatively new, its full potential has yet to be established. However, the major manufacturer of this type, i.e., Barmag, claims that the ring texturing units are superior to other types of twisting methods in that it treats the yarn gently; thus, no filament damage is caused. The production speed could also be increased by about 100 m/min over the stacked disk spindle without loosing any yarn quality.

In the ring twisters, the two rub-surface rings overlap some area with each other and run in opposite directions about their own axes. External pressure is applied to the rib of one ring at the yarn contact zone so that the threadline makes contact on both ring surfaces. This normally applied pressure is the prime source of frictional force needed to twist and forward the yarn movement. The crossing angle of the two rings, the external piston pressure, and the speed ratio ring speed / yarn speed (R/Y ratio) are the determining factors affecting the twist and tension levels of the threadline. The main principle and mechanism of

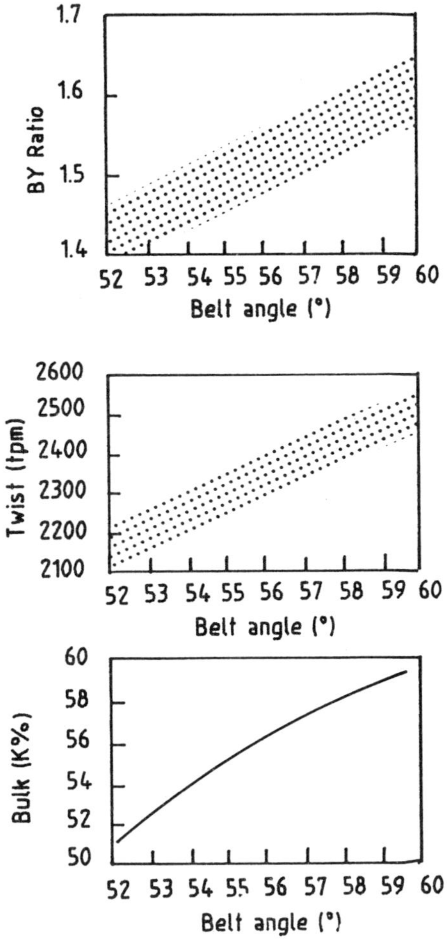

Figure 3.32 Effects of change in belt angle on B/Y ratio, twist inserted, and bulk level of 167f34 polyester yarn

the ring twister is similar to the belt twisters in many aspects, but there are some differences in the operating modes of these two systems because of differences in the geometry of the two systems.[22]

The essential difference between the stacked disk spindle and the ring twisters is that the disk spindle controls the torque and tension level of the yarn by controlling the slippage, whereas the ring spindle controls the threadline torque by adjusting the externally applied pressure to the yarn. The difference affects many practical aspects of friction false-twist texturing. The twister-to-yarn speed ratio (B/Y ratio for belt twister and R/Y ratio for ring twisters) can be reduced to as low as 1.4 from 2.0 of disk circumferential speed / yarn speed (D/Y) ratio of the stacked disk spindles and still give the threadline the same level of twist in the texturing zone.[22]

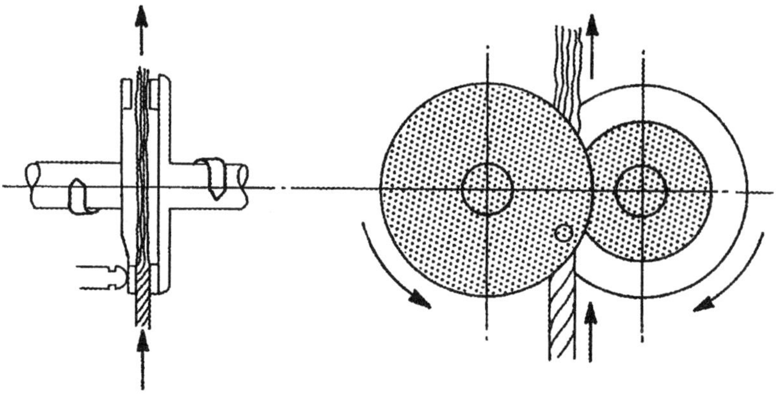

Figure 3.33 Ring false twisters

Theory of ring false twisting. The geometry of the ring twisters as shown in Figure 3.34 dictates that the yarn crossing angles, i.e., the yarn-to-ring contact angle (ϕ), can be changed by adjusting the center-to-center distance of the two rings. The effective contact length S also varies to a small extent with the changes in the ring crossing angle. Similar to the belt-twisting spindles, the effective contact length decreases with the increasing crossing angle; as a result, the frictional force available in the twister will be reduced.[22]

Adopting the same simplifying assumptions to the belt twisting, Kang and El-Shiekh[22] developed the equations regarding the tension variation (T_{in} - T_{out}) and torque generation (Q_{in} - Q_{out}) change across the ring twister as follows:

$$\Delta T = 2\mu NScos\delta$$
$$\Delta Q = 2a\mu NSsin\delta$$

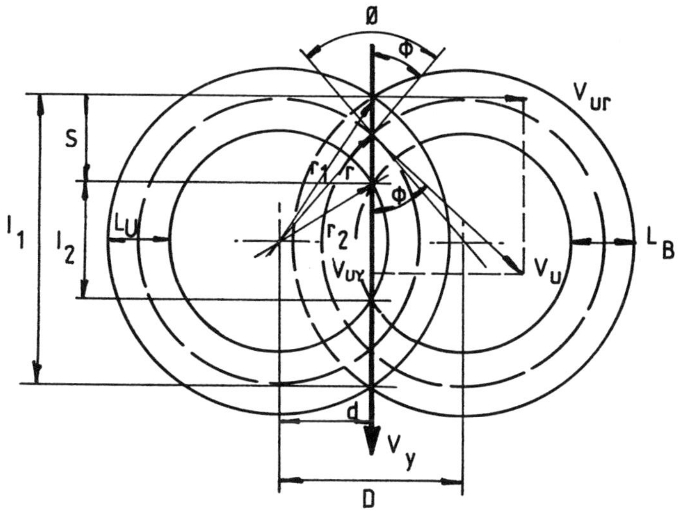

Figure 3.34 Geometry of ring twisters

where
μ = coefficient of friction
N = externally applied pressure
S = contact length
δ = slippage factor

Using the following slippage equation

$$\tan\delta = \frac{\left(R/Y\right)\sin\phi - \tan\alpha}{\left(R/Y\right)\cos\phi - 1}$$

the torque equation can be rearranged as follows:

$$\Delta Q = \frac{(\Delta T)a\left[\left(R/Y\right)\sin\phi - \tan\alpha\right]}{\left(R/Y\right)\cos\phi - 1}$$

These equations give the expressions for the tension variation and torque generation of the threadline in terms of the machine and yarn parameters. The amount of variation in tension and torque levels after passing the spindle changes with the magnitude and the direction of the frictional force generated by the twister. The slippage factor determines the direction of the friction force acting on the threadline.[22]

The overall performance of the ring spindle is determined by the values of the tension variation and torque levels of the threadline. The system is said to be optimum when the threadline tension variation remains zero, i.e., the yarn is neither dragged or pumped by the spindle. In this condition, the total friction force acts perpendicular to the yarn axis and generates pure torque. By adjusting the external pressure *N* and the speed ratio (R/Y ratio), the amount of slippage in the system can be controlled to some extent; thus, the operating speed ratio for inserting the same level of twist into the yarn is lower than that of the stacked disk spindle.[22]

The desirable speed ratio, at which the tension across the spindle remains unchanged, depends primarily on the ring crossing angle (θ). The tension ratio becomes unity when the speed ratio is given by

$$R/Y = \frac{2}{(1+\sec\alpha)\cos\phi}$$

where α is the helix angle and ϕ is the yarn crossing angle to the ring. Thus, the optimum speed ratio increase with the twister angle (θ).

There are four possible modes of operation for a ring twister, i.e., positive grip mode, rotational grip mode, longitudinal grip mode, and slipping mode in both directions. The friction between the yarn and the spindle can be either static or dynamic. It is static when the yarn is running under the positively gripped mode, in which case there is no slippage in either the rotational or elongational direction. It is dynamic when rotational or longitudinal slippage occurs. The actual physical situation of friction is significantly different depending on whether friction is under static or dynamic conditions.[22]

3.2.3.4.3 *Comparison with stacked disk spindle*

Forces acting on the yarn by the ring twister are well balanced because the yarn is clamped between two rings. The externally applied clamping force is the main source of the friction force responsible for tension change and torque generation. In stacked disk spindle, however, the force is applied to only one side of the yarn, and it tends to bend the yarn and shear it. The source of the friction force in the disk spindle comes from the yarn-drawing tension, which cannot be controlled with any machine setting except by changing the yarn draw ratio. Unfortunately, there is no way of correcting the nonuniformity of the drawing tension.[22]

In a stacked disk spindle, the contact angle can be controlled with disk overlap, and the angle of wrap can be changed with the number of disks in each stack. The disk profile changes both the contact and wrap angle, so to produce uniform disk surface-to-yarn relative velocity in both magnitude and direction, the disk profiles and disk overlap should be adjusted in each stack of disks. These will be able to produce a positive grip mode of operation. In ring twisting, however, the positive grip mode is well defined with the function of the speed ratio and ring-crossing angle. Unlike the disk spindle, the externally applied normal force is the main source of friction force, and it adds one more variable to the process settings. The end-to-end uniformity is affected by the variability of the normal force applied to the system, so that end-to-end variability of the product could be larger in the ring-textured yarn than in the stacked disk textured yarn if the ring piston pressure is not precisely controlled for all ends.[22]

3.2.3.5 Other types of false-twist insertion systems

It appears that this type of false-twist insertion devices is seen in only patent literature and has found no practical use so far. Figure 3.35 illustrates four such

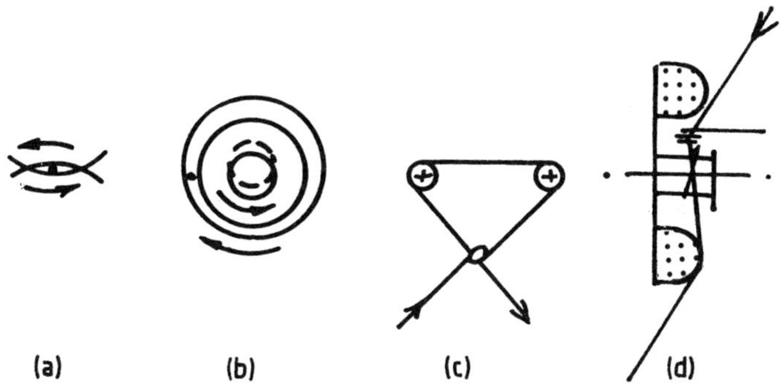

Figure 3.35 Four false-twist insertion means that found no industrial application

approaches, one of which inserts twist to the yarn by direct contact with the inner surface of the element; the other possesses a traveler that is similar to pin twisters. Apart from these thoughts, turbine-driven pin spindles were also developed by Toshiba of Japan. The principle of twist insertion of this turbine-driven pin spindles is shown in Figure 3.36. The advantages of the air-jet twisters are the quite operation and easy adjustment of the twist level by adjusting the air pressure. However, the air-jet-type twist insertion devices are, first, too costly[24] and, second, are unable to work on simultaneous draw texturing machines.

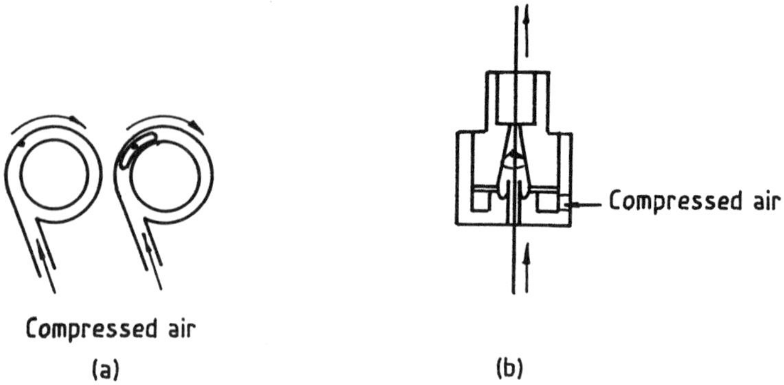

Figure 3.36 False-twist insertion by means of compressed air

3.2.4 Mechanics of Friction False Twisting

Thwaites[25] developed the following theory on the mechanics of friction false twisting by using a bush-type twister.

3.2.4.1 Velocity vectors at friction surfaces

Assume a yarn -running direction perpendicular to the friction surface, as in the case of internal friction twisters (Figure 3.37). V_S will be the velocity vector of the friction surface, V_P is the velocity vector of a point lying on the yarn surface and makes contact with the friction surface. The components of V_P in yarn-running direction and peripheral direction are V_y *and* V_c,, respectively. Between V_P and the surface velocity V_S, there is a relative velocity vector ***R***, which can be resolved in a twist component p and a transport component t. The twist component p is a measure of the twist slippage, which can be defined as

$$\textbf{Slippage} = \frac{p}{V_c + p} \cdot 100$$

In this case, the transport component t acts counter to the yarn-running direction, which means that it is negative or suppressive. This greatly reduces the yarn tension before the twisting element. This and the relative velocity ***R*** have a wear-promoting effect on the yarn and the friction surface, making the internal twisters unsuitable for high-speed texturing.[26]

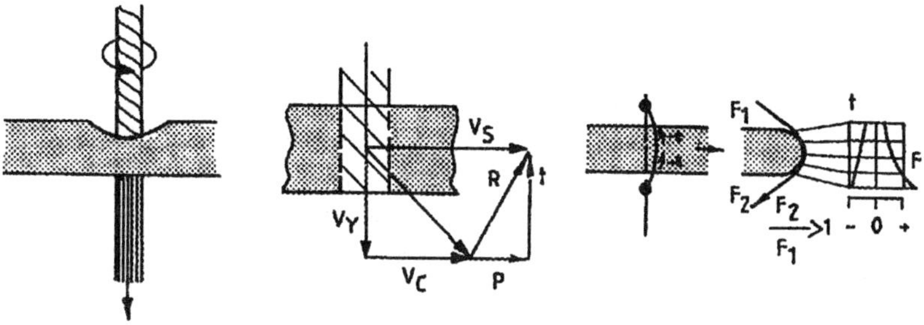

Figure 3.37 Velocities and forces involved on internal (bush) false twister[26]

In the case of outside friction twisters, however (Figure 3.38), it is possible to feed the yarn to the friction surface at an angle. This inclined yarn feed, hence,

Figure 3.38 Velocities and forces involved on external (disk) false twister[26]

reduces the angle between the contact point velocity vector V_p and the friction surface velocity vector V_s with a resultant reduction in the relative velocity ***R*** and in the transport component *t*. In this case, the action of the transport component will be positive. The amount and direction of this transport component *t* can be adjusted within certain limits by means of disk speed and the axial spacing of the three shafts of the multiple disk assembly. Therefore, it is possible with the multiple disk twisting assembly to keep the mechanical stress acting in the yarn-running direction as low as possible.[26] It will be appreciated from Figure 3.38 that one component of the surface speed of the friction surface transmitted to the yarn causes the yarn to twist and the other component causes the yarn to move forward.[27]

It can be shown quite simply that, if no slip occurs, the ratio of the twisting speed to the forwarding speed determines that the twist-helix angle produced at the yarn surface is equal to the angle between the yarn axis and the direction of motion of the friction surface. It also follows that, because the device can forward the yarn, the tension after the friction surface can be made higher or lower than the tension before it, depending on the speeds of the upstream and downstream feed systems relative to the twisting device.[27]

3.2.4.2 Single-disk mechanics

Thwaites[28] presented a simplified model and derived the following equations for yarn-tension ratio and torque generation. Yarn-tension ratio is given in terms of the angle of wrap, $\Delta\theta$, and the coefficient of friction, μ, by a modified capstan equation.

$$ln\frac{T_{out}}{T_{in}} = \mu \cdot \Delta\theta \cdot Cos\psi \cdot Sin\gamma$$

where ψ and γ are the complements of the angle of attack and the angle made by the frictional force with the yarn, respectively (Figure 3.39). The torque generated by the spindle is given in terms of the tension difference by

$$\Delta\mathbf{Q} = (T_{out} - T_{in}) \cdot R \cdot Cot\gamma$$

where R is the yarn radius. Finally, the angle at which friction acts is given in terms of yarn axial velocity, V_y, and yarn angular velocity, Ω, by

$$tan\,\gamma = \frac{V_y - V_s \cdot Sin\psi}{V_s \cdot Cos\psi - \Omega \cdot \mathbf{R}}$$

The simplified model, thus, appears to depend on the constancy over the disk surface of the variables μ, R, and ψ and the ratios Ω/V_y and V_s/V_y..

3.2.4.3 Relation between Twist and Surface-Helix Angle

The following symbols are used in the analysis:

θ = surface-helix angle of the drawn and twisted yarn
N = number of turns of twist per unit untwisted length of drawn yarn
r_0 = equivalent radius of untwisted drawn yarn if 100% filament-packing is assumed (πr_0 x polymer density = drawn-yarn linear density x 10^{-6})

ψ = yarn-contraction ratio, i.e., untwisted yarn length/twisted yarn length
γ = draw ratio in draw texturing
U = undrawn-yarn linear density (dtex)
T = drawn-yarn linear density (dtex)
D = friction disk surface speed
Y = forward speed of untwisted yarn
ρ = polymer density

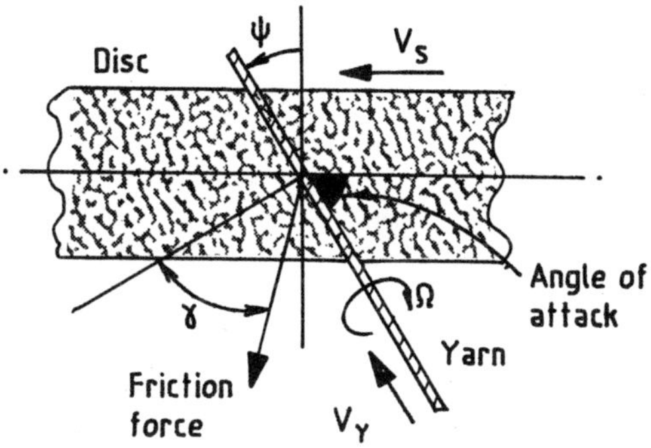

Figure 3.39 Angle of attack and angle made by frictional force with yarn

Two equations relating these parameters are assumed.

$$2 \cdot \pi \cdot r_0 \cdot N \cdot \psi^{3/2} = \tan\theta,$$

$$\psi = \frac{1}{2} \cdot (1 + \sec\theta).$$

The first of these is derived from the yarn geometry. The second is due to Hearle and Morton[29]. From these equations a third can be derived to relate the twist and surface helix angle.

$$N = \frac{2(\sec\theta - 1)}{\pi \cdot r_0 (1 + \sec\theta)}$$

3.2.4.4 Calculation of D/Y ratio

From Figure 3.40, we have

$$\frac{\textbf{twisted yarn surface velocity}}{\textbf{twisted yarn forward velocity}} = \frac{V_R}{V_1} = \sec\theta$$

We also have

$$\mathbf{Y} = \psi V_1$$

and if a no-slip condition is assumed, **D** = V_R, and, hence,

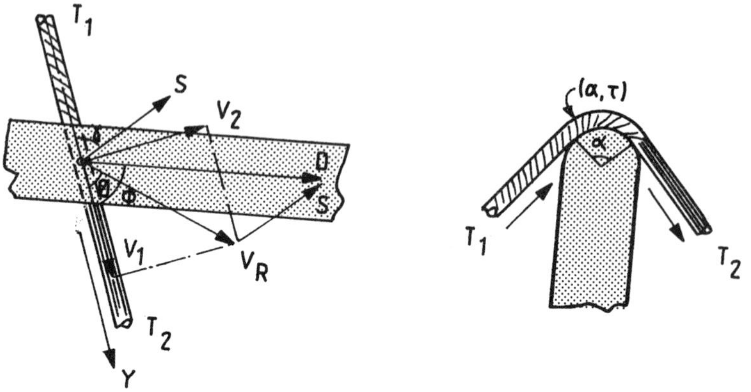

Figure 3.40 Formation of D/Y ratio

$$\frac{\mathbf{D}}{\mathbf{Y}} = \frac{V_R}{\psi \cdot V_1} = \frac{\sec\theta}{\psi}$$

From the preceding equation for ψ,

$$\frac{\mathbf{D}}{\mathbf{Y}} = 2 - \frac{1}{\psi} \quad \textbf{or} \quad \frac{\mathbf{D}}{\mathbf{Y}} = \frac{2}{1 + \cos\theta}$$

Alternatively using the equation for N gives

$$\frac{\mathrm{D}}{\mathrm{Y}} = \frac{3}{2} - \frac{1}{2}\sqrt{\left(1 - 4\pi^2 r_0^2 N^2\right)}$$

In general, the **D/Y** ratio experienced in practice is higher than the previously given theoretical value. This difference is explained by Morris and Denton,[27] at least in part, by the following limitations or deficiencies of the theory:

1. Because the Hearle and Morton[29] geometry is based on a coaxial-helix structure, slow filament migration is implicit in the model, uniform filament packing is also assumed.
2. The calculation of r_0 assumes the yarn density to be equal to the polymer density.
3. A circular yarn cross-section (i.e., one without surface undulations because of individual filaments) is assumed.
4. Yarn kinking at high twist levels is ignored.

3.2.4.5 Curved Friction Surface

The yarn behavior will be different when it rolls over a curved surface than that explainedearlier. Because of the friction, the yarn is deflected and assumes an arcuate path over the surface (Figure 3.38). The effect of this deflection on the relative velocity between a surface point of the yarn and the friction surface as well as on the active transport, and twist components may be calculated for a known deflection. In Figure 3.37 with a vertical yarn path, the transport component is negative and becomes progressively more negative when the yarn runs over the disk. The yarn tension increases constantly. The twist component passes through a maximum. In the case of inclined yarn feed as seen in Figure 3.38, if a tension ratio F_2 / F_1 of unity is secured, then the transport component in the middle would be zero. That is to say, the yarn would neither be retarded nor transported. Because of the deflection of the yarn over the surface, a positive transport component applies when the yarn touches the disk and a negative transport component applies when the yarn leaves the disk. In approximately the middle of the disk, the yarn passes through a point where the transport component is zero. This means that in the upper section of the disk assembly, the yarn tension F_1 is reduced to a certain point, i.e., it reaches a minimum; then, when the transport component becomes negative, the yarn tension builds up again to the value of F_2.[26]

What happens when an asymmetrical friction disk, e.g., a so-called half-disk, is used (Figure 3.38). Because of the different curvature radii of the disk, the yarn passes over a much greater surface at the upper portion of the disk in a more inclined position compared with the lower portion of the disk with its less inclined position.Because experience has shown that not only the angle of wrap is involved in the friction conditions but also the curvature of the friction surface plays an important part, then the positive transport component in the upper portion is effective for a much longer time than the negative transport component applies in the lower portion. Consequently, it is found that under comparable conditions with the asymmetrical disk, the lowest tension ratio F_2/F_1 is reached.[26]

Obviously, a low-yarn tension is desirable in order to protect the yarn. However, to achieve a positive, constant imparting of twist, it is necessary to provide for a certain pressure to be applied by the yarn against the friction surface. For this reason, the yarn tension cannot be reduced at will, without incurring any consequential change in the twist effectiveness and stability of twisting process.[26]

Friction false twisters are basically torque-generating devices. Consider an element of yarn sliding over the surface of a rotating disk as shown in Figure 3.41. If the yarn moves in direction B, then there is a drag force (F_y) acting on the yarn in the opposite direction. The disk surface moving in direction A produces a drag force (F_d). As shown in Figure 3.41b, the resultant of these two forces is (F_R), and this may be resolved into two components, one parallel to and one perpendicular to the yarn element. The component OQ represents the increment of yarn tension, and the component OP produces a torque in the yarn. As the angle α is changed, so the ratio between tension and torque alters. When it reaches a specific value α_2, the increment of tension is zero as shown in Figure 3.41c. If the angle α is further altered to α_3, it is possible for the resultant tension component to act in the same direction as the yarn movement (Figure 3.41a). In other words, the disk "pump" the yarn through the system. Both the forces F_y and F_d depend on the normal force between the yarn and the surface of the disk. Thus, although the magnitude of the torque and tension components change with normal pressure, the ratio does not change appreciably. It follows that the relative positions of the various disks have to be carefully set to give the correct angles so as to give the best torque/tension relationship through the disk stack. Because the amount of twist set into the yarn greatly affects the texture, this is an important parameter.[3]

Several types of disk surfaces are illustrated in Figure 3.42. Claims about them are usually made in terms of twisting effectiveness, crimp level and uniformity, or reduction in slipping speed.[30]

3.2.4.6 Tension ratio over a friction twister unit

In the simplest form, a friction twister unit consists of a rotating disk (Figure 3.43) over which the yarn passes with an angle of wrap α. The yarn is fed into the system by the feed rollers R_1 and is taken out of the system by the delivery rollers. It is a common practice to increase the angle of wrap by inclining the direction of the movement of the disk surface toward the yarn axis by an angle ϕ (disk angle), which is less than 90° (Figure 3.44).

For simplicity, it will be assumed that, despite the existence of a frictional force between the yarn and the twister, there is no sideways deflection of the yarn axis. This means that the sole effect of friction is to rotate the yarn and to cause

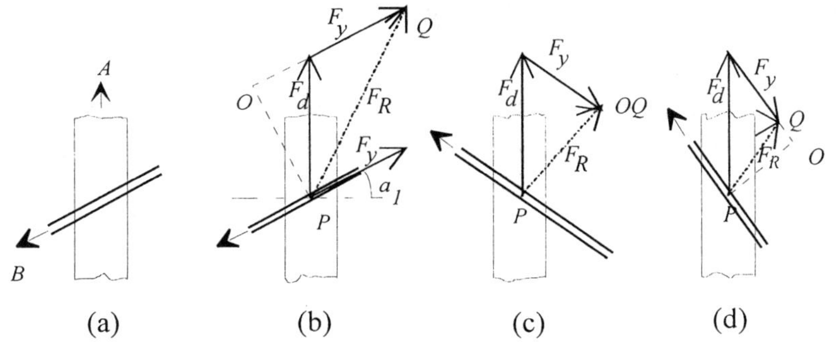

Figure 3.41 Forces acting on yarn at the surface of a single disk twister[3]

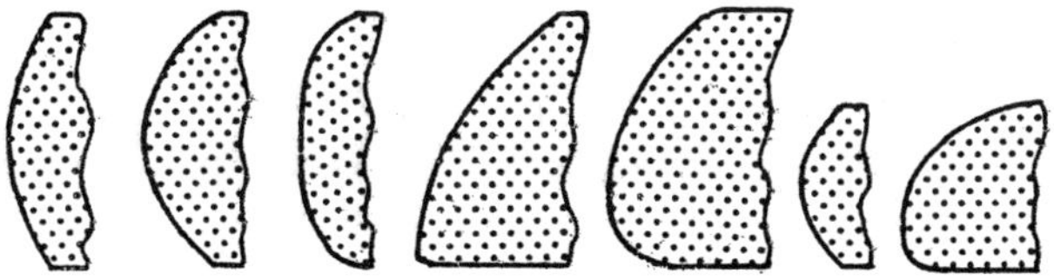

Figure 3.42 Several types of disk surfaces

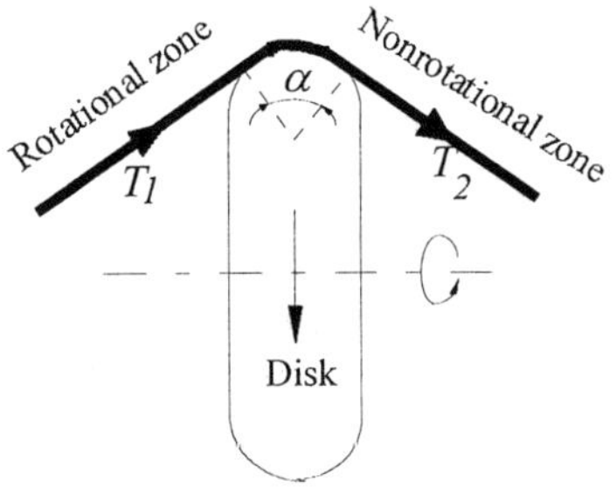

Figure 3.43 Tension ratio over friction-wister unit

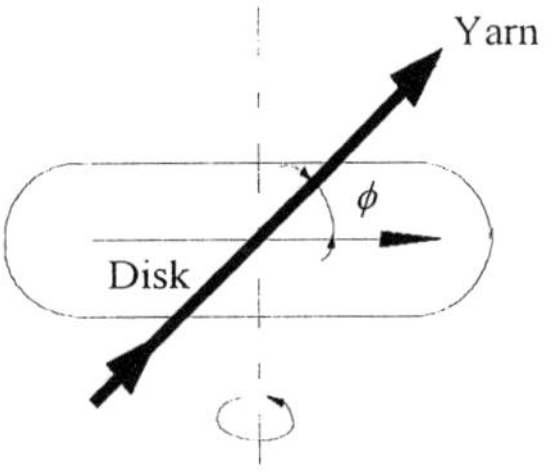

Figure 3.44 Yarn inclination angle with the twist insertion disk

a difference between the yarn tension T_1 before the disk and the yarn tension T_2 after the disk. The phenomenon diskussed here concerns the ratio between these two tensions.[31]

As long as the yarn moves forward but the disk does not rotate, the ratio between the two tensions is given by Amonton's Law which states that

$$\frac{T_2}{T_1} = e^{\mu\alpha}$$

where

μ = the coefficient of friction

This equation, no doubt, indicates that the tension ratio increases with increasing angle of wrap. However, when a friction twister in operation, i.e., when the disk rotates, it is frequently found that the tension ratio decreases with increasing values of α. At first sight, this effect appears to contradict Amonton's Law. This paradox will be explained in the following sections.

Figure 3.40 shows a vector diagram of the velocities operating in slipping friction system, with the yarn oriented at an arbitrary angle to the direction of movement of the friction surface.

The relative velocity between the yarn and the friction surface is

1. Resolved along the yarn axis

$$V_1 \text{ - } D\cos\varphi$$

2. Resolved perpendicular to the yarn axis,

$$D\sin\phi - V_2$$

The "drag angle" γ, which is defined as the angle between the drag-velocity vector, S, and the yarn axis is of considerable importance because it determines the ratio $\frac{T_2}{T_1}$ between the output and input tensions. The drag angle is given by

$$\tan\gamma = \frac{D\sin\varphi - V_R\sin\theta}{V_R\cos\theta - D\cos\varphi}$$

The tension developed at a point in the yarn owing to drag is

$$T = T_1 e^{\mu\sin\varphi\cos\gamma}$$

and the ratio of output tension to input tension is

$$\frac{T_2}{T_1} = e^{\mu\alpha_0 \sin\varphi \cos\gamma}$$

This equation can be regarded as a generalized form of Amonton's Law.

As the friction-surface velocity, D, increases; the drag vector, S, will rotate in a clockwise direction until, when $V_R \cos\theta = D\cos\varphi$, it will point along the direction of V_2, and γ will be 90°. At this point, the drag force has no component along the yarn axis, and $\frac{T_2}{T_1} = 1$. A tension ratio of unity can, thus, be achieved in a geometry where $\theta \neq \varphi$.

As D increases still further, $\tan\gamma$ becomes negative ($\gamma > 90°$), $\cos\gamma$ is also negative, and $\frac{T_2}{T_1} < 1$. The drag component thus acts in a direction to forward the yarn, and output tensions can be less than input tensions for finite values of D provided that φ is less than 90°.

It is generally agreed that disk overlap controls the geometry of the yarn contact with the disks, Given the overlap, the torque and, therefore, the twist generated depends primarily on the number of disks,[26] and the **D/Y** ratio and the machine draw ratio then control their actual values of tension and other variable in the process. The experiments show clearly that only changes in center-spacing affect the angle of attack,[26] but that changes in all of the other parameters, such as center-spacing, draw ratio, and **D/Y** have distinctly different effects on tensions, torque, and twist.[28]

3.2.5 Machines

3.2.5.1 Machine design features

In the early days of false twisting, it was regarded as essential that the yarn should travel the processing zone in a straight line. As long as processing speeds were low, this created no problems because, even with short heaters, the dwell time of the yarn in the heater was long enough to ensure adequate heat setting.[18]

As take-up speeds increased (Figure 3.45) however, the "straight-line" concept increasingly became a stumbling block to progress because heater lengths and, therefore, machine heights had to increase. Some of the high-speed machines have reached heights of 5 m, creating building problems. Studies of this problem revealed that, by proper machine design, the straight yarn path can be changed without detriment to yarn quality or machine efficiency.[18]

Although the twisting elements are regarded as the heart of the false-twist texturing process, other ancillary devices, such as supply yarn package creels, yarn-forwarding rollers and systems, heaters, intermingling devices, oil applicators, and take-up units also constitute an important part for the process.

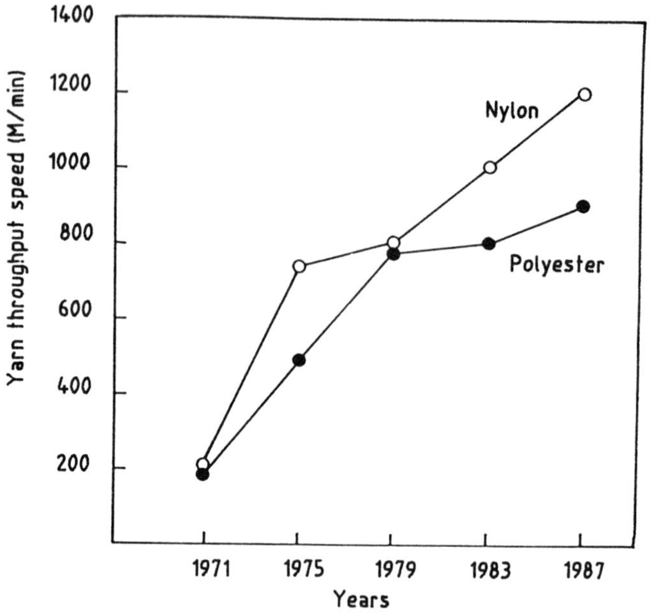

Figure 3.45 Development of false-twist texturing process speeds over the years

Machine design is inevitably determined by the heater length, cooler length and the characteristics of the deflection angle along the yarn path. Depending on the heater arrangement and heater length (up to 2.5 metres), the machines may vary in height from 3 m to a maximum of 5 m.[32] Examples of modern machine layouts from some of the principal manufacturers are shown in Figure 3.46 and in Table 3.1.

Table 3.1 shows the general features of the machines produced by latest machine manufacturers. As seen in the table, all of these machines have rotating or swing-out creels, vapor-phase heaters of up to 2.5 m, setting heater of up to 1.5 m, semi automatic assisted string-up and a traverse stroke of 250 m and their design speeds between 900-1500 m/min.

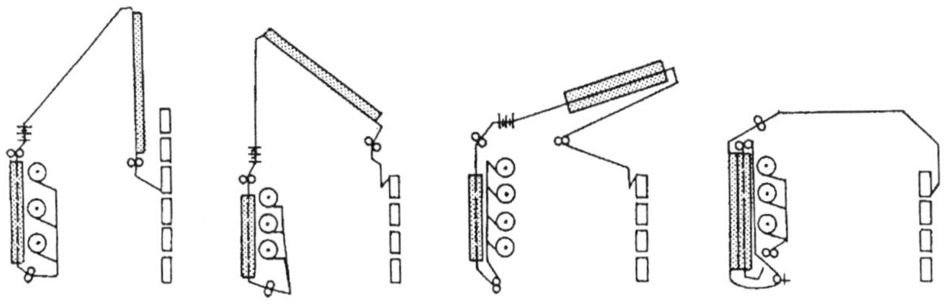

Figure 3.46 Examples of modern machine layouts

Table 3.1 General Features of False Twist Texturing Machines

Manufacturer	Barmag		Rieter-Scragg		Murata	
Machine type	FK6-1000	FK6-V80 -L80 -M80 -D80	SDS1200	DCS1200	No.33H Mach crimper	No.336- 15L, 20L, 20C Mach crimper
Yarn type	SET or HE	SET or HE	SET	HE	SET	HE
False-twist principle	Stacked	Stacked	Positorq	Positorq	Nip twister	Nip twister
Type of friction disk	disk or Ringtex	disk or Ringtex	ceramic	ceramic		
Max. units per machine	240	216	216	216	216	216
Machine length (m)	18.770	17.1	16.176	17.210	16.790	16.715
Section length (m)	1.44	1.44			1.440	1.440
Units/section	24	24	24	24	24	24
Machine width (m)	7.904	7.90	6.390	6.340	7.620	7.385
Machine height (m)	4.890	5.0	5.280	4.010	4.317	4.400
First heater type	Contact	Contact	Contact	Contact	Contact	Contact
Length (m)	2 or 2.5	2.5	2.5	2.5	2.5	1.5 or 2.0
Second heater type	Convection	Convection	convection		semi-cont.	
Length (m)	1.460	1.6	1.46		1.6	
Cooling system type	Plate	Plate	Plate	Plate	Plate	Plate
Length (m)	2.1	2.1	2.95	2.25		1.2 or 2.24
Yarn path	 M-Form	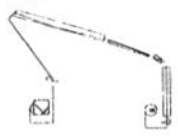 V-Form	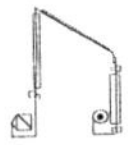 Type B			 Type L
Max. mech. speed (m/min)	1000	1200	1200	1200	1200	1200
Yarn transport system						
Before first heater	Apron	Apron	Nip	Nip	Apron	Apron - nip
After spindle	Nip	Nip	Capstan	Capstan	Apron	Apron - nip
After second heater	Apron	Apron	Nip		Apron	Apron - nip
Maximum total power input (kW)	210.1		176	118	196.3	146.6
Creel Package diameter (mm)	435	435	360	360	430	435
Textured yarn package Diameter * length (mm)	250 * 250	320 * 250	300 * 250	250 * 250	300 * 250	300 * 250
Weight (kg)					7	7
Special features	Autodoff, Unitens, Intermingling,	Autodoff, Unitens, Intermingling,	Intermingling, Qualitor, Introl	Advanced package support, Intermingling,		

Table 3.1 The General Features of False Twist Texturing Machines (Continued)

Manufacturer	RPR		Giudici	ICBT	Teijin-Seiki
Machine type	3SDX/ V or VI	3SDY/ 2 or 2I	TG.20 Super F3 V3	FTF8-E2	HTS-1500
Yarn type	SET	HE	SET	HE - SET	SET
False twist principle	Friction	Friction	Friction	Friction	Friction
Type of friction disk	disks	disks	disks	disks	disks
Max. units per machine	216 - 240	216 - 240	216	96	216
Machine length (m)	16.1 - 17.5	12.6 - 14.0	18.9	10.9	17.39
Section length (m)	1.4	1.4		1.50	
Units/section	9 - 10	9 - 10		16	
Machine width (m)	7.630	7.220	8.8	8.50	10.9
Machine height (m)	4.695	4.315	4.10	3.70	4.125
First heater type	Contact	Contact	Contact	Contact	Convection
Length (m)	2	2	2.5	2.25	1.0
Second heater type	Convection		Convection	Convection	Convection
Length (m)	1.250		1.45	1.20	0.6
Cooling system type	Plate	Plate	Plate	Plate	Plate
Length (m)	1.3	1.3	1.2	1.25	
Yarn path			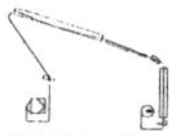 V3 Conf.		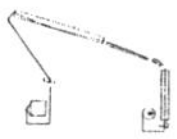
Max. mech. speed (m/min)	1000	1000		800	1500
Yarn transport system					
Before first heater,	Nip	Nip	Nip	Nip	Nip
After spindle,	Capstan	Capstan	Capstan	Nip	Nip
after second heater	Nip		Nip	Nip	Nip
Maximum total power input (kW)	130 - 135	105 - 107	249.1		294.4
Creel package diameter (mm)	250	250		440	
Textured yarn package diameter * length (mm)	250 * 250	250 * 250	250 * 250	300 * 250	300 * 250
weight (kg)	7	7	7		
Special features			Intermingling. Pneumatic threading and doffing		Intermingling. Doff timer. T2 monitor

3.2.5.2 Ergonomics

Consideration of ergonomics is an essential design prerequisite for any texturing machine. Ease of machine operation and operator fatigue levels are vital factors in efficient plant organization. Usual high construction of texturing machines requires extra caution for the operator and routine threadline checks. These should be provided from the floor level wherever possible. Threading times and threading integrity should also be considered during the positioning and detailed design of yarn guides.

3.2.5.3 Creeling

As the size of the supply yarn package increased, the creeling systems gained importance and became an independent unit mostly standing off the texturing machines. These standoff creels have the added advantage of holding the supply yarn further away from the heaters, thus avoiding the problem of fast one-sided ageing of the package. The operator's work is also made slightly easier by the less cramped conditions.[33] Different machinery makers offer different creeling systems, but these are mostly arranged on both sides of the machine. Package diameters up to 435 mm are possible.[34]

Apart from the consideration of supply package size, it is generally true to say that operators prefer standoff creels because of their greater accessibility[35].

Where the texturing machine stands only a short distance from the spinning machine, cyclic operation with truck creel is the best and most economical way of package change-over.

The trucks loaded with supply yarn packages are parked in an air-conditioned intermediate storage room. In this room, the truck creels are joined up to form entire trains. While the empty truck are being run off at one end of the texturing machine, the fully loaded trucks are brought into the working position from the other end. In this way, plants are operated with supply packages up to 28 kg each. The texturing machine must usually not be stopped during doffing. This is achieved by operating the machine at a creep speed, so that the yarns remain in the machine without end breaks. The ends of the new packages can now be tied on, the knotted yarn end being drawn by hand. The machine may then be restarted at creep speed again, saving a great deal of time.[4]

If the transport truck do not act as creels, the supply packages must be pegged onto a fixed creel at the machine. To avoid touching the packages by hand, simply pushing the package over from the creel peg of the truck onto the fixed creel is the usual practice. The positions of the creel pegs on the trucks coincide with those on the texturing machine. So that the full supply yarn packages are easily pushed over.[4]

3.2.5.4 Yarn-forwarding systems

At today's processing speeds, the yarn forwarding systems of the machine must be highly reliable in their control of the fine yarn and able to run for long periods without maintenance.

The yarn-forwarding system, consisting of rollers, in false-twist texturing machines has three main functions to perform. First, it controls the rate of yarn flow through the machine. Second, it controls the tension in the yarn passing over the heater and through the false-twisting unit. Third, the feed roller before the heaters acts as a twist-stopping device.[35]

Choice of positive yarn forwarding devices for the false-twisting machines lies between nip (Figure 3.47a), nip/capstan (Figure 3.47b), capstan (Figure 3.47c), and Casablanca apron-type systems. The apron-type yarn forwarding systems are popularly used on most high-speed texturing machines. Nip systems are some time equipped with diversion rolls to prevent slip.

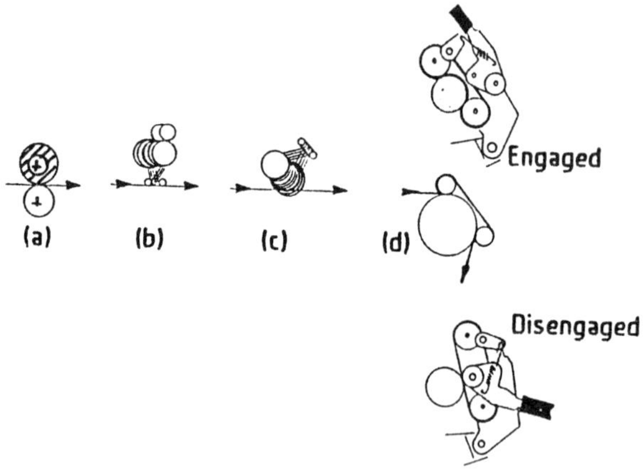

Figure 3.47 Most usual yarn forwarding-systems used on modern texturing machines

Though plain capstans were used on some early machines, this type of feed is no longer used in modern machines because it is believed that this type of feed give rise to filament coring because of relatively low pressure on yarn. The nip feed achieves its grip and twist-stopping action by a high pressure over a relatively short length of yarn and is favored by nearly all manufacturers. Apron feed obtains grip by a moderate pressure over a comparatively long length of yarn. Because it is flexible, the yarn is possibly less likely to give alignment difficulties.Conversely, the nip system gives rise to a more rapid filament migration within the yarn and would appear to have other advantages in simplicity.

Although, in view of the high yarn tension after the spindle, it seems surprising that a single nip or apron on its own can provide sufficient grip at this point, nip/capstan arrangement or multiple wrap nip feed systems are often employed.

3.2.5.5 Heat transfer, heating systems, and heaters

The work of Arthur and Jones[36] is perhaps the first articlethat deals with the heat transfer during the false-twsit texturing operation.

A running yarn is not easy to heat. Because the correct heating of the highly twisted yarn is an essential requirement for a good and uniform yarn quality, the heaters rightly attracted special attention over the years. Essentially, the heater functions to raise the temperature of the yarn, and then to maintain it at the correct temperature for a time long enough to allow the loosening of the intermolecular bonds.

The most important criteria is the yarn temperature, and this will depend on the heater temperature and the speed of the yarn as it passes through the heating zone. Over the years, the production speeds have greatly increased, and consequently heaters have been so designed that the yarn path through the heating zone become progressively longer. Heaters with about 2 m in length are common. Slight variations of heater temperature from spindle to spindle can result in great variations in yarn properties, particularly regarding dye uptake; unfortunately, these are not apparent until the finished fabric stage is reached. Consequently, costly electronic temperature control systems are used in an attempt to maintain temperature to within ±1°C.

The basic requirements of heaters include the following:

1. Reliability
2. Sufficient heating capacity for a range of texturing speeds
3. Constant surface temperature, independent of heat consumption
4. Maximum temperature uniformity from one heater to another
5. Collective setting of related value for the entire machine
6. Simple and effective temperature control
7. Applicability for conventional and draw texturing operations
8. Economy in operation.[33]

For the false-twist texturing process, several types of heaters, based on any one or a combination of the three modes of heat transmission –conduction, convection and radiation– are used. Of these, conduction-type heaters have been the most popular; however, the convection-type hearters known as high-temperature heaters have also become important lately.[37]

The contact or residence time of yarn in these heaters is approximately in the range from 0.2 to 0.5 sec depending on the speed of the yarn and the length of the heater. During this short time, the temperature of the yarn has to be raized to about 200°C. While the yarn is in contact with the heater, it rotates at a high speed because of twisting. Furthermore, in addition to heating by contact, the yarn is also heated by radiation from the surrounding surfaces of the groove in which it rides, and by the hot air in the immediate vicinity of the heater surface. Because of these factors, the yarn is heated uniformly all around its circumference.

According to Newton's law, the rate of the increase in the surface temperature will be proportional to the difference between the heater and surface temperatures, T_h and T_s, respectively. Thus,

$$\frac{dT_s}{dt} = s(T_h - T_s)$$

where s is the proportionality constant. Solving the differential equation with the initial condition that at $t = 0$, $T_s = T_R$,

$$T_s = T_h - (T_h - T_R)e^{-st}.$$

This is an expression for the surface temperature of the yarn, and s is a constant that depends on the material and the linear density of the yarn. It is a measure of the rate of increase of the temperature of the yarn. The higher the value of s, the faster the yarn temperature rises. Polyester yarns were found to be heated faster than nylon yarns, and consequently the parameter s was higher for the former. This larger increase in the temperature of polyester yarns was due to its lower heat capacity. The heat capacity of the polyester yarns had been found to be 0.45 and that for nylon to be 0.79.[37]

High-linear-density yarns were found to have larger values of the parameter s than lower-linear-density yarns; hence, the higher-linear-density yarns are heated slower. This is due to the increase in the ratio of surface area to the mass of the yarn with increase in the linear density.

As the yarn heated in the heating zone, heat flow from the surface toward the center of the yarn; hence, there is a temperature gradient between the surface and the center of the yarn. Using the basic laws of the heat transfer and assuming that the highly twisted yarn behaves as a solid rod (Figure 3.48) and making further simplifications by omitting some negligible terms the following equation is reached.[37]

$$T(r,t) = T_h - (T_h - T_R)\frac{2J_0(\frac{\lambda r}{a})}{\lambda_n J_1(\lambda_n)} e^{(-\frac{\alpha \lambda_n^2}{a^2})t}$$

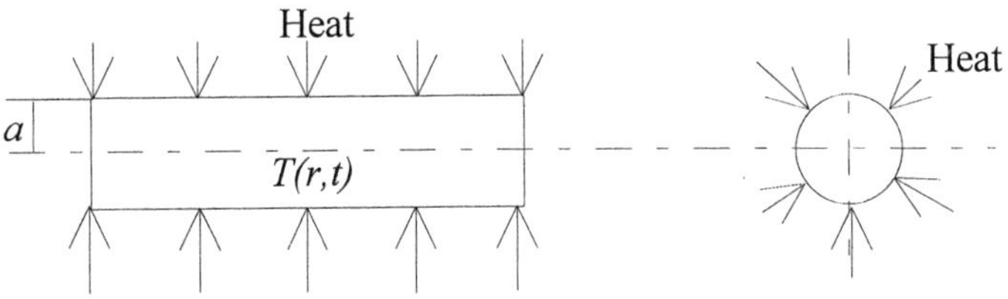

Figure 3.48 Heat transfer to a highly twisted yarn

From this expression, the temperature in the cross-section of the yarn can be calculated. The temperature difference across the cross section of the yarn in the texturing process was found to be small for the range of linear density and yarn speeds used conventionally.[37]

The yarn may be heated by direct contact with a plate or tube, or by hot air and radiation in an enclosure. Direct action of pressure steam has also been used. Furthermore, the plates, tubes, or enclosures may be heated individually by electricity or by a circulating fluid system.

The heaters may be heated by various means, such as direct electric wire, hot air, and dowtherm vapor. The most common type, however, is electrically heated vapor-phase closed condensation heater, which relies on the condensation of the dowtherm liquid vapor on the back surface of the yarn-heater tracks or tubes to transmit heat. And most of the heaters are controlled either individually or in blocks of four channel.

Steam in direct contact with the yarn is also suggested for heating the yarn, but relevant problems, such as steam losses and string-up difficulties, hinder the wide- spread use of such a method.

Important features of heaters are the length, a flat temperature profile along the length, energy consumption, string-up, and cleaning facilities. Around these parameters, several different designs have been developed and actually are being used in industry today. There seems to be no preferential design for the time being.

It has been established by Coll-Tortosa[38] and Moussa[39] that heat is transferred essentially through convection, even in the case of contact heaters; therefore, a particular importance attaches to the rotational speed of the yarn in the heater. The heat flux, Q per unit length can be calculated from the following formula:

$$Q = \alpha \cdot 2\pi \cdot R(\upsilon_L - \upsilon_F) \cdot 10^{-6} \quad [\text{W/m}]$$

where

υ_L = surrounding air temperature, i.e., heater temperature

υ_F = yarn temperature

R = yarn radius

$$\alpha = \frac{\lambda_L}{\delta_L} \cdot 10^3 \quad \left[\text{W/m}^2 \cdot {}^0\text{C}\right]$$

where λ_L is the heat transfer coefficient of the air and δ_L is the hypothetical air-film thickness that the rotating yarn creates in the heater.

The air-film thickness decreases with increasing rotational speed (i.e., increases) and so does the heat flux Q. It is seen from the formula that the heat flux is influenced to a large extent by the difference between the yarn and heater temperatures. This nature, therefore, implies the use of heaters with stepped temperature profile, the so-called shock heaters (Figure 3.49). In these shock heaters, the yarn first passes through a zone with highly overheated temperature influenced to a large extent by the difference between the yarn and heater temperatures. This nature, therefore, implies the use of heaters with stepped temperature profile, the so called shock heaters (Figure 3.49). In these shock heaters, the yarn first passes through a zone with highly overheated temperature and then through one with normal texturing temperature.[15]

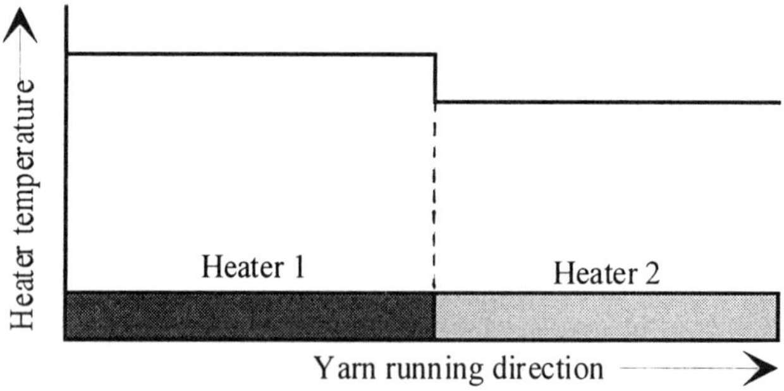

Figure 3.49 Temperature profile of shock heater

Hot air or steam, saturated steam, radiation, laser beam, high frequency, open flame and solid-state or liquid-state bath-heating systems are proposed as the alternatives to the conventional heating systems. The use of hot-air and hot-steam heaters, and also saturated steam heating involves all the problems of balloon influence on the heat transmission. The question of heater sealing and yarn string-up is a critical factor especially in the case of saturated steam heaters.[15]

The heat setting of the twisted filaments can also be achieved by laser heating 20 to 40 times faster than ordinary heaters. However, the structural changes in the individual filaments as a result of laser and ordinary heat treatment are fundamentally different.[40] The structure and properties of yarn textured with laser heating mainly depend on the power density and exposure time of laser (as on the temperature and contact time of an ordinary heater). The cross-section of filaments are deformed from round to polyside with both ordinary heating and laser heating. But the amount of deformation by ordinary heating is higher than laser heating because of the much longer treatment time in ordinary heating than in laser heating. Regarding the filament surface structure, after texturing, both filament surfaces appear to be crude and contain cracks. The filament surface of yarn textured by laser heating is cruder than that of yarn treated by ordinary heating, because the treatment intensity of laser is much higher than that of an ordinary heater.

Use of laser technology also involves difficulties with respect to adjustment and focusing of laser beams for many texturing positions.[15] In summary, it seems apparent that the vapor-phase-type heaters are dominant in all modern texturing machines.

3.2.5.6 Cooling systems

Setting does not only occur in heating but also in cooling, and it follows that inadequate cooling before untwisting will affect the efficiency of the process.

To ensure proper heat setting of the yarn in the twisted state, the length of the cooling zone extending from the exit end of the heater to the false-twisting element should be consistent with heater length and yarn throughput speed.[33] Strictly speaking, setting occurs in cooling, not heating, and an efficient heater is useless unless it is supported by efficient cooling to a degree that the yarn temperature is below 100°C before the yarn is untwisted at the spindle[33].

With low-speed machines, it is possible to rely on natural cooling to reduce the yarn temperature below T_g, but with high-speed machines other means of cooling become necessary. One possibility is to use a cooling block that absorbs heat from the yarn and reradiates the energy to its surrounding medium by virtue of its large surface area. The use of cooling air or gas is also possible. The main

objective is to get the temperature down quickly so that the yarn path is not unduly extended.[3]
Trials with different cooling conditions, such as the use of a cooling bar or extending the cooling zone through deflected yarn paths, under otherwise identical conditions, showed that the twist level in the heater zone was reduced as a result of the additional friction points. The crimp factor, however, tended to display opposite trend, i.e., despite lower twist level in the heating zone, superior or identical crimp values were found.[15]

3.2.5.7 Spindles: false-twisting devices

Magnetic pin-spindle units are still keeping their small market share. The quoted theoretical speeds are up to 900,000 and 1,000,000 rpm, permitting production speeds up to 360 m/min.

Triple-stack friction disk devices, however, dominate the current false-twist texturing scene. The number and material of the disks are so optimized that good yarn qualities are obtained even at high speeds. Improved bearing systems also contributed to this development.
The disk material can be divided into two classes. The first group is soft materials, such as polyurethane, which have a high inherent grip on the yarn resulting partly from a high coefficient of friction and perhaps partly from local indentation into the disk surface. However, because of many problems encountered with polyurethane materials, including wear, surface damage, deposits, sensitivity to temperature and chemical agents, the use of such materials is limited.[41] The second group is hard materials, of which ceramic-coated metal is typical. These rely for their grip on the mechanical keying of the roughness of the ceramic surface into the surface filaments of the yarn.[27] Ceramics have the advantage of being hard, inert, nonelastic, and insensitive to wear, heat, and chemical agents. Although the initial work was conducted with solid ceramics, ceramic coating later become widely used. Polyurethane disks become commodity with service life of 6 to 12 months. Yarns textured with polyurethane disks have higher strength, elongation, and bulk, and there is little snow formation. Polyurethane disks are often combined with disks of other materials.[34] Although disks of the solid ceramic type are less critical regarding their own wear life, there may be a greater danger of yarn damage, and the grip on the yarn may be less reliable. A rough ceramic surface will cause significant yarn damage. A smooth ceramic surface will provide inadequate grip.[27]

Although common disk thickness is about 9 mm, disks with 12 mm thickness are also on offer from Rieter-Scragg Ltd. of the United Kingdom.[34] It is claimed that by increasing the surface area of the disks, the disk life is increased because of the reduced pressure and, hence, abrasive damage.[42]

Most machines have endless belts to drive the individual positions. However, individually driven friction disk units are also on offer. Narrow speed variation of ±1% ensures that yarn is uniform from one package to another. The noise levels are lowered, and energy savings of up to 15% are achieved with these individual spindle drives.[43]

Although the state-of-art twisting spindles are triple stack friction units, the belt and ring twisters are also being exploited by different machinery makers. These so- called third-generation twist-imparting systems claim to assure virtually slip-free thread run, exact twist-level adjustment, better yarn quality, and higher bulk. Murata of Japan and Barmag of Germany are the two machinery makers that install belt and ring twisters, respectively, on their machinery.[34] The universal acceptance of these type twisters are, however, not yet accomplished.

3.2.5.8 Oiling systems

Oil is generally applied between the delivery speed and the take-up of the texturing machines as seen in Figure 3.50.

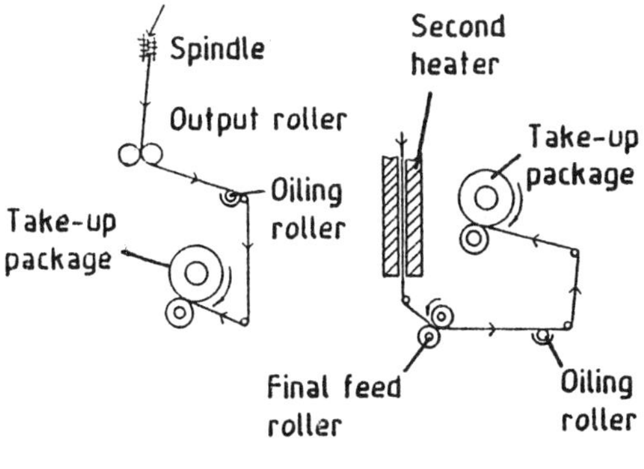

Figure 3.50 Oiling systems for conventional and draw-texturing processes[44]

The trend in oil application is towards precise application of oil to the yarn at the last possible moment before the take-up. Therefore, most of the modern oiling systems operates on a pressure-circulated and filtered-return principle in which oil is circulated and cleaned.

As the texturing speeds have increased over the years, the following three basic problems associated with oil and oiling systems in false-twist texturing have appeared.

Splashing of yarn lubricant This problem causes hazards to operators especially if the oiling unit is situated at eye level. There is also splashing from the floor-level oiling units, but the oil is thrown directly onto the floor and not into the operator's face. Also, the floor-level oiling unit enables some splash guards to be fitted that do reduce hazards of splashing to some extent.

Splashing can be reduced considerably by the use of suitable reduced- splash oils. Also drag and splash can be reduced by the use of serrated oil rollers.

Low oil pick-up High speed movement of especially high linear density yarns reduces the amount of oil picked-up by the yarn.

Overfeed before the oil roller This is thought to be caused by the drag imposed on the yarn as it passes over the oil roller and guides. The phenomenon shows itself by what appears to be overfeed between the feed roller and the oil roller, this eventually causing yarn stoppage because of yarn wrapping around the feed rollers.
To overcome the overfeed, there are several mechanical means. The first alteration that can be considered is to reduce drag by reducing the angle of contact of yarn with oil roller. Roller speed needs to be increased to maintain oil application. The second alternative is to increase the overall tension in the system.[44]

3.2.5.9 Take-up systems, package formation, and unwinding

Increasing throughput speeds inevitably forced to design and develop improved take-up units. Furthermore, a stable package is also desirable for further processing, such as knitting and weaving. Economical constraints in modern industry also induces larger and automatically handled take-up packages.

To overcome many difficulties, such as winding tension, edge and pattern formation, and physical demands of the yarn, development in design of cradle system dynamics, traverse mechanism, and the incorporation of sophisticated package edge and pattern disturber mechanism have been brought forward by mainly texturing machine manufacturers.

Rieter-Scragg, for example, has developed an Advanced Package Support (APS) which, by its design ensures that packages are controlled precisely throughout their build cycle (Figure 3.51). Vibration is minimized, resulting in the ability to construct a package more accurately. In addition to the advantages in package unwinding performance, the APS also offers advantages in operation, with ease of access for doffing, and is suited for automatic package handling systems.[46]

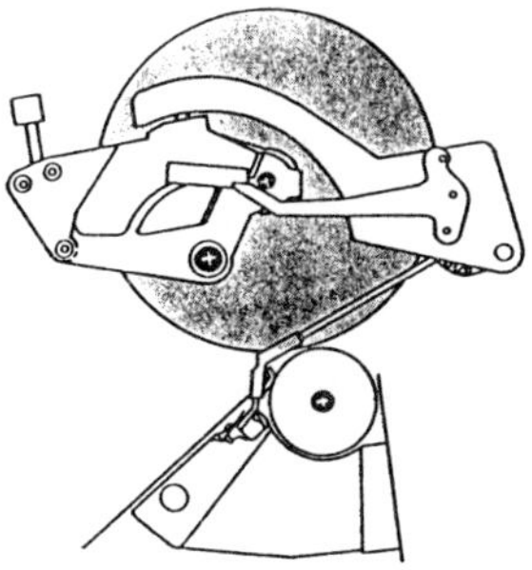

Figure 3.51 Advanced package support of Rieter-Scragg machines[45]

The take-up systems are closely related to the package formation. Therefore, structure and properties of packages, and unwinding problems are also be diskussed. For design and cost reasons, virtually all texturing machines are equipped with winding systems operating on the "random wind" principle. When this system is used, there is a risk that more or less distinct patterns will develop over the package structure. The most common method of preventing such patterns is to modulate the double-lift throughout the package build-up. The need to break up patterns depends highly on the sensitivity of the yarns being wound, which involves count, E modulus, and mechanical processability. Edge builder and traverse cycle systems are of critical importance to package structure.[47]

Ribbon winding, frequently known as patterning has an important influence on the yarn unwinding. It is a known fact that in random winding, patterning occurs when the speed of package rotation **n** is an integral of the number of double-traverses per minute **DH** or it represents a simple fraction, i.e., when B = **n/DH** = 1/4, 1/3, 1/2, 2/3, 3/4, 1, 4/3, 3/2, 5/3, 2, ... The diameter d_B at which patterning occurs because of such a fraction is calculated as follows:

$$d_B = \frac{2H}{\pi \tan\beta \cdot B}$$

where
H = traverse,

β = lead angle

The characteristic feature of patterning is that at the particular diameter the yarn is no longer distributed uniformly over the surface, but the coils are repeatedly laid on the same part and are superimposed on each other forming ridges. These ridges easily slip in and create unwinding problems.[48]

The detrimental effect of patterning is pronounced where the yarn coils causing the patterning cross over each other less frequently in each double-traverse. However, a significant influence is also exerted by the package diameter at which patterning occurs. The disturbing influence is greatest in the middle-diameter range. Figure 3.52 shows in geometrical form, characteristic patterning in a package. It may be seen that disturbing patterning occurs mostly at $B = 1$. In this case, the yarn coils producing patterning do not cross at all within a double traverse. Patterning $B = 1$ is practically nonexistent in textured yarn packages. At $B = 2$ there is one crossing point per double-traverse.[48]

Figure 3.53 shows the diameter d_B of the severest patterning in relation to lead angle β for a typical package of spinning industry; the results, however, are applicable to texturing industry. The practical package range is shown with cross-hatching. With increasing traverse and reducing lead angle, patterning shifts to the larger diameters. This also enables risky patterning to be removed from the winding range.[48]

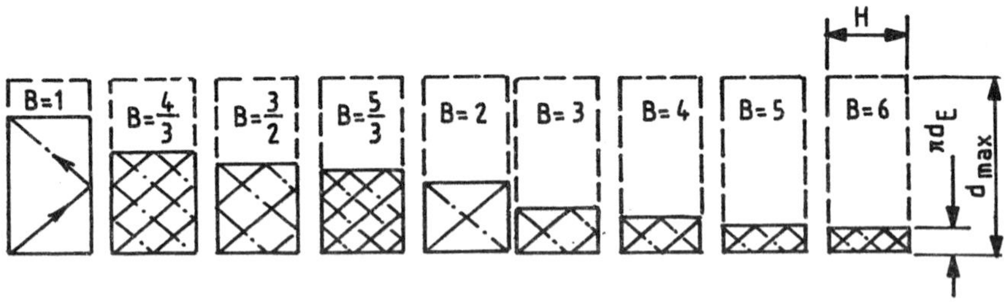

Figure 3.52 Some types of patterning that may occur on a package[48]

Modern winding-up systems incorporate antipatterning means or pattern breakers. This involves periodic changes in the number of double-traverses per minute. Figure 3.54 demonstrates how the number of double-traverses per minute can be varied in accordance with various functions. The practical significance of these functions must, however, be questioned as the inertia of the reciprocating roller prevents it following the sudden frequency changes imposed by the controller at a high frequency of antipatterning. When using a grooved drum, the reciprocating system has even greater inertia.[48]

The purpose of such antipatterning is to interrupt the ridged build-up of yarn coils repeatedly by a network of windings uniformly distributed over the package surface. The amplitude must be great enough to form a network, and the higher the frequency of pattern interruption attainable, the greater the number of networks that can be interposed [48]

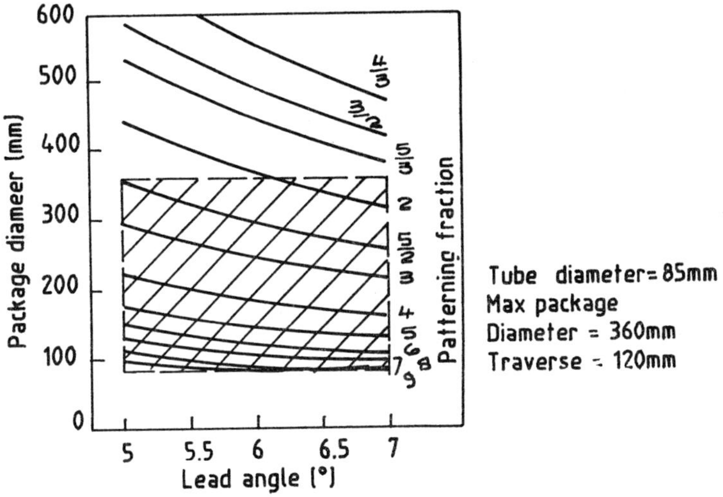

Figure 3.53 Diameter at which patterning occurs d_B in relation to lead angle β[48]

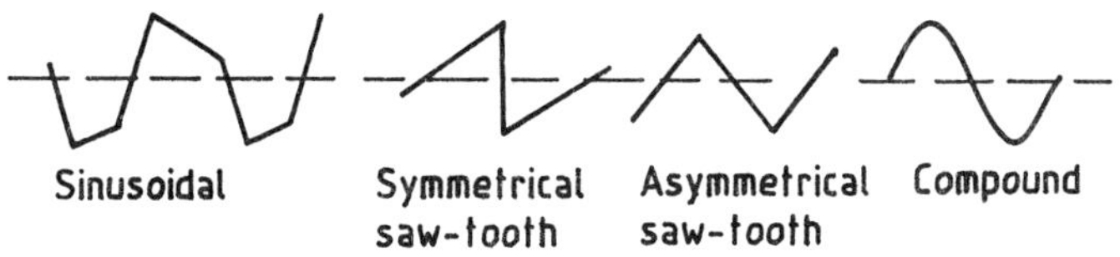

Figure 3.54 Various breaking functions for antipatterning[48]

Lower-traversing speeds and, hence, smaller crossing angles generally result in a distinct improvement of unwinding conditions of textured yarn packages. This is mainly attributable to improved hardness distribution across the package traverse and to a lesser extent to reduce yarn tension at the yarn guide reversal points. The relationships between yarn unwinding properties and the crossing angle were closely examined in extensive test series under production conditions. When the crossing angle is reduced, unwinding problems are distinctly reduced. The traversing speed can be further reduced by optimizing friction roller wrap from the yarn deposit to the package.[47]

The package density greatly affects yarn unwinding properties. As the package density increases, the unwinding becomes easier. But there are limits to package density because of deformation of the cardboard tubes, felting of the package edges, package appearance, filament breaks resulting from excessive winding tension, and uneven dying caused by tension peaks during winding. To compensate the bulging of dense packages, a system has been developed that controls the package ends during the winding process, assuring that the package ends are theoretically made slightly concave. This over compensation results in packages that have practically straight-end surfaces.[47]

In general, the textured yarn is taken up on friction-driven winders with reversing thread shafts, pattern breaking, and package contraction. As an alternative, Barmag offers ribbon free random take-up systems, with the traversing speed changed over individually. The approach to patterning is detected digitally on each winding position, and the ribbon is skipped with optimal timing, so that the traversing speed is varied abruptly.[34]

Filament intermingling also contributes to improved package build-up and as a result, unwinding.

3.2.5.10 Automation

International competition forces texturing machinery manufacturers to lower production costs by means of automation and rationalization. Depending on the end use, the automation can be divided into two main areas: (1) covering the process and production machine development (e.g., continuous processes and automatic manipulators)and (2) covering the machine supervision and control, quality control, material flow, and planning.[49]

Filament yarn-texturing efficiency is highly dependent on the stability of the overall process.Because any intervention in the process involves a certain amount of risk, reducing such intervention is frequently a goal for automation.

Through the use of microprocessors, virtually automatic setting up and scanning of all machine parameters is made possible by some machine manufacturers. A keyboard is provided to feed in the nominal and tolerance values, and a visual display unit displays these values.

It is a common feature that all the machines are equipped with automatic temperature controls, combined with warning and fault indication systems as well as process intervention. In addition, the shaft speeds, both rotational and peripheral, are indicated, and also the setting and display of the traverse speed to the wind-up.[50]

Increased textured yarn package size and weight lead to the development of automatic and semiautomatic doffing systems. The automatic doffer first cuts the yarn ends and feeds these ends into yarn aspirators. The free full packages are then released from the take-up unit and collected on to the doffer. Then an empty package is fed to the package holder, placed against the rotating drive roller, and the sucked yarn is fed to the package for tailing,and, finally, winding.

In general, automation in texturing covers transporting feed packages to the creel, doffing the textured packages with the yarn running, removing the textured packages into intermediate storage facilities, controlling process and quality, packaging, warehousing, and controlling dynamic production. Quality grading is also a subject of automation in texturing. Threading up, defect elimination, cleaning, and maintenance are still manual. However, there has been development work going on to thread up the texturing machine automatically to reduce labor requirements.

Automatic doffing and package handling is the most advanced automation task in the texturing industry. To deliver exact metered packages, the doff cycles should have been automated. With the increase of quality consciousness, automatic quality determining and grading features are all added to this system. In addition to the machine-fixed automation systems, automated guided vehicles or overhead rail transport systems are used for both supply and textured yarn package transport. As a natural consequence of this development, fully automated texturing plants that incorporate all the preceding automation features are in reality. In this computer-integrated production system process, control plays an important role. Continuous monitoring and evaluation of the process parameters brings on-line quality control.

Automation system provides the following benefits to the texturing industry:

1. Less labor intensive
2. Improved quality through "no-touch"
3. Less waste
4. Metered packages
5. Continuous data tracking from creel to packaging
6. Easy adaptation of the line to different requirements
7. Increased production

3.2.5.11 Machine environment

Because of the presence of heaters and fume emission from these heaters, the texturing room requires extensive air conditioning. It is a known fact that only a small proportion of the heat from the heaters of a texturing machine is absorbed by the yarn, and on most machines the excess heat is lost through the fume extraction system.[51]

Apart from the heat and fumes, the noise is also another environmental issue for the texturing plant. The noise emitted from the driving motors, yarn transport shafts, and the yarn suction devices amount to considerable levels. In the presence of intermingling jets, the noise becomes an important issue. Therefore, the machine operators are usually requested to wear ear plugs in the texturing hall.

3.2.6 Properties of False Ttwist Textured Yarns

3.2.6.1 Structural geometry and mechanics of false twist textured yarns

During the development of the false twist texturing process, the structural mechanics of the yarns were also examined by several investigators.[5, 52, 53] In the early days, Arthur[53] described the false twist textured yarn as "a twist-free bundle of twist-lively filaments, which has considerable retractive power derived from the snarling tendency of each filament." This description basically explains the bulk and stretch formation resulting from the snarling of the individual filaments. However, although snarls do appear in false twist textured yarns, they do so in only special conditions, for example, when low twists are used in the texturing process. Other investigators explained the bulk and stretch formation resulting from the setting of filaments in a constant radius helical configuration with several helical turns equal to the yarn twist. It can be demonstrated that, when such helices are untwisted by the same number of turns, one half of the coils becomes reversed as in Figure 3.55, and the textured yarn may be regarded as a system of helical springs in series or parallel, with one half of the springs being coiled in the direction of texturing twist and one half in the opposite direction.

In practice, however, the filaments in most false twist textured yarns are set in the twisted yarn, neither in a completely straight form nor in helices of constant radius, but they migrate between the yarn center and surface. The pitch of the migrating helices may sometimes stay constant but their radii increase or decrease randomly. However, to make some approach to the structural mechanics of these helices, the pitch of the migration will be assumed to be constant, and the decrease or increase of the radii will be assumed to be changing steadily. Figure 3.55c illustrates one section of such a migrating helix.

It can be practically demonstrated that the larger-diameter section of such a helix tends to reverse, whereas the smaller-diameter section keeps the helical configuration in the direction of the twist originally inserted. Because it is complex to develop a mathematical approach on this model, a further-simplified model with a two-radius helical system is used as seen in Figure 3.55d.

The expressions for the tension and torque in a stretched helical spring have been derived by Love.[54] For each side of the helix, these may be written as

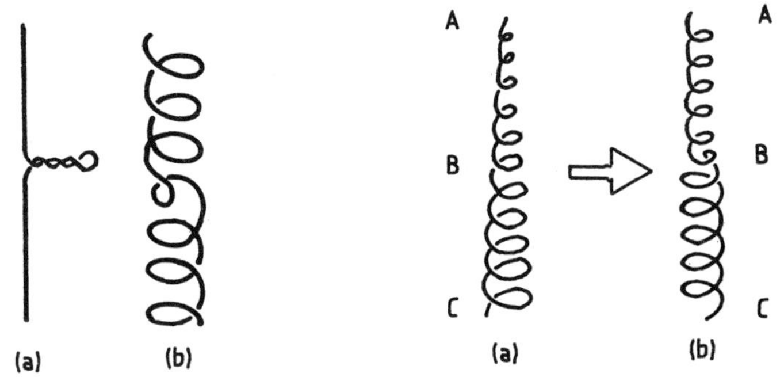

Figure 3.55 Various types of snarl formed in false twist textured yarn[52]

$$F_1 = T\left\{\frac{\sin\alpha_1}{r_1}\left(\frac{\sin\alpha_1\cos\alpha_1}{r_1} - \frac{\sin\alpha_{01}\cos\alpha_{01}}{r_{01}}\right) - \frac{E}{2\upsilon}\frac{\cos\alpha_1}{r_1}\left(\frac{\sin^2\alpha_1}{r_1} - \frac{\sin^2\alpha_{01}}{r_{01}}\right)\right\}$$

and

$$K_1 = T\left\{\cos\alpha_1\left(\frac{\sin\alpha_1\cos\alpha_1}{r_1} - \frac{\sin\alpha_{01}\cos\alpha_{01}}{r_{01}}\right) + \frac{E}{2\upsilon}\sin\alpha_1\left(\frac{\sin^2\alpha_1}{r_1} - \frac{\sin^2\alpha_{01}}{r_{01}}\right)\right\}$$

where T is the torsional rigidity of the filament, and E/ν is the ratio of Young's modulus in bending to shear modulus. In each case, the first half of the expression represents the contribution to tension or torque of the torsional stress in the filament and the second one-half the contribution of bending stress.

The preceding four expressions can be written in the abbreviated form:

$$F_1 = (T/r_0^2)(F_{T1} + F_{B1}) \qquad F_2 = (T/r_0^2)(F_{T2} + F_{B2})$$
$$K_1 = (T/r_0)(K_{T1} + K_{B1}) \qquad K_2 = (T/r_0)(K_{T2} + K_{B2})$$

In these expressions, F_{T1}, F_{T2}, K_{T1}, K_{T2} represent the relative contributions of torsional stresses to the retractive force and torque of the filament; F_{B1}, F_{B2}, K_{B1}, K_{B2} represent the relative contributions of bending stresses. Both torsional and bending stresses make some contribution in each of the preceding four cases. It is concluded that over the practical range of yarn relaxations, i.e., up to 20% relaxation from the full extended condition, it is primarily the bending forces in the filaments that determine the total retractive force. This conclusion is contrary to the common belief that the torsion within the filaments make the greater contribution to the elastic properties. Further investigations suggest that the retractive force of a false twist textured yarn is proportional to the filament linear density and independent of the yarn linear density.

The preceding theoretical approach to the structural mechanics of the false twist textured yarns explains the structure with the reversing helix configuration. One consequence of this approach is that the crimp frequency reduces as the filament extended. This is a consequence of the reversal point between the left-hand and right-hand helices rotating around the helix axis as the filament is stretched and relaxed.

The preceding approach serves to explain, at least in a qualitative manner, many of the observed characteristics of the false-twist yarns and is of assistance in promoting an improved understanding of their properties.

3.2.6.2 Crimp

The purpose of texturing is to set the filament into crimped shapes so that the yarn and fabric produced by such filaments have a substantial degree of stretch and bulk. The crimps produced by false twist texturing process are considered as three-dimensional crimps, which consists of a helices twisted in the S and Z directions, and connected by a reversal zone.[55]

3.2.7 End Uses of False Twist Textured Yarns

Stretch yarns are used in woven and knitted structures for a wide range of end uses; swim suits, half-hose, tights, briefs, stretch covers for chairs, underwear, girdles, ski pants, gloves, stockings, surgical stockings, and many other types of stretch-to-fit garments. Because the yarn bulk is more readily developed in nylon than in polyester, most stretch fabrics contain textured nylon yarns.[14]

False twist textured polyester yarns are probably unique in their development. Within a 5-year period from 1965 to 1970, their consumption has grown by a factor of 12. During the subsequent 10-year period, it again grew rapidly by a factor of 4, reaching more than 1.4 million ton and by 1992 growth to about 2 million ton is reached.

Excellent stretch properties of false twist textured yarn coupled with high bulk rendered their use in mostly tight-fit textile products, such as hosiery, pants, swim suits and underwear. Most (80%) false twist textured yarn is converted into textile goods by circular knitting.

To a great extent, the development of textured yarns has shaped the weft-knitting industries in the United States and Europe. Despite the successful use of textured

yarns in both woven and warp-knitted products, weft knitting is the largest user of textured continuous filament yarns.[56]
Although the false twist textured yarns are both used in weaving and knitting, the fabrics knitted from false twist textured yarns are not free from problems. One of the severest and widely prevalent problems found in dyed knitted fabrics of textured filament yarns of nylon and polyester is known as barré or barriness. It is the term used for lighter or deeper horizontal lines than the rest of the fabric occurring at regular intervals across the fabrics and coinciding with knitting courses. The cause of barriness can vary from filament production to knitting operation. However, considerable amount of barriness is caused by the texturing process. Variations in heater temperature, heater dwell time, spindle speed, and take-up speed could all lead to barriness effect in the final knitted fabric. With the development in all these aspects of the texturing operation, the bariness problem has also been partly eliminated.

3.2.7.1 Market structure of false twist textured yarns

There has been, in the past, a continuous growth in the consumption of textiles, and it is a justified assumption that the demand for clothing will continue to rise, because of the rising world population and rising standard of living. This increase in demand will have to be covered through the synthetic fibers, because sheep rearing and cotton cultivation have their natural limits in the available agricultural land.[57]

False twist textured polyester yarns fill about 60% of the textured yarn market, and most of them are treated by the post-texturing setting. The crispness and hard-wearing properties of the polyester coupled with the covering power and character derived from the texturing process produce inexpensive easy-care fabrics with lasting appeal. They are used in most outerwear applications: ladies' dresses, trouser suits, jackets and skirts, and in men's sport trousers, suits and general leisure wear.[14]

3.2.8 Relations between process parameters and textured yarn properties

During texturing, the thermoplastic yarn is subjected to mechanical stresses at high temperatures. As a result, the texturing occurs because of the changes within the macromolecular structure of the supply yarn. The degree of these changes are closely related with the process parameter as well as the properties of supply yarns. It was observed that while crystallinity of the supply yarn increases on texturing, there is a decrease in the orientation of the crystalline and amorphous regions. The following subsections will investigate further effects of the process parameters on the textured yarn properties.

3.2.8.1 Heater temperature

It is obvious that there should be an optimum setting temperature for every supply yarn depending on its glass transition temperature, melting temperature, and molecular structure. This optimum temperature is either suggested by the supplier of the feed yarn or established by the texturizer through experiments. However, a knowledge of the effects of varying heater temperatures on the textured yarn properties facilitates the easy establishment of the desired heater temperature.

Yarn with 100 dtex, 20 filament nylon 6 yarns, having negligible producer twist, has been textured[58] under varying heater temperatures, while keeping all other process parameters constant. As the heater temperature is increased from 135°C to 210°C with 15°C increments, the breaking load of the textured yarn decreased (Figure 3.56). This drop in breaking load is a function of the structural changes in the polymer during the process. It should be noted that the tenacity of the textured yarns are always, irrespective of the heater temperature, lower than that of the supply yarn. This drop in tenacity largely depends on the structural changes occurring in the polymer, particularly disorientation of the macromolecular chains, during the texturing process.

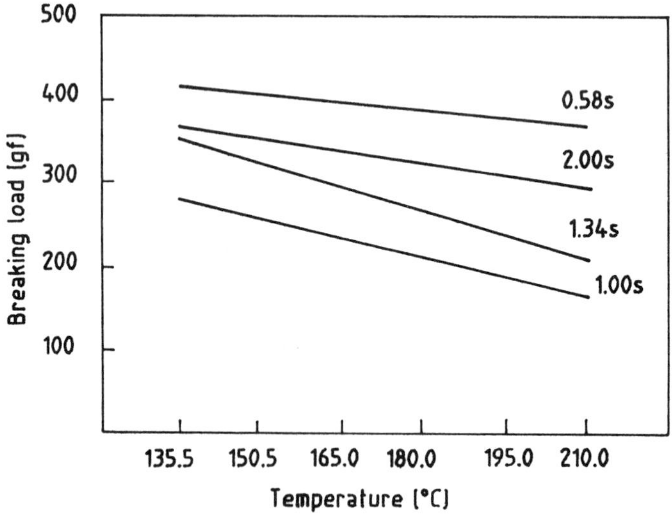

Figure 3.56 Effects of heater temperature on breaking load

Despite the increase in crystallinity of textured yarns, the tenacity and tensile

modulus decrease. However, it is found that the crystallite orientation decreases as the heater temperature is increased. Thus, it would appear that these mechanical properties are more influenced by molecular orientation than by the increase in crystalline content.[59]
In another example,[60] 84 dtex 24 filament polyester yarn was textured on a pin-spindle-texturing machine and processed at varying heater temperatures of 185, 200°C, 215°C and 230°C. It was found that the degree of crystallinity of textured yarn is higher than that of the supply yarn, and the degree of crystallinity increases as the heater temperature and the contact time increases.

The higher crystallinity of the textured yarns, compared with the supply yarns, is due to the recrystallization at elevated temperatures. It is known that when the fibers are subjected to a heat-setting treatment, partial melting and recrystallization occurs, and this results in an increase in the size of the crystals.

In considering the effect of heater temperature on the degree of crystallinity, the following points should be remembered:

1. The temperature of the heater is not the actual temperature of the yarn. In fact, the temperature within the yarn will have a gradient, the surface being hotter than the core. At a fixed contact time the average temperature of the yarn is generally less than the heater temperature. The factors that govern the actual yarn temperature are the heater temperature, the residence time of the yarn in the heater, and the thermal properties of the yarn.
2. For polyester yarn, the rate of crystallization is around 180°C to 190°C

These two points suggest that the rate of increase of crystallinity would be around 200°C which roughly corresponds to the optimum temperature of crystallization for polyester yarn. The experimental studies of Gupta and Kumar[60] also confirm this conclusion.

The supply yarn is cold drawn, resulting in high-molecular orientation. During texturing, this yarn is heat set in a twisted state at temperatures between 185°C to 230°C, after which it is cooled and untwisted. As a result of this heat setting, the molecular orientation is higher in the supply yarn than the textured yarn. There arc two factors affecting the crystallinity orientation. These are the torsional stresses that help to disorient the molecular chains, and the tensile stresses that orient the crystallinity and the amorphous secions of the chain molecules further. As the heater temperature increased, the effect of these factors also increasesbecause of the increase mobility of molecules in the fiber. However, the tensile stresses appear to be the dominating factor; hence, the molecular orientation increases as the heater temperature increases.

Birefringence is a measure of the average molecular orientation of the fiber. The factors that influence birefringence are the nature and amount of crystalline and amorphous material present in the fiber and their respective orientations. The studies[60] have shown that the birefringence of the supply yarn is higher than that of the textured yarn. Because the supply yarn is cold drawn before it is textured, the molecules in both the crystalline and amorphous regions are highly oriented; therefore, its birefringence is relatively high. The textured yarns have lower birefringence because their amorphous content is low, and the orientation of the molecules in amorphous regions is also low. As the heater temperature increases, the birefringence of the yarn decreases because of the decrease of amorphous content and orientation with increasing heater temperature.

Because the heater temperature affects the molecular structure of the fibers as explained earlier, one of the important properties of textured yarn, dye uptake of the textured yarn, is thus affected by the heater temperature. It is generally accepted that dye diffusion is mainly in the amorphous phase of the fiber and is substantially altered by the degree of orientation of this phase. The experimental studies[61] with polyester yarn have shown that as the heater temperature increases, the dye uptake of the textured yarns first decreases and then increases. The approximate minima is observed between 200 to 215°C heater temperatures. This event could be explained in light of the preceding diskussions. It is already demonstrated that as the heater temperature increases, the amorphous phase and amorphous orientation progressively decreases, resulting in a decrease in dye uptake because the total available space into which the dye can go will be reduced. Conversely, the decreased amorphous orientation will encourage the dye uptake further. When the effects of these two causes are combined the observed behavior, first a decrease and later an increase, can be understood.

The yarn tensions before and after the false-twisting unit decrease slightly with increasing heater temperatures. High temperatures soften the thermoplastic filaments more than low temperatures and, hence, causes low-yarn tensions.[62]

To get a good crimp, one needs high heater temperatures to set the yarn properly. However, if a textured yarn with good strength is desired, then low heater temperatures are required. Crimp stability increases with higher heater temperatures, and there seems to be no influence of the heater temperature on the uniformity of the crimp.[62]

The physical properties of textured yarns, such as tenacity, breaking elongation, and crimp rigidity decrease as the heater temperature increases.[63] Figure 3.57 compares the effect of heater temperature on the mechanical properties of textured and flat polyester yarns. For the textured yarns, there is little decrease in mechanical properties as the temperature of the heater is raised. The changes for the textured yarn are generally less than for the corresponding flat yarns. The impression is gained that texturing tends to form a more homogeneous morphology, which is only moderately sensitive to changes in heater temperature.[64]

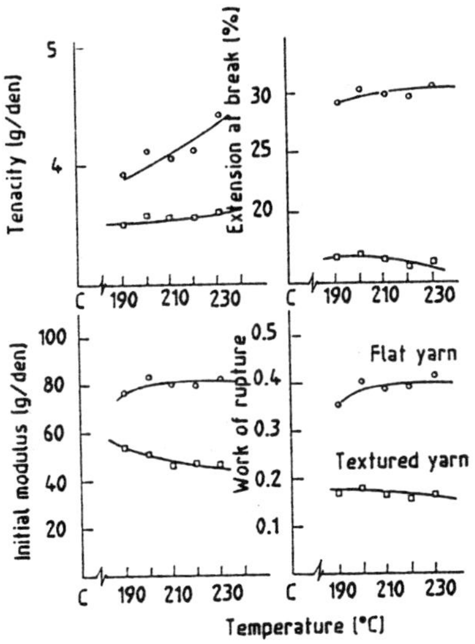

Figure 3.57 Effects of heater temperature on mechanical properties

The effect of heater residence time (at the indicated heater temperatures) on texturing is shown in Figure 3.58, where for two cases (191°C and 228°C heater temperatures), it is shown that crimp rigidity increases from the low levels at short residence times to a constant value from about 0.75 sec. residence time and above.[10]

3.2.8.2 D/Y ratio: ratio of disk surface speed and yarn speed

The ratio of the disk surface speed to the yarn speed usually referred to as D/Y ratio falls typically in the region of 1.4 to 3.0 for most existing systems. The higher the D/Y, the higher the disk speed in relation to the yarn speed, and in general, the more torque is transferred to the yarn. This, in turn, leads both to more twist in the yarn and more slippage between yarn and disk resulting after a point in increasing instability of texturing tension and twist insertion. The careful choice of D/Y is, therefore, clearly important in setting up optimum processing conditions.[65]

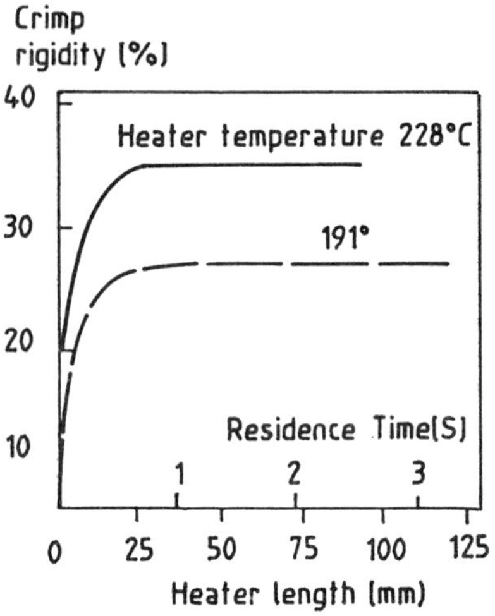

Figure 3.58 Influence of heater temperature and residence time on crimp rigidity

One advantage of friction twisters is the possibility of varying the ratio between the output and input tensions (T_2/T_1) over a wide range on either side of unity so that the output tension can be higher or lower than the input tension or equal to it.

The method for adjusting the tension ratio is to vary the ratio between the twister and yarn speeds (the D/Y ratio). The higher the D/Y ratio, the lower is the tension ratio.[18]

Figure 3.59 shows the relationship between T_1 and T_2 for a typical ceramic-coated disk aggregate at a range of D/Y ratios.[65] At the lowest D/Y, the yarn is being partially pulled through the disks, and stretched between the aggregate and the output nip. This gives a high T_2 tension. As D/Y increases, the forwarding action of the aggregate increases and the tension ratio, defined as T_2/T_1, moves through unity to less than one, where yarn is effectively being overfed by the aggregate.[65]

3.2.8.3 Process speed

Figure 3.60 illustrates the development of texturing speeds during the period 1950 to 1990.[15]

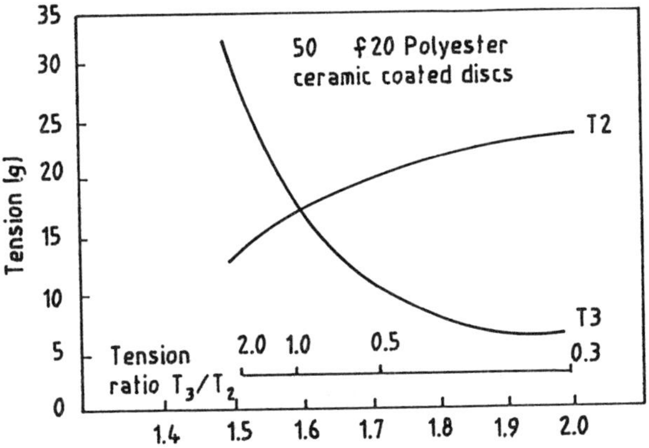

Figure 3.59 Relationship between T_1 and T_2 for a typical ceramic-coated disk aggregate at a range of D/Y ratios[65]

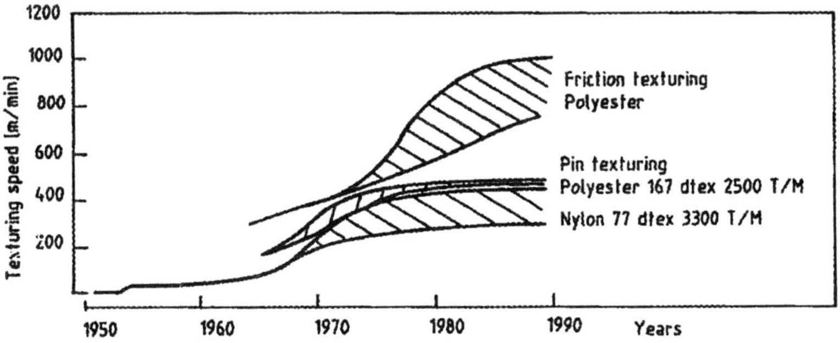

Figure 3.60 Development of texturing speeds during 1950 to 1990[15]

The process speed is inversely proportional to the contact time of the yarn in the heater. It is thought to be appropriate to study the effect of contact time under the title of process speed. By varying the process speed, the contact time was varied. Consequently, the cooling time also varied accordingly, with the length of the heater being fixed. As the contact time is increased from 0.58 second to 1, 1.34 and 2 sec, it is observed that the breaking load first decreases, then increases and, finally, decreases again as seen in Figure 3.61. This rather peculiar behavior of the textured yarn cannot be easily explained, because it is normally expected that the increased contact time means increased heat transfer from the heater to the filaments and, thus, increased texturing, i.e., decreased tenacity. However, it is observed that the effect of higher thermal input because of the increase in contact time is overshadowed by the variation in actual twist imparted to the yarn. Figure 3.61 also includes the twist levels recorded from the heater. This similarity in trend explains the peculiar effect of the contact time. Conversely, the variation in the twist imparted is thought to be due to the slip in friction false-twisting assembly.[58]

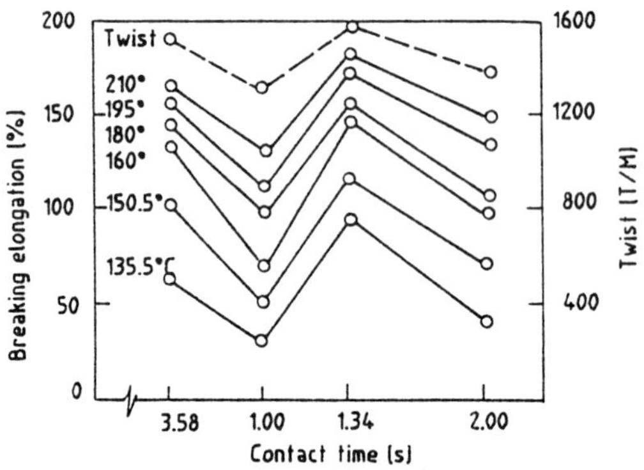

Figure 3.61 Relationship between contact time and breaking elongation at various heater temperatures[58]

As machine speed increases the minimum texturing tension required to suppress tension instability and surging increases. This can be taken as the most significant fact in the interaction of friction false-twist texturing and machine capability.

3.2.8.4 Yarn tension

The yarn tension level required for stable yarn travel in the texturing zone rises with the texturing speed. From the relationship between spinning and drawing speed and the related yarn tension in the texturing zone, one can determine a minimum yarn tension as a function of the texturing speed that is required to avoid surging. The level and course of this minimum yarn tension are independent of the spinning speed for a given material and yarn linear density.[47]

Optimization of yarn tensions before and after the texturing spindle is an essential factor for satisfactory texturing. A ratio of 1/2 to 1/3 is found to be giving desired texturing conditions. To adjust the tension ratio, draw ratio, twist level, and the heater temperature are the parameters to be used.

The effect of tension level on the structural properties of 76 dtex 24 filament polyester textured yarns were investigated by Gupta and Amirtharaj.[66] The researchers found that the crystallinity of all textured samples is higher than that of their supply yarns because during heat setting at 215°C, partial melting and recrystallization occurs; hence, the degree of crystallinity increases. It is also found that at low tensions the crystallinity is high because of the fact that under low tension, large and well-developed crystals are allowed to grow and crystalline and noncrystalline phases become distinctly separate. The tension during texturing reduces the possible degree of disorientation. Thus, the higher the tension in texturing zone, the greater the orientation of the crystallites.[66]

During texturing, filaments undergo considerable deformation, and their cross-section is no longer circular. There is, therefore, short-term variation of birefringence in the filaments. However, because of the importance of the birefringence as a structural parameters, their approximate, average values are measured at varying tension levels and found that as the tension increases birefringence first increases; however, at high tensions the birefringence slightly decreases. This is thought to be due to the increased twist levels at high tensions.[66]

As the formation of large crystals are encouraged at low tensions, the dye uptake of the filaments will be enhanced. However, the increased tensions do not allow the formation of large crystallites; hence, the dye uptake decreases. Therefore it could be concluded that the tension is the main factor controlling dye uptake.[66]

3.2.9 Relations between Supply Yarn and Textured Yarn Properties

The retractive force of a textured yarn can be written in the following way:

$$\frac{F}{D} = \frac{C_1 E d}{A^2}$$

where

F = retractive force of the yarn

D = total linear of the yarn bundle

E = filament yarn elastic modulus

d = filament linear density

A = diameter of the helix

C_1 = constant

It can be seen that the retractive force is directly proportional to the modulus and filament linear density. Because $E_{N6} < E_{N66} < E_{PES}$, the retractive force for a given total linear density and filament count will increase in the same way, i.e., $F_{N6} < F_{N66} < F_{PES}$.[67]

Because the square of the helix diameter is directly proportional to the total linear density, the preceding equation becomes

$$F = C_2 Ed$$

Therefore, one sees that for yarns of different linear densities, but twisted to same twist angle, the specific retractive force depends on only the filament linear density. For a given filament linear density, the retractive force per decitex is inversely proportional to the total linear density. This also explains the known fact that, for a given linear density, the stretch and recovery is better for yarns with lower numbers of filament.[67]

3.2.9.1 Yarn material

The main supply yarn materials for the false twist texturing process are polyester and polyamides (nylon 6 and 6.6), with the former being dominant. Only 25% of the total polyamide filament production is textured, and this is predominantly for fine hosiery yarns. The use of polypropylene filament yarns is somehow inhibited because of the poor texture development in these yarns. The reason for this may lie in the fact that polypropylene supply yarn has a much higher crystallinity and a poorer thermal conductivity than polyester and polyamides.[68]

3.2.9.2 Total yarn linear density

Because the false twist texturing process requires a high twist insertion followed by the removal of this high twist, the total yarn linear density is seriously restricted to fine yarns. The range only varied from as fine as 8 dtex to a practical maximum of 330 dtex. Although the fine yarns up to 167 dtex are dominantly textured by friction disks, the coarse yarns between 167 to 330 dtex are usually processed by either ring or crossed-belt-type false twister.

As the total yarn linear density increases, the energy required to twist increases rapidly. As a result, either the process speeds need lowering, or some other methods of texturing may be needed.

3.2.9.3 Filament linear density

The torsional rigidity of a circular filament is proportional with the fourth power of the filament diameter. Therefore, as the filament diameter increases, the twisting becomes increasingly difficult. The difficulty does not only stem from the increased torsional rigidity of the filament, but the heat transfer to the core of the filament also becomes difficult, making the texturing process unsuitable.

Consequently, fine linear density filaments (<2 dtex per filament) are usually preferred for an effective texturing. However, the microfilaments (< 1 dpf) are also gaining increasing importance in the texturing industry. Because of the low tenacity of the microfilaments, the texturing of these filaments requires special attention. The yarn's path as well as the friction disks should be specifically adopted for the texturing of microfilament yarns.

3.2.9.4 Filament cross-section

The visual impression and handle of the fabric can be varied within certain limits by means of the cross-sectional shape of the filament. Texturing is associated with a deformation of the cross-section, which, in some cases, leads to hexagonal structures giving rise to flat surfaces on the filaments that can produce reflections and, hence, glitter. Filament yarns with profiled cross-sections (i.e., trilobal, pentalobal, and octalobal) are likewise found to be deformed after texturing. However, this deformation is less than with round-starting cross-sections, so that the formation of these flat surfaces is likewise less pronounced, and glitter effects are, thus, avoided.[57]

Yarns produced from slit films can have certain advantages over the conventional multifilament yarns produced by spinning and drawing, particularly in terms of economics. The films apparently have biaxial orientation to begin with, as opposed to the uniaxial orientation of conventional yarns, so compared with the conventional yarn, slit-film yarn will differ in structure and properties.[69]
When such slit-film yarns are textured with a pin twister, it is observed that the initial flat "filaments" are considerably distorted on texturing. However, the textured yarns have properties similar to those obtained for conventional multi-filament textured yarns.[69]

3.2.9.5 Supply yarn spin finish

The function of the spin finish is to give desired frictional properties and control the static electrification at all stages of filament processing. For most of the filament yarn processes, such as spinning, drawing, and winding, low filament to contact surface is required. However, for the texturing process itself, a relatively high filament to contact surface friction is required. It is neither economical nor practical to have two different spin finishes at two different stages of filament yarn processing. Therefore, a single spin finish is required to meet these two contradicting demands.

For the false twist texturing operation, heat stability of the spin finish is the deciding factor controlling the selection of the finishes. A heat-stable spin-finish component should posses the following properties:

1. Minimum color development on heating
2. Minimum formation of insoluble or only partly soluble hard and resinous decomposition products at the temperature of the heater surfaces
3. A flash point appropriate to the temperature of the heater surface
4. Appropriate volatility characteristics such as the degree to which volatility is used to meet friction requirements, and environmental conditions (i.e., odor, toxic fumes, and condensation of volatile substances) on the machine parts[70]

Spin finishes on the filament surface are subject to high temperatures as they pass through the heater and it is necessary to deal with the fumes produced. Also spin finish and particles of polymer, monomer and oligomer can be deposited as a white powder (snow). Snow accumulation on sensitive working surfaces (e.g. godets) can cause mechanical errors which show up in the yarn[3].

Too smooth a lubricant will cause loss of friction against the twisting element and result in insufficient or irregular crimp. Too rough a lubricant increases polymer scuff at the twist element, and filament breaks because of excessive friction. Too volatile a lubricant means evaporation loss, excessive application means lubricant scuff, and too low an application means polymer scuff and tight spots. These lubrication factors are amplified at higher texturing speeds. There is also a connection between the application amount and crimp, depending on the lubricant type. Tight spots are also dependent on the yarn tension ratio before, between, and after the spindles at medium lubricant levels. Because this tension ratio is directly dependent on lubricant friction, this relationship is also a question of lubrication.[47]

3.2.9.6 POY and FOY effects

When a POY is used as supply yarn to the false twist texturing process, there may be a difference between the outer and the core filaments. The combination of torque and tension applies a greater elongational strain on the outer filaments of the bundle, and these are also likely to be heated to a higher temperature.

However, the work done on the filament bundle appears as heat inside the bundle and this helps to overcome the heat-transfer problem. Nevertheless, it is possible for sheath and core filaments to have quite different mechanical and thermal stress histories by the time that the process is completed. This, hence, affects the consistency of dye uptake.

A further complication is that by drawing the twisted bundle in the softened state, the individual filaments develop flat surfaces, such that the filaments becomes roughly pentagonal or hexagonal in cross-section. This produces a sparkle in the finished product and also affects the hand.

During the simultaneous texturing POY feed stock, polymer ageing becomes an important phenomenon toconsider. The drawing characteristics of the polymer alter markedly as the polymer approaches its shelf life and working with overaged feed stock can cause considerable operational difficulty. Thus great care must be taken to manage the stock of feed yarn properly.[3]

Textured POY has a higher skein shrinkage, lower torque, improved dye uniformity potential, and slightly (5% to 10%) reduced tensile strength compared with normally crimped yarn. The greater crimp development tends to produce a double-knit fabric with a bulkier and drier hand than a fabric produced from textured fully drawn yarn.[71]

Among the advantages of using POY are more uniform dying, better processing performance, more tolerant texturing process, lower torque values in the textured yarn, easier high-speed processing, and larger package sizes.[71]

Table 3.2 shows the properties of false twist textured yarn made with a single heater process from fully drawn and POY supply yarn. Yarn tenacity is somewhat lower for POY supply, but the reduction is not significant in subsequent processing or use. High-crimp contraction under the relatively heavy 5 mg/denier load is a significant difference for the POY product, leading to an increase in stretch potential of woven stretch fabrics.

Table 3.2
Properties of Textured Polyester Yarns[72]

	POY Supply	Fully Drawn Supply Yarn
Linear density, denier	157	158
Tenacity, g/denier	3.9	4.2
Elongation, %	22	26
Shrinkage on boil-off, %	3	3
Crimp contraction at 5 mg/denier load, %	14	10

The underlying advantage of POY as supply yarn is that the texturing starts from a low crystallinity base, and yarn does not have to undergo the crystal melting and reformation of a fully drawn supply yarn. Therefore, POY textured yarn has better yarn-to-yarn crystalline uniformity, and this results in superior results in pressure dying.[72]

3.2.10 Process and Quality Control

Process production efficiency is easy to monitor as it is based on the presence or absence of the yarn. Each threadline-processing position of false-twist machines is equipped with detectors immediately before the take-up that trigger a cutter after the creel if the thread breaks. By scanning the detectors and cutters, it is possible to determine the threadline number down, time down, time back up, and, hence, the duration of the fault. Management information (e.g., calculations of machine, shift, operative and plant efficiencies, and identification of persistently faulty threadline) can be derived. This information can then be used to assist in planning of maintenance, highlighting of training needs, and as a basis for bonus payment schemes. In addition, these data are considered by management to represent a reflection of supply yarn quality.[73]

The process and quality control efforts are mainly concerned with rationalization of yarn feed, textured yarn package quality, textured yarn package handling, and textured yarn quality control.

Operational data collection is an important topic in the acquisition of texturing machines. Most modern draw-texturing machines are, therefore, equipped with means that measure and compare heater temperatures, delivery speeds, rotational speed of the twisters, crossing-angle in the package, and D/Y ratio with the nominal value. End breaks are optically signaled via the stop motions, specifying the location and machine.

Convenient data collection and processing via microcomputer and data acquisition terminals with trouble printout and direct connection to central computer are also possible.

3.2.10.1 On-line yarn-tension measurement

A possible location for a tension transducer in the texturing zone would be between the cooling plate and the false-twisting unit, in the zone where the yarn is subjected to the most sensitive and critical heating, cooling, and twisting operations. However, measurement in this zone would affect the process by reduction in passage of twist by the yarn guides of the transducer. Moreover, some possible tension effects resulting from the false-twisting device may not show up. A more satisfactory location for a transducer is in the T_2 zone immediately after the false-twisting unit, where its presence will have minimal effect on the twist and tension balance of the process.[65]

Measurement of the yarn tension is a suitable method for detecting faults during texturing, as many faults affect the yarn tension curve. A stable process would give a stable tension trace just after the false-twisting unit (Figure 3.62). Examination of tension faults show that there are three broad classes of fault.

1. Continuous instability, as could be caused by worn machine components; incorrect processing specification; incorrect threading of the machine; or a gross change in supply yarn, which affects the process specification
2. A change of mean, as would happen if the heater temperature control failed; or a progressive change in supply yarn linear density occurred
3. A tension transient, resulting from a sudden tension flick that occurs randomly; and for no clearly understood reason[73]

Simulation of known and likely faults during texturing makes it possible to determine a relationship between the yarn tension signal and the cause of the fault, so that the point at which the fault occurs can also be detected.[74] Intensive efforts are being made to develop a yarn tension sensor integral with the twisters. The OLT sensor as seen in Figure 3.63 is such a device that measures the yarn tension directly after the twister for each position on the machine for on-line quality control. It is possible to continuously register tension peaks as well as the long-term drift from the mean value. Built-in evaluator emits visible signals when the tension exceeds the higher or lower limits of the preset values of the peak and mean tension.[19]

The On Line Tensor (OLT) developed by FAG and the Unitens yarn tension sensor developed by Barmag are two major on-line yarn tension measuring instruments that facilitate the previously mentioned measurements.[75]

3.2.10.2 On-line yarn bulk measurement

Yarn volume may also be selected as the criterion to monitor the textured yarn quality. The "Qualitor" developed by Rieter-Scragg is the device used for measurement of yarn volume. It can be used on only double-heater machines, where it is located immediately after the second heater. Because of overfeed, yarn passes through this hollow device in bulked state. The rate of measurement is at 10 digital values/sec. It is argued that this device reacts most sensitively to disturbances and changes during processing. The collected data are transferred to a microprocessor-controlled system "Introl" for setting, control, and display of texturing data.[19]

3.2.10.3 On-line twist measurement

To maintain a constant twist level during friction twisting, a feedback control system has been proposed. The twist level in the running threadline was measured through, for example, the helix angle of the twisted yarn. A control signal was then fed back to the friction twister, which adjusted the friction twisting unit to keep the twist level constant.[71]

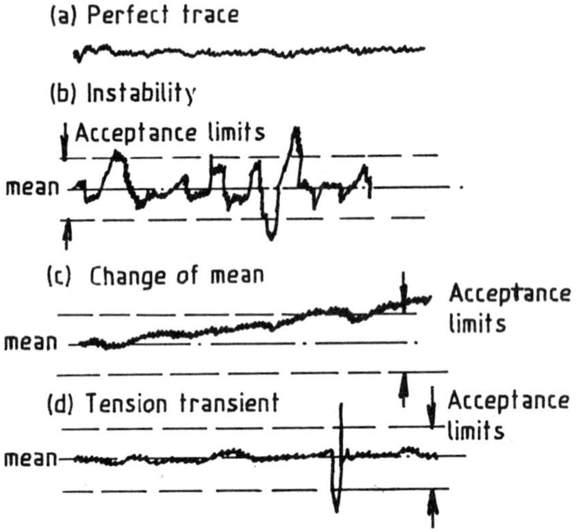

Figure 3.62 Types of tension faults observed in T_2

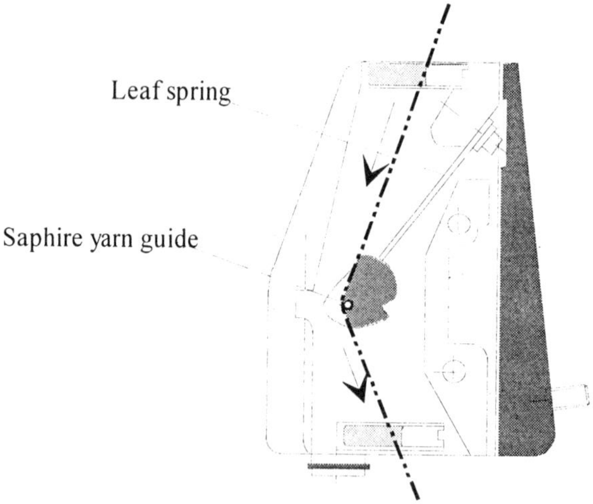

Figure 3.63 On-line tension sensor

Inside the twisting zone of a false twist texturing machine, the yarn has both rotational and longitudinal movements. A point on the surface of this yarn has a speed component in the circumferential direction of V_u and in the axial direction of V_{ax} (Figure 3.64). The resultant value of the two components is the surface speed of the yarn V_O, whose direction in relation to the yarn axis is determined by the twist angle α. The formula for the yarn twist can then be obtained as $T = \dfrac{\tan\alpha}{\pi d}$. It is clear that when the helix angle α and yarn diameter d are measured, the twist can be calculated.[76]

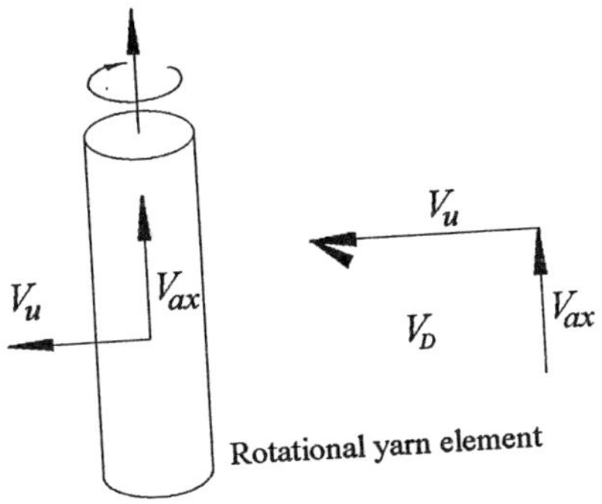

Figure 3.64 Circumferential and axial components of yarn surface speed

To measure the helix angle, either a razor blade or, even better, a tiny ceramic pin that ensures low friction and reduced wear are used. For the yarn diameter measurement, conversely, a small sensor with laser light has been developed.[76]

The operating principle of the on-line twist-measuring system is shown in Figure 3.65.

Several devices had been developed to measure the helix angle, but the response would have to be rapid. Although these devices would probably prove to be invaluable as twist-measuring instruments for day-to-day quality control in the commercial process, the feedback system appears to be complicated to implement on all commercial machines.[71]

Other textured yarn faults, such as thin and thick places, snarls, and loops can also be detected on-line using feeler heads and appropriate electronic control devices. Such a feeler head is developed by Textechno under the trade name of

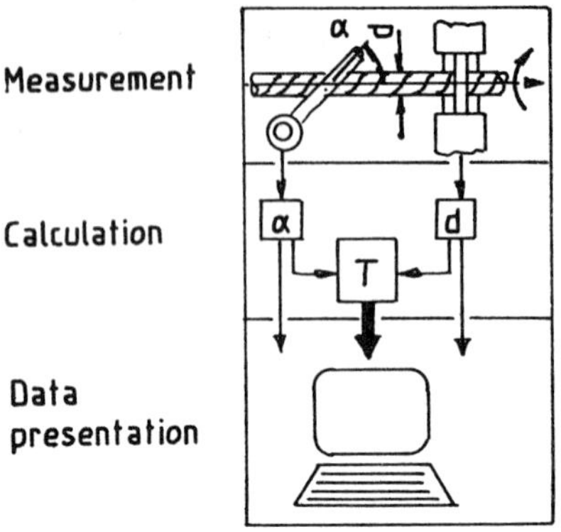

Figure 3.65 The operating principle of the on-line twist-measuring system

ELKOTASTER[77] (Figure 3.66), which continuously monitors the thickness of the yarn and send signals to the control unit. If the yarn thickness is more than the predetermined level, either the operator is warned or the unit is stopped.

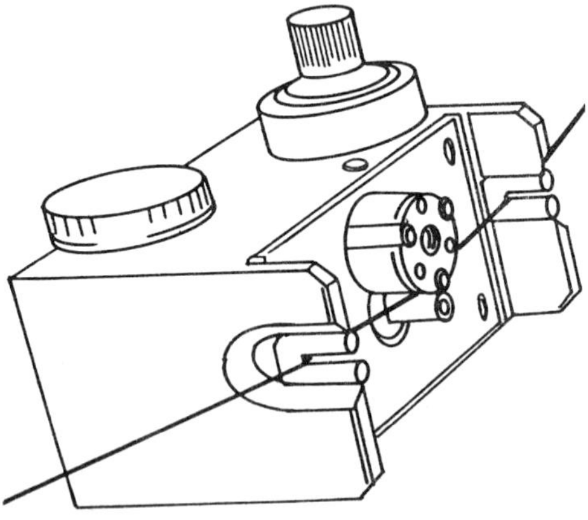

Figure 3.66 ELKOTASTER[77] continuously monitors yarn thickness

3.2.11 Textured yarn tests and instruments

The important factors for yarn-quality consideration are linear density, tenacity and elongation, crimp properties (crimp contraction, rigidity, retraction, stability, and module), package density, broken filaments, uncrimped portion, partially drawn portion, bulk variation, snarling tendency, and spin-finish content.[78]

3.2.11.1 Linear density variation

Linear density variation is a long-term variation and is not common. It is mostly the responsibility of the filament manufacturer to ensure that a yarn suitable for false twist texturing operation has less than 3% inear density variation.

However, draw texturing creates a short-length variation in the linear density of individual filaments basically because of the drawing operation. This short term variation may only be measured by using an Uster Evenness Tester having carefully washd the filament, since the spin-finish residues cause false readings.

3.2.11.2 Tensile properties

Tensile properties, tenacity, and breaking elongation, are again directly related to the supply yarn used. Therefore, there is little that texturizer can do. For different end uses, different tenacity and breaking elongation values are recommended. A lower tenacity than recommended may cause filament breakages; similarly, a high elongation may lead to defects, for example, in a woven fabric as tight or shiny ends as the yarn is susceptible to tension variations during warping, weaving, or weft unwinding. Tensile properties of textured yarns are tested on any tensile tester. The evaluation of tensile properties, however, are not only made by the tenacity and breaking elongation values, but also the shape of the stress-strain curves.

Rothschild, a test equipment manufacturerer, has a range of testing apparatus for the dynamic measurement of textured yarn elongation-force/extension properties. One piece of Rothschild equipment, Tensomodul, has been one of the standard instruments for this purpose. Another instrument, the Dynamic Elongation Force Tester R-3075, operates on the same principle as Tensomodul, in which a running yarn passes between two feed-roller systems with preset underfeeds or yarn extensions variable between 0% and 1,000%, with the yarn tension being continually recorded. The instrument can be used in the following three modes:

1. Elongation-force uniformity: The extension is held constant; the diagram shows periodic or random variations in elastic properties.
2. Dynamic force-elongation properties can be obtained between two preselected elongation limits.

3. Force-extension hysterisis diagrams are obtained by cycling the running yarn between two elongation limits.[23]

3.2.11.3 Crimp properties

3.2.11.3.1 *HATRA Crimp Rigidity Test*

This test[79] consists of suspending a hank of yarn in water at 20°C under a load of 0.1 g/denier. After 2 min, the length of the hank is measured (L_1). The load is then reduced to 0.002 g/denier, and the length again measured after a further period of 2 min (L_2).

The percentage crimp rigidity is defined as $\dfrac{(L_1 - L_2)}{L_1} \times 100$.

Crimp Retraction

A modified HATRA Crimp Rigidity Test procedure is used to define the Crimp Retraction.[80] Textured yarn sample of 90 cm length under a pretension of 0.002 g/denier is subjected to a total pretension of 0.1 g/denier at room temperature, and after 2 min its length (L_1) is measured. The load is reduced to 0.002 g/denier and the length again measured after a further period of 2 min (L_2).

The percentage crimp retraction is then taken to be $\dfrac{(L_1 - L_2)}{L_1} \times 100$.

Rothschild Crimp Retraction Tester is an established instrument. It continually monitors the crimp-retraction force in textured yarns as the yarn passes through a crimp-developing heater with a preset level of overfeed. It is capable of showing cyclic variations of crimp level resulting, for example, from spindle speed variations or surging. The effects of slipping or worn-out feed systems are easily detected.[23] The other instrument for the crimp contraction tests is Dynafil M of Textechno.[81]

3.2.11.3.2 *False-twist textured yarn characterisation (DIN 53840)*

The quality of textured yarn is measured by its crimp properties as these have a profound influence on fabric quality. The German standard DIN 53840[82] for the measurement of crimp properties is usually the method followed by major textured yarn manufactures. This standard specifies a method of determining the crimp contraction, crimp module, and mechanical crimp retention. These three characterize the crimp properties of a yarn.

According to DIN 53840, the crimp of a textured filament yarn in the form of a skein of specified linear density (2,500 dtex) is developed by treatment with hot air (120°C for polyester and polyamide) for 10 min while the specimen is subjected to a pretensioning load (0.001 cN/dtex). Having conditioned for 24 hours, the straightened length l_g of the specimen is measured 10 sec after a high tensile force (0.2 cN/dtex) has been applied. After a specified recovery time (10 min), the contracted length l_z of the filament yarn, shortened by the effect of its crimp, is measured under a low pretensioning. Next, a specific tensile force (0.01 cN/dtex) is applied and 10 sec after the application of this load, the resulting length l_f, which serves to characterize the elongation behavior of a textured yarn, is measured. To determine the mechanical crimp retention, the specimen is subjected temporarily (10 sec) to a high tensile force (1 cN/dtex) the contracted length l_b subsequently being measured under a low pretensioning load (0.001 cN/dtex) after a specified recovery time (10 min) has elapsed (see Figure 3.67).

The crimp contraction E% or KK is then the reduction in length of a textured filament yarn as a result of its crimped structure when the crimp is developed. It is expressed as the ratio of the difference between the straightened length l_g and the contracted length l_z to the straightened length l_g.

$$E\% = \frac{l_g - l_z}{l_g} \times 100$$

The Texturmat M of Textechno is an automatic crimp contraction apparatus in which the retractive power of small hanks of textured yarn is automatically measured between two predetermined loading conditions. This is simply an automation of the standard, manual Crimp Contraction or *KK* test. Crimp stability can also be assessed by this tester.

The crimp module K% characterize the elongation behavior of a textured yarn in the range of crimp elasticity. The crimp module K is the ratio of the difference between the straightened length l_g and the length l_f of the yarn under a specific tensile force, to the straightened length l_g

$$K\% = \frac{l_g - l_f}{l_g} \times 100$$

The mechanical crimp retention *B%* indicates the crimp contraction of a textured filament yarn following a defined tensile strain.

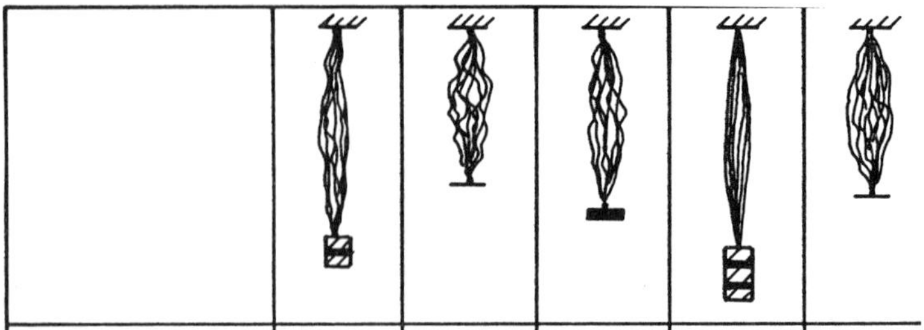

Load g/tex	2	0.01	0.1	10 or 20	0.01
Duration of load	10 sec	10 min	10 sec	10sec	20 min
Length of hank	1g	1z	1f		1b

Figure 3.67 Operation steps of DIN 53840

To determine the value of the mechanical crimp retention, the specimen is subjected temporarily to a high-tensile force, F_{KB}. Then the length l_b of the filament yarn, contracted as a result of its crimp, is measured.

The mechanical crimp retention $B\%$ is expressed as the ratio of the crimp contraction calculated on the basis of l_b following the application of a tensile force, to the crimp contraction E.

$$B\% = \frac{l_g - l_b}{l_g - l_z} \times 100$$

Further details of these measurements may be found in DIN 53840.[82]

The crimp stability is defined as the percentage ratio of the retraction of a hank of yarn after and before a load of 1.5 gf/dtex has been applied to it for 30 sec.

It is well established that crimp values, although of great value for quality control, do not define the behavior of yarn in fabric form. For example, a certain value of retraction in the tube test or Hoechst K test can be obtained in a set yarn by a certain difference in temperature between the first and second heaters. If the temperatures of both heaters are reduced, with their difference still maintained, the crimp will stay at about the same value, but the stability of that crimp to high tensions will be reduced. This is well introduced and explained by previous workers.[83]

One can obtain a similar effect by increasing speed. An increase in speed on a double-heater machine usually increases the crimp value because of the smaller setting effect of the second heater. The effects of reduced heating times on the crimp stability of the single-heater draw-textured yarns are shown in Figure 3.68.[83] High speeds clearly reduced the crimp stability, especially at lower processing temperatures.

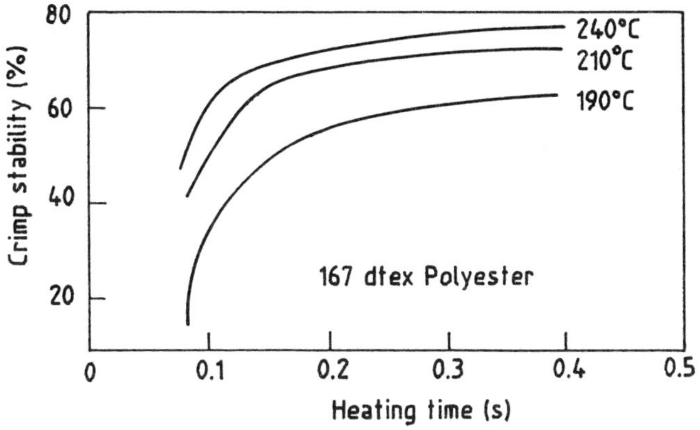

Figure 3.68 Effects of reduced heating times on crimp stability of single-heater draw-textured yarns[83]

Continuously measured crimp contraction: Textured Yarn Tester (TYT) To monitor textured yarn bulk and dyeability characteristics, a continuous two-zone measurement concept was developed and implemented on TYT (Figure 3.69). The most important advantage of the TYT is that it eliminates a test time of more than 20 hours required for conventional Knit-Dye-Grade-Cull (ship) testing (TYT takes 3 hours) and the human error attending all of the handling involved.[84]

As Figure 3.69 illustrates, the yarn enters the instrument through a precision tensioner to remove the soft crimp and allow accurate length measurement by the constant-speed input rollers. After the input rollers, the yarn passes through a low-pressure relay jet and is then delivered to the sensor of the first zone of the TYT. The calibrated sensor maintains the tension in this zone to an accuracy of ± 1 mg with a set point as low as 50 mg. From this sensor, the yarn passes vertically through a convection-type noncontact heater that develops the crimp and shrinks the yarn. The yarn then passes onto the bottom rollers. The speed of the bottom (intermediate) rollers is controlled by the sensor to maintain a constant low tension in the heater. The computer uses the roller speed differential between the input and intermediate rollers to calculate the total contraction (total recovery).

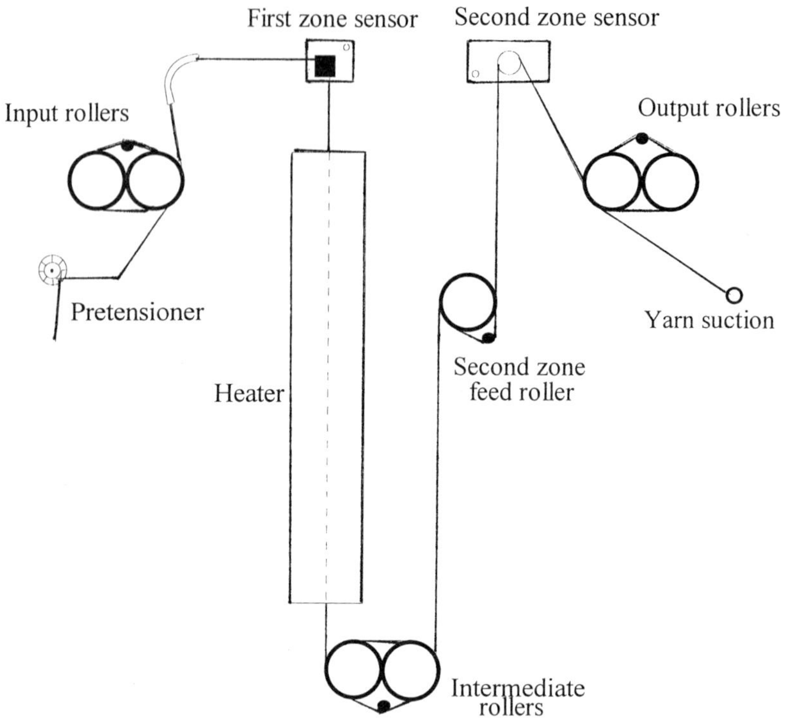

Figure 3.69 TYT apparatus of Lawson-Hemphil

The second zone is similar in principle, but utilizes a heavier sensor to control the output rollers of the TYT. The tension level is usually set to match the input level and remove the crimp developed in the heater. If the crimp is removed, the only change in the yarn left to measure is caused from shrinkage. The computer uses the roller speed differential between the input and output rollers to calculate the filament shrinkage.

Textured yarn packages are placed on a six-position rotating turntable for testing in sequence. Packages are knotted into a continuous threadline, then they go to a storage feeder. Yarn is accepted after "soft crimp" has been removed and fed to the first zone sensor, then through a heater tube to an output roll system. A computer compares the speed of the output rolls with the input rolls, calculates the value of total recovery, and then displays and prints it out. A sensor at the top of a second (heater) zone applies a tension comparable with initial tension entering the TYT. By comparing the speed of the final output rolls (after second zone) relative to the constant speed of the input rolls, the computer calculates, displays, and prints filament shrinkage (zone two shrinkage). The yarn enters an aspirator and is collected in a waste box. Before entering the final feed rolls, the yarn passes through a knot sensor to enable automatic sequencing and identification of the packages.[84]

Computer then continuously calculates the total recovery, filament shrinkage, and true filament crimp using the speed data provided by the TYT. The computer also compares the results with the predetermined limits. If the results are outside of these limits, the computer then prints these results in red and activates a flag indicating the faulty package that can easily be separated from the accepted packages by the operator.

Another commercial equipment for continuous crimp contraction force measurement is Textechno's DYNAFIL M[81] (Figure 3.70). The measurement principle of this device is also given in Figure 3.71.

Yarn coming from the bobbin that is fed in runs through a tensioner where a constant tension (0.2 g/dtex) is given the incoming yarn. The yarn is driven by the first roller at a constant speed. Then it is transferred over a weight loaded compensating roller, which keeps a defined preliminary tension force that remains constant during the testing. The second drive roller is designed as a two-step galette, where the rear galette feeds the yarn into the measurement zone while the front, replaceable galette removes the yarn from the measurement zone. The rotational speed of this stepped roller may be adjusted precisely. Yarn may either be wound onto a spool or sucked into a discard bin.

From the rear galette of the second drive roller, the yarn goes through the heater downward, to the measurement roller, which is connected to the force measurement head, and then runs outside the heater upward to the front galette of the second drive roller. The force transducer measures the yarn tension, which is created by over- feeding or under-feeding of the yarn and the heater temperature.

A recorder records the variations in this tension over a preselected interval.

3.2.11.4 Hardness of textured yarn packages

The shore hardness of textured yarn packages should be 65° as measured by shore hardness tester. Too high package density cause defective yarn by creating crimp loss, and a soft package causes unwinding problems and tension peaks during unwinding resulting in defective fabric.[78]

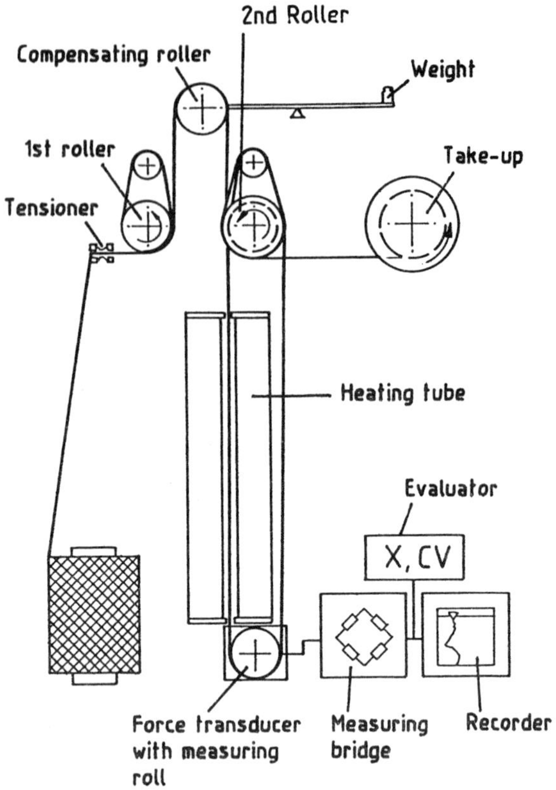

Figure 3.70 DYNAFIL M[81]

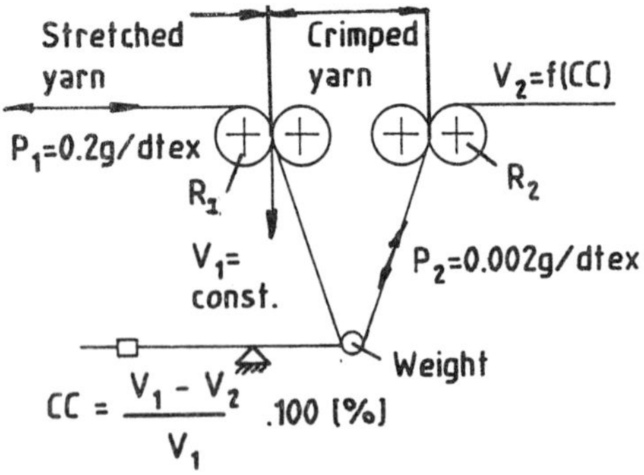

Figure 3.71 Principle of continuous crimp measurement

3.2.11.5 Unwinding properties

The most important question that requires an answer regarding the unwinding properties of textured yarn packages is that what causes the yarn to break during unwinding.

In investigating this point in detail, it can be concluded that unwinding breaks (and some fabric faults) are the result of tension peaks during unwinding, which are multiplied through guides on weaving looms/knitting machines to a level that causes either break, a short pick, or tight course. These tension peaks are in turn caused by loops, broken filaments, knots, and slubs snagging during unwinding. Because it is tension peaks that cause unwinding breaks and some fabric faults, one of the means of assessment of package performance is the measurement of unwinding tension.[85]

A method has been developed for measuring the unwinding properties of textured yarn packages in which tension peaks during unwinding are recorded and computer processed. The instrument, Package Performance Analyser of Rieter Scragg, allows for unwinding speeds of up to 2,000 m/min and is equipped with a stop motion cutter system that acts to prevent lapping if suction fails. The system is also used to determine the effective number of stops within the test period. Extensive measurement series have shown that determining tension peaks over a period of 3 hours is sufficient for reliable deductions regarding the running properties to be expected.[47]

Any filament breakages cause entanglement and lead to stoppages during further processing. Therefore, packages should be carefully inspected against such faults.

The R-3060 Automatic Bobbin Analyser, a Rothschild instrument, analyzes the ribbon-breaking cycles, traverse variations, and edge formations of cylindrical packages. The instrument simply monitors the position of the yarn coming off the unwinding package, and the traverse pattern of the package is printed out in several modes of presentation that indicates the particular characteristic being investigated.[23]

3.2.12 Common yarn faults

3.2.12.1 Uneven dye uptake

Although most of the filament migration occurs during the twisting stage, some occur during the untwisting stage too. If those core filaments that have been heated less and perhaps drawn less come to the surface, then there are differences in dye uptake between these and the nonmigrated surface filaments. It is the thermal history of the surface filaments in the final textured yarn that determines the color shade. Therefore, an even heating and cooling of filaments is of great importance.[3]

The wear of twisting disks shows up as mechanical faults in the yarn and produces strict periodicities in the final product, which give rise to moiré patterns[3].

3.2.12.2 Tight spots

Tight spots are short regions of highly coherent twisted yarn, the twist direction being the same as that in texturing zone (Figure 3.72). The occurrence of these tight spots is a reflection of the nonsteady-state operation of the texturing. Another probable explanation for the formation of tight spots is the variations in the supply yarn structure along the threadline.[86 - 88]

The level of twist in a false-twist textured yarn affects the yarn characteristics and, hence, those of the fabrics from them. During the friction false twisting, if the torque gets too high, or the tension too low, tight spots are produced. Twist slips through the disks and a concentration of real twist appears in the final product as seen in Figure 3.72. The fact that it is cancelled by a longer length at a low and opposite twist does not alter the fact that a tight spot is present. At the point of high local twist the bulk is unable to develop and the surface appearance changes. Usually tight spots are a warning of impending surging and such instability usually causes severe filamentation or end breakages[3].

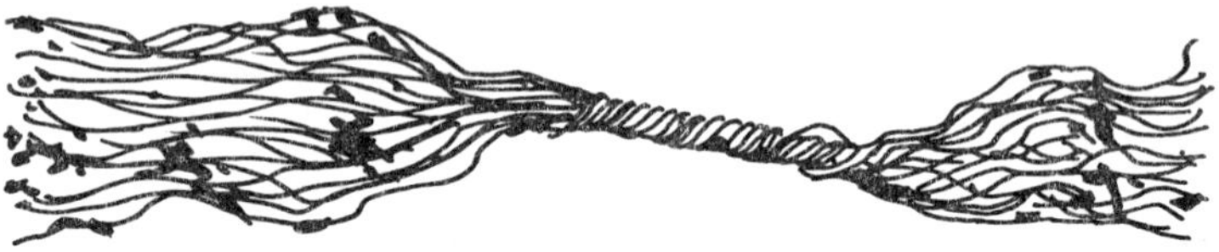

Figure 3.72 Tight spot; a short region of highly coherent twisted yarn

If the overfeed in the first zone is too high, the tension falls and tight spots and poor set are produced. If the tension drops too low, there are excessive end breaks.

An increased tendency to tight spots is generally encountered in simultaneous draw-textured yarns as compared with conventionally textured yarns. This may be explained by a higher initial untwisting torque in the yarn arising from a greater filament adhesion, due in turn, to a closer moulding together of the filaments during the combined drawing and twisting stage of the process. This may, perhaps, be coupled with a greater variation along the yarn length due to slight draw-point oscillation.[89]

The tight spots can be removed (i.e., the twist can be redistributed) by running the yarn over a sharp edge under high tension[87] close to the exit of the false-twisting unit. It is believed that this guide or "tight-spot eliminator" fulfills its function by holding back and accumulating any negative twist in the yarns, which then provides extra torque for untwisting the positively twisted potential tight spot as it appears at the exit of the twisting unit. It has also been observed that, in a situation in which tight spots do occur, their incidence is often reduced when the supply package to the texturing machine is unwound in such a way that the small twist generated by unwinding overend is opposite to the direction of false twist, which, thus, gives rise to twist in the negative direction immediately after the twisting unit.[89]

For similar reasons, any positive twist occurring after the false-twisting unit, which is due to small changes in spindle speed, will encourage the formation of tight spots.

It should also be mentioned that, in a two-heater process, any tight spots present in the threadline after the spindle that pass into the second heater are set permanently into the yarn by the second heater. Therefore, it is impossible to remove them completely, and they remain visible even in the final fabric. For this reason, a tight-spot level of more than one per 20 m of yarn is regarded as commercially unacceptable.[89]

3.2.12.3 Filamentation

The fundamental cause of filamentation is the stress breakage of individual filaments in the twist zone. The inherent filament to filament uniformity at melt spinning is clearly important here, and fiber manufacturers are continually striving to upgrade their quality in this respect. Disk and heater parameters also play a part in the onset of the filamentation.[65]

3.2.12.4 Kinks and snarls

Because of the setting of filament molecules in a helical configuration, filaments tend to take up this form physically. This tendency, on the one hand, creates the bulkiness of the yarn; on the other hand, it creates snarling of the individual filaments and, hence, the yarn. There exists, therefore, a certain balance between the desired bulking effect and undesirable snarling. During a well-adjusted, steady operation, this balance is normally established. However, because of certain problems in twist insertion or heating, there may be a local true twist accumulation creating a tight spot or a long-term pseudo-true-twist accumulation creating snarls and kinks.

3.2.13 Common Process Problems

3.2.13.1 Surging

The phenomenon of threadline instability in false-twist texturing, i.e., surging, was reported as early as 1973 by Klein and Trummer[90] in a study of draw texturing by friction twisting. This undesirable phenomenon occurs as the texturing speed rises. Surging may also be described as threadline tension instability, i.e., periodic yarn tension fluctuations of varying amplitude.[47]

The yarn tension fluctuations cause several more or less sinusoidal variations in yarn properties. A yarn textured under surging conditions has a typical structure in which twisted, i.e., untextured, and nearly untwisted, i.e., textured, places alternate. The frequency of these variations depend on the spinning and texturing conditions, such as yarn linear density, texturing speed, and draw ratio.[47]

Figure 3.73 shows the various tension profiles observed during an experiment on a commercial texturing machine (Rieter-Scragg's SDS800B machine) with a draw ratio of 1.582 and varying delivery roller speeds from 710 to 830 m/min. It is obvious from this figure that the tension goes through five patterns with the increase in yarn speed:

1. Almost constant, small random fluctuations at lower tension
2. Intermittent surging
3. Continuous surging
4. Intermittent surging again
5. Sawtooth-like curve at higher tension

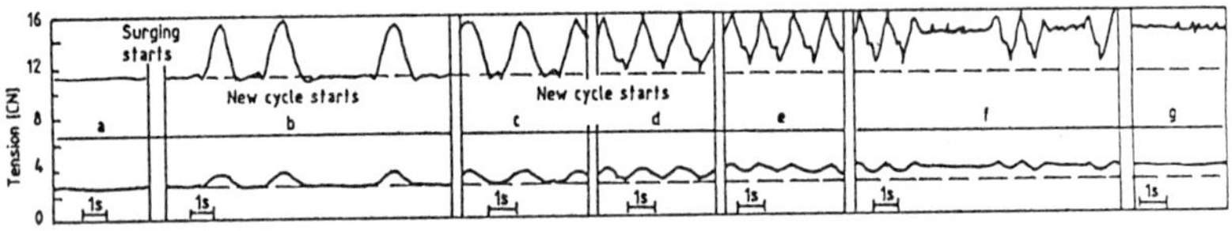

Figure 3.73 Various tension profiles observed during an experiment on a commercial texturing machine (Rieter-Scragg's SDS800B)

It is also clear from the figure that as the yarn speed increases, the length of one surging cycle becomes shorter, and the tension amplitude becomes smaller[91].

The five observed patters are commonly assigned to three modes of operation;

1. Presurging mode: pattern 1

2. Surging mode: patterns 3, together with the transitional zones 2 and 3
3. Postsurging mode: pattern 5 [91]

Control of surging can be accomplished by increasing texturing tensions and the draw ratio, but this may reduce bulk and cause broken filaments and yarns. The main parameters that influence instability, i.e., surging, are texturing speed, draw ratio, D/Y ratio, yarn linear density, and twist level. One explanation of threadline instability may be based on the idea of ballooning. When a disruption in the yarn allows a deflection from the correct path, tension must be counted on to restore the threadline. Because the threadline is being twisted quickly, there is a (centrifugal) force acting to throw the yarn out and expand the original deflection. This balloon-forming force is equal to the product of the mass of the loop, the average distance from the center line, and the rotational speed. An expression for the restoring force from tension can be obtained from the trigonometry of the loop. At the conditions where the threadline is barely under control, the two forces have the same magnitude.[84]

The draw ratio, D/Y ratio, and texturing speed have an appreciable effect on the level and variation of the yarn tension in the texturing zone.[92] With a high draw ratio, there seems to be no longer any surging even at high texturing speeds (up to 1,200 m/min). As the draw ratio decreases, the surging effect is displaced from high texturing speeds toward the low ones. The susceptibility to surging also decreases at low draw ratios and high texturing speeds.[92]

At sufficiently high draw ratio (1.3 or 1.4 depending on the supply yarn), no surging occurs, irrespective of the D/Y ratio. The yarn tension in front of the twisting unit increases with a rise in the D/Y ratio, the tension after the twisting unit decreases. At a lower draw ratio (1.2 - 1.3), the likelihood of surging increases with a rise in the D/Y ratio. These trends apply to all texturing speeds. If a still lower draw ratio (1.1 - 1.2) is selected, the likelihood of surging increases with a rise in draw ratio. This trend is also encountered at all texturing speeds. Yarns textured at low draw ratios, and low D/Y ratios have, as a rule, no adequate crimp properties without any surging occurring.[92]

Surging results in a deterioration in the properties of the textured yarns. Periodic yarn tension variations during texturing produce alterations in Z and S twist in the yarn. This effect is attributable to the fact that the yarn slips through the twisting unit, and the twist is not released, as in the case of uniform yarn tension.[92]

3.2.13.2 Heater cross-over

Heater cross-over is a common process problem whereby a portion of threadline will "climb" out of a groove of the primary heater and cross-over into an adjacent groove occupied by another threadline. Heater cross-over is a serious problem if it occurs frequently, because either one or both threadlines sharing a common heater groove usually break; hence, texturing performance suffers.[93]

Familant[93] observed that cross-over actually occurs in stages. Initially a segment of a threadline will begin to oscillate on the heater. The amplitude of the oscillation increases in time; finally, a cross-over occurs when the oscillating section of the threadline reaches the top of the heater groove. It has also been observed that the threadline always crosses over in the twist direction, i.e., on the right side of the heater groove for S twist and on the left side of the groove for Z twist.

The theoretical study of Familant[93] considering the force balance of a yarn in the heater groove, has shown that the yarn oscillates on the side of the heater groove; as the friction coefficient between the yarn and the heater wall increases with time, the yarn climbs further and results in a cross-over. The reason for the increase of coefficient of friction is mainly the contamination of the heater wallbecause of the spin finish from the filament surface. Therefore, it could be concluded that low friction spin finishes reduce the possibility of cross-over phenomenon. The steep wall design for the heater groove could also reduce the cross-over. It is also suspected that dynamic factors, such as surging, could lead to cross-over too.

3.2.13.3 "Snow" formation

Snow deposits are in general a housekeeping problem. However, they also lead to problems with texturing machine drive belts, oiling systems, intermingling jets, and area exhaust systems. Snow deposits are commonly vacuum cleaned as often as once a day while the machine is running. Even with this frequent attention, it is often difficult to prevent excessive build-up.[94]

Snow is made up of small polymer particles loosely held together by spin finish. Snow consists of about 60% polymer from the filament and 40% spin finish material.[94] It is generated by abrasion of the yarn during friction twisting and can be a considerable problem on draw-texturing machines, for it is deposited on drive belts and interferes with the correct working of other parts of the process, such as oil application and intermingling. It is also deposited on the friction surfaces themselves, and its build-up has been associated with process instability.

Mechanism of snow formation. A basic model for yarn/friction disk interaction is illustrated in Figure 3.74, which shows a cross-section of the yarn and disk in the contact region. The items of interest are the disk (which acts as the source of motion), the yarn, and the thin layer of finish (which transmits the force from disk to yarn). In reality, the yarn/disk contact zone will be made up of several small contact points. If there is some direct contact between the disk and yarn, all force will not be transmitted through the finish as indicated. However, the major effect of finish on friction twist responses justifies the assumption that force is transmitted from the disk to yarn through a finish film.[94]

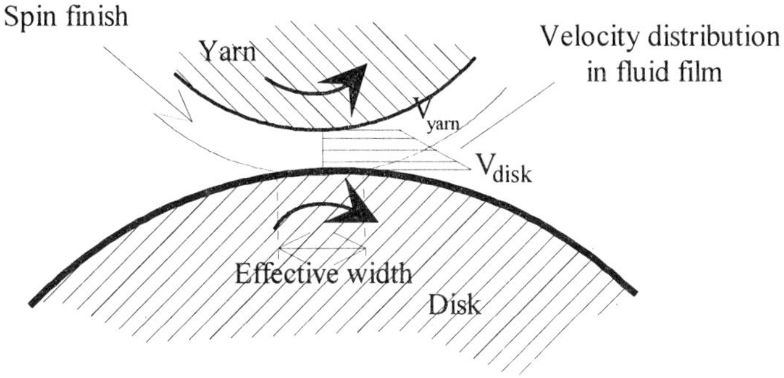

Figure 3.74 Model of yarn-disk interaction

The mechanism of snow generation suggested by the model is "local yarn heating caused by viscous heat generation in the thin finish film softens the yarn locally and causes small pieces to be wiped off during the process." The model predicts that the snow generation could be reduced by increasing finish viscosity, effective contact width, wrap length, and fraction of yarn heated while increasing the torsional rigidity, finish level, D/Y ratio, and fractional heat flow to the yarn. Although it seems that the surface characteristics of the friction disks would be an important contributor to the snow gereration, the tension ratio before and after the false-twist unit has been found to be a major determining parameter for the snow generation (Figure 3.75).[94]

3.2.13.4 Noise

The false twist texturing machines show some unusual features in their noise emission. Nevertheless, they share many of the characteristics of other types of yarn-processing equipment. The machines are long and narrow, usually two-sided, with power, gearing, and control facilities in consoles at the end of the machines. They have rotating "spindles" driven by belts running the length of the machine. Yarn-feed and take-up rollers are invariably driven by machine-length drive shafts, and the yarn is usually taken up onto packages with associated winding rollers and mechanisms.

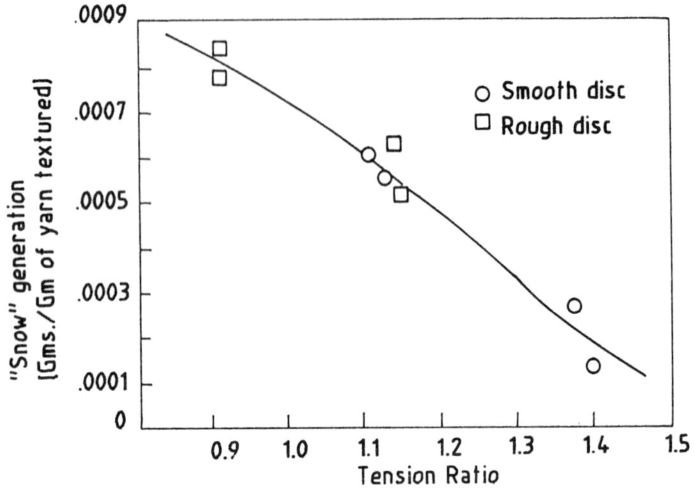

Figure 3.75 Effects of tension ratio and surface characteristics of the friction disks on snow generation

The relative contribution of the principal sources of noise to the overall level varies considerably for the different makes or generations of machines, but these principal noise sources are generally recognized as the following:

- False-twist insertion unit and its drive systems
- Yarn-feed rollers and transmission elements
- Yarn take-up rollers and transmission elements
- Yarn take-up traverse units
- Main-drive motor(s) and power distribution elements
- Ancillary systems (including intermingling jets, suction doff, fume extract, etc.)[95]

The main source of noise in the pin-type false twist texturing machines is the spindle and associated drive equipment. Noise-shielding equipment is frequently used to reduce the amount of noise emitted to the environment. Improved quietness is one of the advantages of the friction twisting units.

3.2.14 Error Sources in False Twist Texturing

The first error source is the raw material. The supply packages should be acclimated before creeling. This is accomplished either by air conditioning the raw materials in the warehouse or by unpacking the raw materials at the machine 1 to 2 days before the processing.[96]

3.2.15 Modifications to the Basic (Conventional) Process

3.2.15.1 Draw Texturing

Until the mid-1970s, the starting material for any texturing process was a fully drawn continuous filament yarn. Increased texturing speeds by friction false-twisting elements, however, made it possible to combine the drawing process with the texturing process; hence, either completely undrawn or partially drawn yarns are textured on one machine to produce false twist textured yarns.

Although, in the mid-1970s, there have been difficulties and concerns about the possibility of draw texturing because of the inadequacy of friction aggregates and lack of supply yarns with long enough shelf life, the processing quality of friction false twisting elements reached such a state that nearly all the present false twist texturing machines operate on the draw-texturing principles; the development of partially oriented supply yarns with necessary storage life also contributed to this development.

3.2.15.1.1 *Two modes of operation*

The draw-texturing can be carried out in two ways.

1. Drawing and texturing in two successive zones (sequential)
2. Drawing and texturing in one zone (simultaneous) texturing

In sequential texturing, the partially oriented supply yarn is drawn before texturing operation on the same machine. Therefore, it requires extra draw zone with feed rollers, hot pin, or hot plate. In simultaneous texturing, however, the drawing occurs between the feed rollers and friction false-twisting elements simultaneously with twisting, heating, and cooling, making the process more economical. Hence, the simultaneous operation today is dominant.

Sequential texturing operation produces similar yarns with characters similar to that of the yarns produced by the basic process. However, to facilitate drawing before the texturing operation, an extra set of feed rollers and, in the case of polyester, temperature-controlled drawing pins or plates are normally required. This method, therefore, becomes more expensive in machine cost than simultaneous texturing which, also has certain drawbacks –especially in yarn quality.

During a simultaneous texturing operation, it is likely that some filaments will remain partly undrawn, whereas some others are fully drawn during heat setting because of the filament migration. As seen in Figure 3.76, a yarn is both twisted and stretched at the same time during the simultaneous texturing operation. In this case the outer layers on the filament bundle sustain a higher tension and the inner layers a lower tension. The filaments at the center of the bundle sustain only the tension from the stretching force applied to the yarn plus a superposed torsional stress about their own axis. If the residual draw of the POY is now drawn out by the tensile force, the outer filaments will be overdrawn, whereas the filaments inside the bundle will be incompletely drawn.

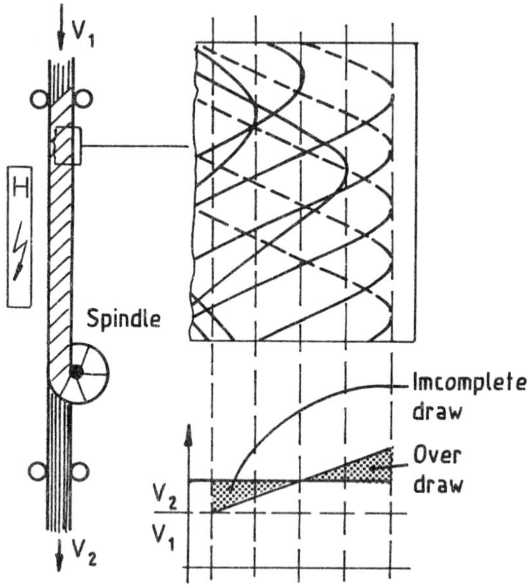

Figure 3.76 Draw distribution in a yarn

When such incompletely drawn filaments reach the false-twisting element, they are drawn at this stage; hence, the crimp imposed on them is removed by this drawing action. The filaments having no crimp then protrude from the body of the yarn and form so-called daisies with stronger dye uptake. If the tensile force is increased, to draw the inner filament layers adequately, the number of broken filaments increases, because in the outer layers of the yarn the drawing exceeds the breaking point.[4]

Another special feature of simultaneously draw-textured yarns is their distorted cross-sections (Figure 3.77 shows the comparion of the cross-sections of simultaneous draw-textured and conventional textured filaments). This distortion can give rise to slightly higher values of crimp, –hence, a crisper hand– but, more important, to a glitter effect, which may be attractive in some end uses but less desirable in many others. A small loss of tenacity may also result because of the distortion of individual filament cross-sections.

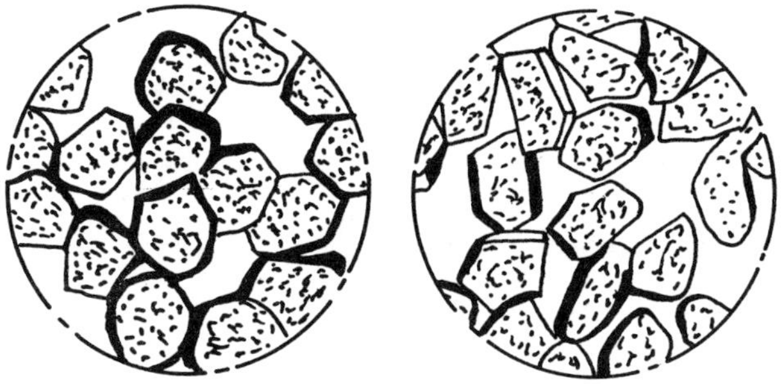

Figure 3.77 Comparison of cross-sections of simultaneous draw-textured yarn

Dye uniformity is generally improved in the draw-textured yarn compared with the yarns textured with the basic false-twist texturing. This is probably due to, first, the elimination of the separate draw-wind or draw-twist stage and the subsequent storage on package or bobbins, and, secon, to the large distortions that occur at the molecular level, swamping any minor differences that may have occurred in the undrawn or partially drawn yarn. An alternative explanation for the improved dye uniformity is that the drawing during the simultaneous texturing process occurs at a lower rate and, hence, increased heating time.[97] The increased dye uniformity is also associated with the increased light reflectance of the simultaneous draw textured yarns.[98] Typical spectra of the intensity of light reflected from the two types of yarn are shown in Figure 3.78.[98] The draw-textured yarn has a much greater intensity, and it is also distributed differently in wavelength. It is argued by Thwaities[98] that this approximately 10-fold increase is responsible for the improved dye uniformity of the draw-textured yarns. Furthermore, it is argued that this variation of light reflectance is due to the variation of yarn and filament linear density during a draw-texturing process.

The roots of these advantages of the draw-texturing process undoubtedly lie at the molecular level. In the first place, the large draw effectively breaks down the intermolecular bonds, which, in basic process, require a time delay at high temperature for loosening, particularly with nylon. This more rapid breakdown of bonds is partly attributable to the mechanical action of the draw and partly to the heat of draw, suggesting shorter heaters or lower heater temperatures. The simultaneous texturing process also has the advantage that the molecular orientation is immediately in the direction of crimp.[97]

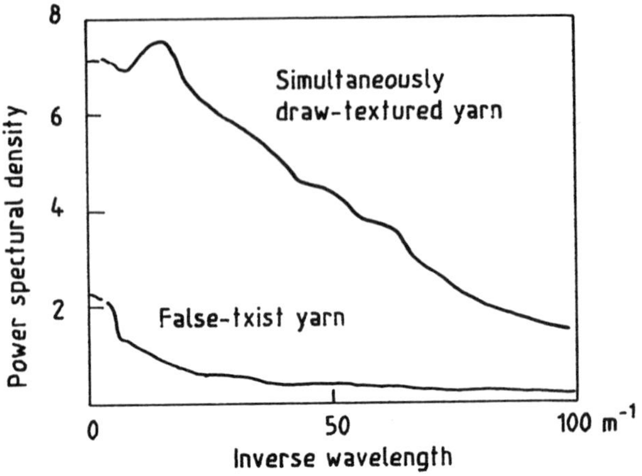

Figure 3.78 Typical spectra of intensity of light reflected from two types of yarn[97]

It is generally recognized that the draw ratio must be carefully controlled and will be less in simultaneous draw texturing than in the basic processing. This change in draw is presumably due to the fact that the outer parts of filaments in torsional distortion and outer filaments in the twisted yarn will be more stretched than the rest because of the filament migration. This means that if these parts of yarns or filaments are at the normal draw level of the basic process, the average draw will be lower.[97]

Because the level of tension during the simultaneous texturing operation is set by drawing tension, it may not be the optimum tension for efficient spindle performance. During the sequential texturing process, however, the yarn tensions in both drawing and texturing zones can be independently adjusted by the speed ratios; hence, the optimum tension level can be obtained.

The relative advantages of simultaneous and sequential draw texturing are outlined in Table 3.3.

It is generally accepted that POY is the supply yarn for draw texturing rather than undrawn yarn. This is basically because of the short life-span of the undrawn yarn, which is expressed in terms of only days, whereas POY has a life span of months, not weeks.

Table 3.3
Comparison of Draw Texturing Techniques

Simultaneous	Sequential
Machine and Process. Simpler machine design leading to lower machine and maintenance cost, to easy threading, high speed and efficiency, lower torque required to twist yarn	*Machine and Process.* Supply yarn has normal linear density, suitable for nylon because of cold draw of nylon
Yarn. Improved dye uptake and regularity, lower twist liveliness, better crimp contraction (even better than the basic yarn), poor crimp stability	*Yarn.* No tenacity decrease, comparatively little cross-sectional distortion of the filaments, textured yarn properties such as crimp contraction and crimp stability, are similar to basic product
Fabric. clearer structure, especially in knitted fabrics, good handle, more brilliant	

In simultaneous draw texturing, the speed ratio between the input rollers and the intermediate rollers (on a double-heater machine as seen in Figure 3.79) or between the input and output rollers (on a single-heater machine as seen in Figure 3.80) determines the degree of drawing. To calculate the starting-point draw ratio, the following formula is used:

$$Draw\ Ratio = \frac{Supply\ Yarn\ Linear\ Density}{Draw\text{-}textured\ Yarn\ Linear\ Density}$$

This starting draw ratio is used to texture the yarn, the textured yarn is then sampled and tested. According to the test results, the draw ratio is adjusted. For example, if the yarn strength is low, several broken filaments exist, and the running efficiency is low, this is an indication that the draw ratio is too high and should be decreased.[33] It is reported from industrial experience[71] that increasing the draw ratio reduces the depth of shade obtained in dying but increases the yarn tenacity.

Increasing the twist level also improved some POY products, permitted lower tension to be obtained, reduced the number of broken filaments, and gave a definite improvement in dyeability.

3.2.15.1.2 *Supply yarns (POY) for draw texturing process*

Basic structural characteristics of a POY, compared with undrawn and fully drawn yarns, are shown in Table 3.4.

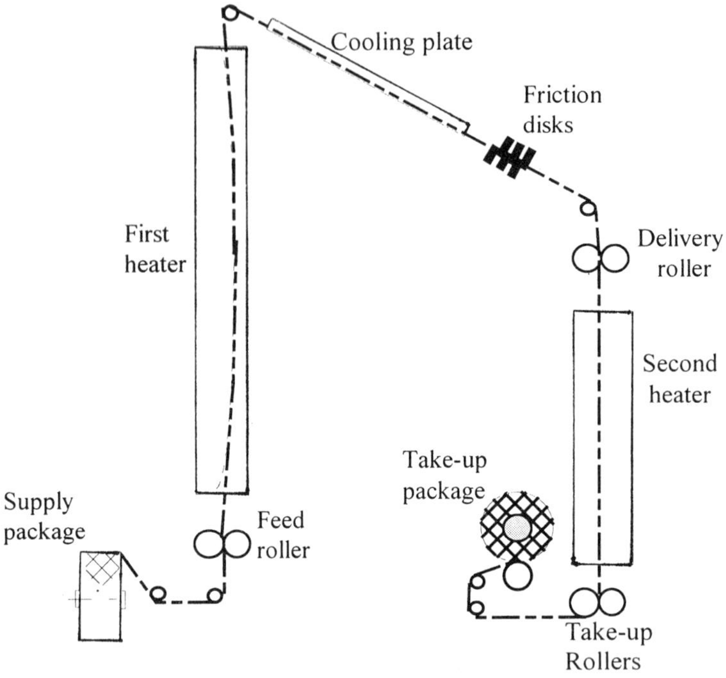

Figure 3.79 Schematic illustration of a double-heater draw texturing machine

POY yarn is distinguished particularly by a higher level of orientation than undrawn yarn, while retaining a low level of crystallinity. X-ray and density data suggest somewhat higher crystalline organization for POY compared with undrawn yarn, which explains the better thermal and age stability of POY.

Table 3.5 shows typical tensile properties of the tnree yarns. The relative levels of orientation in POY and undrawn yarn account for differences in tenacity, elongation and modulus. In POY, the high shrinkage is due to the combination of relatively high orientation and low crystallinity[72].

POY is thought to be successful in simultaneous texturing primarily because of its long shelf life and the lower tension ratio at the false-twist spindle.[99]

However, undrawn supply yarn is also of some interest to many yarn producers, provided that they are textured within 8 to 10 days of extrusion.

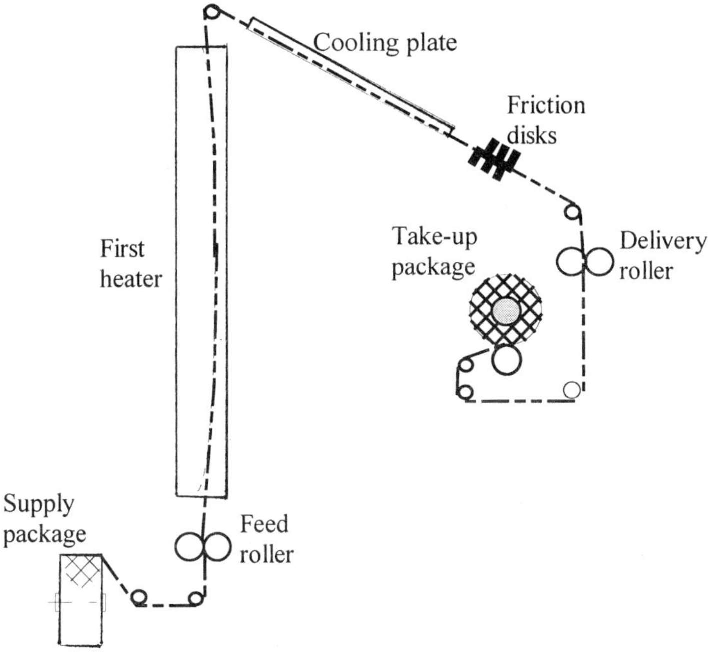

Figure 3.80 Schematic illustration of single-heater draw texturing machine

Table 3.4
Structure Properties of POY, Undrawn and Drawn Yarns[72]

	POY	Undrawn Yarn	Drawn Yarn
Orientation			
Birefringence, Δn	0.038	0.005	0.18
Sonic velocity, Km/s	1.83	1.56	3.5
Natural draw ratio, X	1.45	3.00	–
Crystallinity			
X-ray, %	< 5	0	30
Density, g/cm^3	1.341	1.338	1.380

Table 3.5
Tensile Properties of POY, Undrawn, and Fully Drawn Yarns [72]

	POY	Undrawn Yarn	Drawn Yarn
Linear density, denier	245	496	150
Tenacity, g/denier	2.2	1.2	4.3
Elongation, %	130	400	30
Modulus, g/denier	30	22	110
Shrinkage, %			
Boil-off	60	44	8
160°C dry heat	71	55	13

It is generally agreed that when undrawn yarn is simultaneously textured, the operations of stretching and twist setting occurs at the same time, and this results in some "looped filaments".[52,90] Stretching in the twisted state is the cause of looped filaments, and in POY, where the stretching is less than that in undrawn yarn, this effect is less conspicuous. This problem of looped filament formation with undrawn supply yarns can be avoided if the yarn is twist set at a temperature above its T_g. This can be achieved by heating the undrawn yarn above its T_g before it reaches the noncontact or convection-type heater. When the heater is of the contact type, the initial 20 to 25 cm of contact can be avoided by setting a guide, so that the yarn is preheated to its T_g before the actual twist setting occurs[99].

3.2.15.1.3 *Influence of process variables on the textured yarn properties*

Heater temperature. The relationship between the yarn dye uptake and heater temperature is shown in Figure 3.81. The dye uptake decreases with temperature, starting at 195°C, and then passes through a minimum at approximately 218°C. It starts to rise again with further increase in temperature. It is suggested that with the rise in heater temperature at a given yarn speed, the amorphous portion in the fiber structure decreases because of the formation of crystallites, but the orientation in the amorphous regions also decreases. The decrease in amorphous portion will reduce the available space for the dye molecule, leading to decrease in dye uptake of the yarn. On the contrary, the decrease in amorphous orientation will increase segmental motion. Therefore, the dye molecules will easily penetrate the fiber when the chain segments move to produce greater space for the dye, leading to an increase in dye uptake.[100]

Yarn-crimp contraction increases with the heater temperature as shown in Figure 3.82. The phenomenon could be explained as follows. There are two types of deformation imposed on the twisted continuous filament yarns. One is a

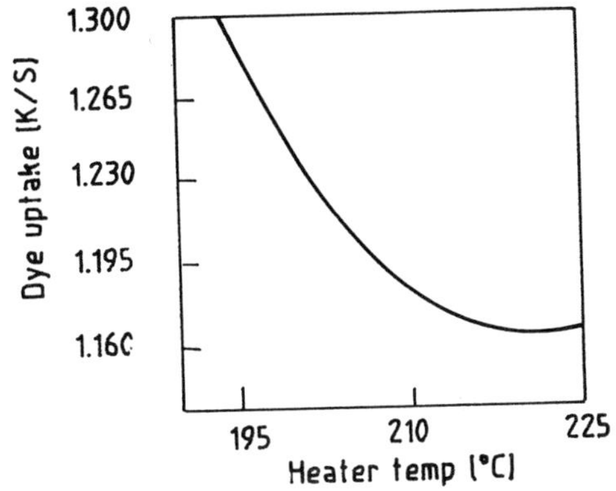

Figure 3.81 Relationship between the yarn dye uptake and heater temperature

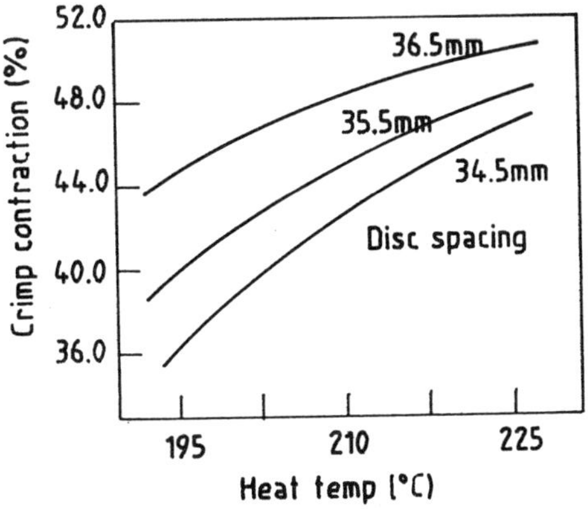

Figure 3.82 Crimp contraction increases with the heater temperature

twist along the length of the filaments, and the other involves bending of some of the filaments so that they follow a helical path about the yarn axis. In addition, filaments in the twisted yarn migrate in a helix from the yarn surface to the yarn center and back again. Therefore, filaments are not evenly deformed along their length. Those parts of the filaments that stay on the yarn axis will be relatively straight and are deformed in a purely torsional way. But those on the yarn surface will have a helical as well as a torsional deformation. The radius of the helix will be equal to the distance from the yarn axis. As the yarn is heated in the twisted configuration to set these deformations into the filaments, the twisted yarn becomes torque free. However, as the yarn is detwisted, the filaments are again restored to a torque-lively condition. The retractive power of the yarn is created mainly by these torsional forces when they are relaxed fully, but the effect of the bending forces in the filaments are predominant in the practical relaxation range of the yarns.

By increasing heater temperature, the helical coils are expected to reduce in diameter and also produce higher recovery force in the yarn, because the retractive power of a stretched helical spring is inversely proportional to the square of its diameter. Crimp frequency also increases as a result of the diminishing diameter of the helices, because the coils increase in number as the yarn is untwisted. The yarn crimp contraction and bulk will depend on the retractive power as well as crimp frequency.[100]

Preheating. In the false-twist draw-texturing process, only one heater (main heater) is employed in the texturing zone. In the early stages of development of draw-texturing processes, when undrawn yarns (UDY) were used as a supply for the process, an electric-type contact heater, having a temperature profile with a low-temperature input,[101] or a low temperature preliminary heater[102] was used to prevent undrawn filaments from coming into immediate direct contact with a high-temperature surface. From the experimental studies,[103] it is observed that the use of a preliminary heater in draw-texturing polyester yarn has some effect on the characteristics of the draw-textured yarns even when POY rather that UDY is used as the supply yarn,[103]

Experimental studies[103] have shown that as the temperature in the preliminary heater is increased from room temperature, the yarn tension in the cooling zone steadily increases until the preheating temperature approaches 82°C, the T_g of polyester yarn. It then increases rapidly and levels off as the temperature is further increased. When the other properties of the draw-textured yarns with and without preheating are compared, it is concluded that the characteristics of draw-textured yarns with preheating more closely resemble those of conventional false twist textured yarns.[103]

Draw ratio. The yarn tension before the false-twisting unit cannot be influenced by the overfeeding of the yarn as in conventional texturing. The draw ratio determines the yarn tension. Also, the yarn tension in the draw texturing cannot be increased as much as is needed, because of the different path a single filament makes. The distribution of draw ratio in the filaments at draw texturing is, therefore, different from the drawing operation. The experiments have shown that as the draw ratio is increased, the tension before and after the false-twisting unit (pin) is increased. To get a sufficient yarn strength, a high draw ratio needs to be selected. But high draw ratio means high yarn tension; this, in turn, may cause filament or even yarn breakages.

When the stress-strain curves of the simultaneously drawn and textured yarns are analyzed, it will be clearly observed that a high draw ratio leads to steeper curve than a low draw ratio. The steepest curve is obtained from the conventional textured yarns. The draw ratio, however, is not the only factor that influences the stress-strain behavior of simultaneous textured yarns. The preorientation of the supply material also influences the tensile behavior. For example, texturing preoriented yarn leads to better yarn strength.

A crimp contraction above 50%, as can be achieved by simultaneous texturing, is better than the crimp contraction of conventional yarns. Crimp contraction and elastic pull (crimp module) decrease with increasing draw ratio. This is contrary to crimp stability (mechanical crimp retention), which increases with the draw ratio. Low draw ratios lead to a crimp stability that cannot be accepted. A simultaneous yarn, textured at low draw ratios, will be drawn to a certain extent while testing the crimp stability. It looses its crimp. Such yarns are not fully drawn. They still have a little drawability, which means plastic elongation. The crimp stability of simultaneous yarns is below that of conventional yarns.

The dye take-up decreases linearly with increasing draw ratio as seen in Figure 3.83. A higher orientation of polymer resulting from higher draw ratio leads to higher molecular density. The amorphous regions in which the dyestuff migrates are getting smaller. This is the reason for smaller dye take-up. Yet, the dye take-up of simultaneous textured yarn is much higher than that of the conventionally textured yarn.

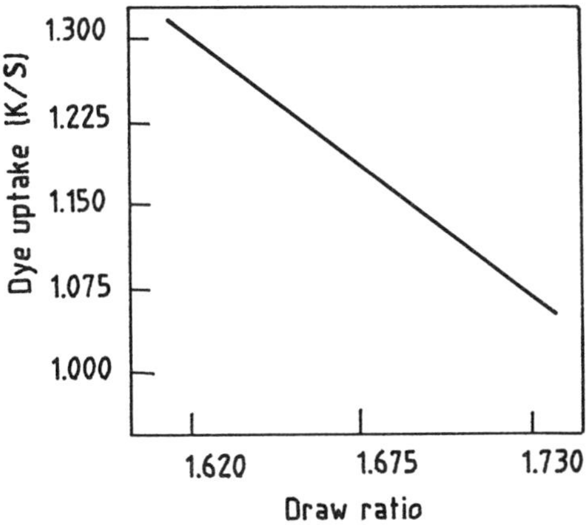

Figure 3.83 Dye take-up decreases linearly with increasing draw ratio

The crimp characteristics could be appreciably improved by increasing the draw ratio. This applies particularly to the crimp rigidity, which is between 80% and 90% at the high draw ratio. The crimp contraction and the nominal crimp decrease by approximately 5% in the speed range 800 - 1,200 m/min. This loss can be partly offset by increasing the heater temperature. In the case of the yarns textured with low draw ratios, surging occurs as from a texturing speed of 1,000 m/min. The highly twisted, uncrimped points cause the crimp characteristics to drop. The crimp rigidity is generally lower at low draw ratios.[92]

The tenacity can be considerably improved by increasing the draw ratio. Like the breaking elongation, it does not depend on the texturing speed. The elongation loss associated with the increase in draw ratio can be looked as a disadvantage.[92]

The snarling tendency also shows a dependence on the draw ratio. As the draw ratio increases, the snarling tendency falls and, in the case of uniform yarn tension, is not dependent on texturing speed.[92]

D/Y ratio. The effect of D/Y ratio on the yarn dye uptake is not pronounced. Only about 4% decreases in K/S value, which is the ratio of the light absorption the light scattering from the sample is observed as the D/Y ratio is changed from 1.872 to 2.022. As an indicator of dye uptake, the K/S ratio may also be calculated as

$$K/S = \frac{(1-R)^2}{2R}$$

R is the degree of reflectance.

The yarn twist level increases with D/Y ratio at lower levels and passes through a maximum[2]. However, further increase in D/Y ratio causes a rapid fall in twist level until it is stabilized. This relatively low twist level may result in the decreased dye uptake. As discussed, the lower twist level would have a less disorienting effect in the amorphous phase and a less distorting effect on the crystallites at the temperature of texturing. Therefore, lower dye uptake at high D/Y ratio can be expected.[100] There is little effect of the D/Y ratio on the crimp contraction of the textured yarns.[100]

3.2.15.1.4 *Filament deformations in simultaneous draw texturing*

Torsional strain along the filament axis. For each turn of yarn twist, the filaments are subjected to a turn of twist about their own axis. Processing twist levels of about 2,450 turns/m extended length correspond to 1 turn per 400 μm extended filament length, i.e., 0.9° per μm. For a filament of 20 μm diameter (i.e., of approximately 5.5 dtex polyester), this corresponds to a surface strain of about 9°, falling to zero at the axis of the filament. The molecular chain axes in the final draw-textured filament might be spread over this angular range, depending on the extent of strain relaxation in the twist-setting zone of the heater. This deformation occurs in conventional textured yarns as well as draw-textured yarns and would reduce the conventionally textured filament birefringence in comparison with that of the supply yarn.[104]

Radial compressional stress-strain. Photographs of cross-sections of yarn samples snatched off the primary heater show that all the filaments are tightly close-packed, with no visible spaces between them. The circular cross-sections of the supply filaments suffer varying degrees of distortion. The radial stresses are generated by the torque and tension on the yarn. In a POY filament, and even more so in a fully drawn filament, the radial yield stress is much lower than the tensile yield stress, which would draw the filament further.[104]

The strains that convert a circle into a regular hexagon of equal area will occur at fairly low radial stresses in a yarn bundle. Filament deformations into cross-sections of high aspect ratio would require greater stresses, but, as the twisted yarn moves up to the heater, the yarn temperature will rise, and the yield stresses become lower. The constant torque in the threadline will, therefore, be able to deform the filaments more as the twist increases[104].

Inspection of the cross-sectional shapes of the draw-textured circular supply filaments show that the least distorted shape is usually a regular hexagon (aspect ratio = 1.15). This means that, before any significant twist develops just beyond the feed roller, the cylindrical filaments close-pack with hexagonal symmetry.[104]

The ideal hexagonal close-packing of Figure 3.84 shows that no single filament lies on the axis of the 167 dtex 30 filament yarn, and the envelope of the filament bundle is an elongated hexagon. Perhaps a more probable packing for a 30 filament yarn is sketched in Figure 3.84, which shows a 1-6-10-13-filament arrangement, with the center filament on the yarn axis and the envelope roughly circular in section. The radially compressed filaments are shown in Figure 3.84 with five pentagonal filaments in the bundle of 30.

Parallel hexagonally close-packed cylinders cannot be twisted about the bundle axis without either extensive movement of filaments from the center outward to reduce progressively the population of the concentric shells or considerable distortion of the cross-sections by elongation along the radial direction of the bundle (yarn)[104].

Cross-sectional deformations resulting from twist. Hexagonal close-packing fills all available space in the untwisted state. The existence of so many hexagonal cross-sections in the textured yarn shows that initially, just beyond the nip point of the feed rollers, filaments pack predominantly hexagonally.

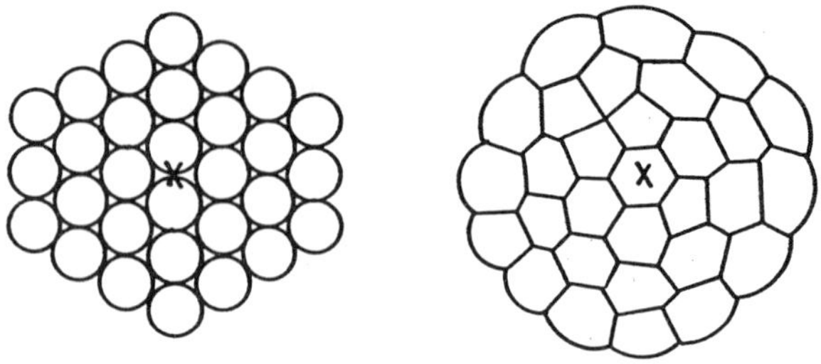

Figure 3.84 Ideal hexagonal close-packing of 30 circular filament

The filament migration pattern to maintain the equilength condition must also be established here. A center filament with the six nearest neighbor seems the most probable packing for the core of the yarn, but, as soon as the twist build-up occurs, the filament cross-section has to become thinner in the yarn tangential direction and longer in the yarn radial direction.

This effect is illustrated in Figure 3.85, where parallel circular-section filaments in any given concentric shell become elliptical in cross-section when twisted to a helix angle θ in that shell. Filaments that compress radially into regular hexagons in the yarn cross-section will deform into elongated hexagons in filament cross-sections at a helix angle θ.[104]

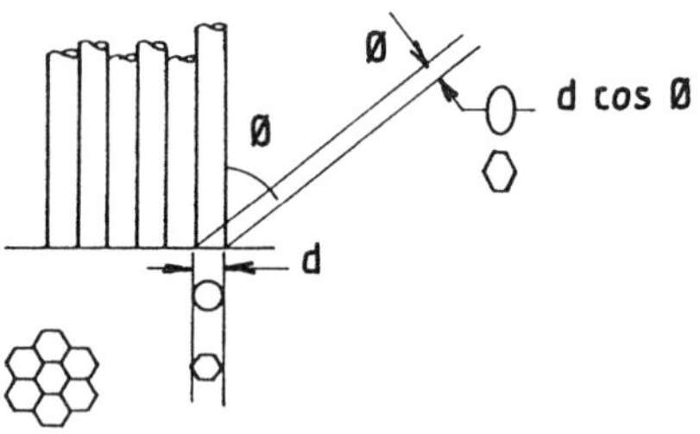

Figure 3.85 Filament cross-section becomes thinner in the yarn tangential direction and longer in the yarn radial direction

3.2.15.1.5 *Effects of draw texturing on filament linear density*

It has been observed by Thwaities and Hooper[105] that there are significant and observable differences in the variation in linear density between filaments from draw-textured yarns and those from conventionally false twist textured yarns. The differences are much the same as those detectable by reflectance measurements for the two types of yarn. This supports the argument that linear density variations may be the reason for the more uniform dyed appearance of fabrics made from simultaneously draw-textured yarn.

The experiments also show that there is no essential difference in linear density variation between conventionally textured filaments and their supply filaments. This indicates that there is little or no differential shrinkage between filaments in this process. The distribution characteristic of draw-textured filaments is, conversely, different. It appears to be accurately described by the differential-drawing model, provided that only the neck region is considered.

3.2.15.1.6 *Heat transfer during simultaneous draw texturing process*

During the draw texturing process, it is likely that some stress-induced crystallization occurs simultaneously with drawing, so the following equations are used to obtain estimates for the temperature rise of the yarn from the latent heats of drawing and crystallization, respectively:

$$\Delta T_{drawing} = \frac{9\times10^{5}(R-1)T_0 p \sec\alpha}{DC_p}$$

$$\Delta T_{crystallisation} = \frac{\Delta H}{C_p}$$

where

R = draw ratio
T_0 = drawing tension across the heater
α = helix angle resulting from yarn twist
D = linear density of supply yarn
C_p = specific heat of the yarn material
(≅0.45 cal/g°C (1.88 J/g°C) for polyester)
p = conversion factor, g cm to calories = 2.34×10^{-5}
ΔH = net heat of crystallization

For a particular case of 289 dtex, 34 filament polyester POY, draw ratio of 1.8, texturing speed of 365.8 m/min, 2,087 tpm which corresponds to a helix angle of 55°, with 1.52 m long vertical heater and 42 gf yarn tension in the heater, together with 32.2 J/g of net heat of crystallization during texturing process, it was found[106] that $\Delta T_{drawing} \cong 11°C$ and $\Delta T_{crystallisation} \cong 17°C$.

Theoretical equations for calculating the yarn-temperature profile on the primary heater in the simultaneous draw texturing operation, with the temperature rise of the yarn from latent heat of drawing and crystallization considered are not available. Jones and Mason[107] derived a simple empirical expression for the yarn temperature as a function of the contact time on the heater for the conventional texturing process, which is as follows:

$$\ln\left(\frac{T_H - T_L}{T_H - T}\right) = \frac{703t_n}{C_p D_n}$$

where T_H is the heater temperature, T_L the yarn temperature at the entry to the heater, T the local temperature of the yarn corresponding to a nominal contact time on the heater of t_n sec, C_p the specific heat of the yarn, and D_n the nominal untwisted yarn linear density.

The experimental and theoretical yarn temperature data are plotted against the contact distance on the heater in Figure 3.86[106]. The experimental and theoretical curves coincide exactly at the 180°C heater set temperature. At the higher heater set temperature conditions, the theoretical curves were displaced somewhat above the experimental curves. The initial temperature of the yarn at the entry to the heater was 72°C to 78°C, depending on the heater set temperature condition. The ambient temperature was approximately 27°C.

In the conventional texturing operation on which the theoretical yarn temperature calculations are based, the feeder yarn is drawn and crystalline, so that latent heats of drawing and crystallization are not significant. If the drawing occurred within the heater of a simultaneous texturing process, it would, therefore, be expected that the experimental temperature curves would be displaced above the theoretical curves, but this was not observed. It is therefore concluded that the drawing occurred not on the heater but possibly at the entry to the heater, where it is believed that the heat of drawing contributed to the high initial temperature of the yarn. The difference between the initial temperature of the yarn and the ambient temperature was 45°C to 50°C, which is higher than the estimated total heats of drawing and crystallisation for this process. It is believed that the yarn was prewarmed by radiation from the heater surface or by convection just upstream from the entry point to the heater and that this prewarming effect also contributed to the high initial temperature of the yarn.[106]

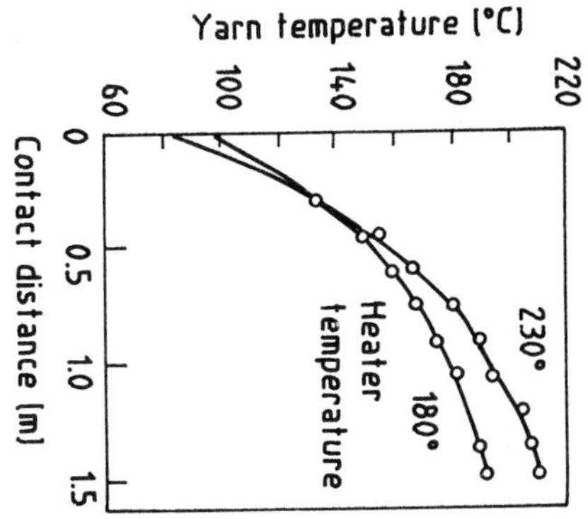

Figure 3.86 Experimental and theoretical yarn temperature against the contact distance on the heater[106]

3.2.15.2 Post-texturing setting

Many items of apparel, especially outerwear, need the bulk, opacity, and warmth provided by this process but not excessive stretch. Wool yarns have properties suited for most outerwear applications and they stretch by less than 10% compared with many times this value for the stretch textured nylon yarns. By reheating the stretch yarn in a partly contracted condition (in which much of its bulk is allowed to form) and, hence, setting it in this condition, the force of contraction is virtually removed while most of the bulk is retained. One can, thus, produce a yarn that will stretch by only about 10% from its fully relaxed condition, making it suitable for applications where wool was formerly used.[14]

The purpose of the setting operation is to reduce the net extensibility and twist liveliness of the stretch textured yarn without significantly affecting its bulk. In essence, the setting process serves to relax the residual stress that the yarn carries over from the primary texturing process. Further, the setting operation serves to guard against snarls developing in the yarn during knitting by reducing its residual torque.[108]

This second heat treatment was first performed by overfeeding the stretch yarn by about 15% on to a loose package followed by a heat-setting treatment in steam or a package dying process.

The whole process was soon made continuous by overfeeding the stretch yarn (allowing 15% to 20% relaxation) through a second heater fitted to the texturing machine. These yarns are called “stabilized” or “SET” yarns.

3.2.15.2.1 *Mechanism of post-texturing setting*

The pretensioned, straightened, stretch textured yarn possesses a small residual torque. The threadline torque in the setting zone is much lower than this residual torque, as reflected by the known reduction in residual yarn torque resulting from passage across the setting zone from entry to exit. Thus, a stretch yarn is fed into the cold segment of the setting zone; the requirement of torsional equilibrium along threadline leads to its immediate setting.

3.2.15.2.2 *Pseudo false twisting in the second heater*

The behavior of a torque stretch yarn as it moves through the resetting heater is dictated by its thermomechanical history. Despite the absence of a second twisting device, the yarn undergoes temporary untwisting over the first portion of the resetting heater. It then untwists and remains at zero twist over the latter part of the heater.[108] The mechanism of this pseudo false twisting is as follows.

During the steady-state operation of the resetting zone, the entering yarn twist equals that of the exiting from the primary texturing process and is generally zero. But the torque of the entering yarn is not zero. Rather, it is small (compared with the torque that "drives" the primary process), and it is negative in sign (relative to the direction of the original texturing twist). This corresponds to exiting conditions from the primary or first heater texturing zone. When this yarn subsequently emerges from the resetting zone, its twist is still zero, but its torque is even smaller and still negative. Using a noncontact heater for the resetting process, constant torque and tension throughout the setting zone can be assumed, neglecting inertia effects. The tension level is dominated by the degree of overfeed in the setting zone and to a lesser extent by the temperature of the heater.[108]

When the yarn enters the resetting zone, it up-twists to a low level in the direction of the original texturing twist. As the yarn then moves into the heater, it up-twists significantly. A short time later, while still in the heating sub-zone, the yarn untwists completely, and subsequently remains at zero twist as it moves away from the heater, and then exits the apparatus (Figure 3.87) In effect, the textured yarn passing through the second heater resetting zone is, thus, subjected to false-twisting action. The degree of this false twisting in setting zone varies directly with the heater temperature, the residence time in the heater, and the degree of overfeed in the setting zone.[108]

To reduce the effect of this pseudo false twisting effect and eliminate some of its side effects, such as tight spots, a snub pin is placed in the resetting zone, preferably after the secondary heater. The presence of the snub pin provides for consistent, significant increase in residual crimp. The effect of the snub pin on shrinkage is negligible.[108]

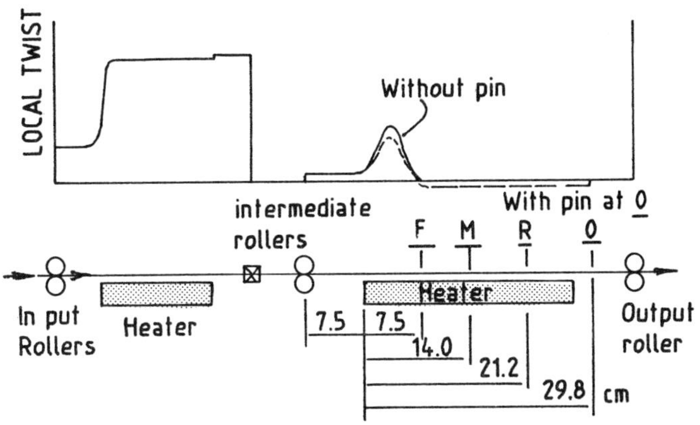

Figure 3.87 Twist distribution along the false twist texturing process with post-texturing setting

The secondary heater accepts the textured yarn and suppresses its contractive power and twist liveliness, while retaining as much as possible of the filament crimp. Heat-setting efficiency depends on similar factors to those applying in the first stage-setting zone, with the heating time being the most important of these.[33]

The secondary heaters are mostly noncontact type convection heaters because of the fact that the yarn is textured, and, thus, more open and bulky. The low-yarn tension is another reason for both use of convection heaters and downward yarn delivery through the secondary heaters.[33]

The second heater is usually a noncontact heater because of the following reasons:

1. As the yarn is under low tension in the second heater, small frictional differences between one contact heater plate and another could have some influence on yarn relaxation on the heater and, hence, on yarn properties.
2. As the yarn will also be bulky and will not be rotating, only a few filaments could be heated by contact. Most of filaments will be heated by radiation or convection.

Because of the low-yarn tension conditions, a vertical heater and downward yarn path would be preferred in these heaters. Experimental investigations of Lünenschloss et al[109] showed that crimp contraction decreased for increased setting zone temperature, or for decreased resetting zone overfeed. This behavior is not unexpected, because the higher setting temperatures should reset the yarn more effectively to the geometry dictated by the zone overfeed, i.e., at crimp levels much less than that of the fully relaxed torque stretch, single-heater processed yarn. Further, as the zone overfeed decreased, a geometry with lower crimp was set in the yarn. Lünenschloss et al[109] showed that yarn bulk responded to the resetting temperatures and overfeed in parallel with their effects on crimp development potential.[110]

3.2.15.3 Twist liveliness of the false twist textured yarns and its elimination: Detorqueing

Because the filament molecules are oriented and set into a migrating helix configuration, the extended filaments create a torque to go back to the original helical configuration. This latent torque of the molecules will result in the twist liveliness of the false twist textured yarn because of the ends of a yarn (or, indeed, of a single filament) are fixed, and the tension is lowered by bringing the two ends together; the yarn or filament will assume the minimum energy level by a snarl (Figure 3.88a) while the ends of a torque-free yarn stay separate (Figure 3.88b).

The snarling tendency, i.e., twist liveliness of the high elasticity (HE) yarn is also high. This high twist liveliness of the basic false twist textured yarn sometimes creates problems in later process stages of textile manufacture. Especially for the applications where the high stretch and bulk are not absolutely essential, the bulk and, hence, the twist liveliness of the yarn is reduced in a setting operation. The HE yarn is fed into a noncontact-type heater at a higher rate than it is taken away; thus, the stresses are relaxed, and the pitch of the helix introduced in the first heater is increased. This, consequently, results in a decrease in bulk, stretch, and twist liveliness of the textured yarn.

The twist liveliness causes problems both during the further processing of a twist-lively yarn and at the final fabric. During the conversion sequences from yarn to fabric, a twist-lively yarn tends to balance the internal torque by forming snarl. Especially at the process stages where the yarn tension is low, such as low-speed yarn wind-off as seen in warp knitting, a snarl is formed (Figure 3.89).

This snarl may not be fully opened at the later stages of the operation and appears as a fault in the final fabric, or such a snarl may get into the neighboring machine parts and cause a yarn break and an operation stop. When a twist-lively yarn is converted into a knitted fabric, the twist liveliness creates fabric deformation in the form of fabric spiralitiy for circular knit goods or course inclination for flat knit fabrics. Such defects may sometime require further treatment.

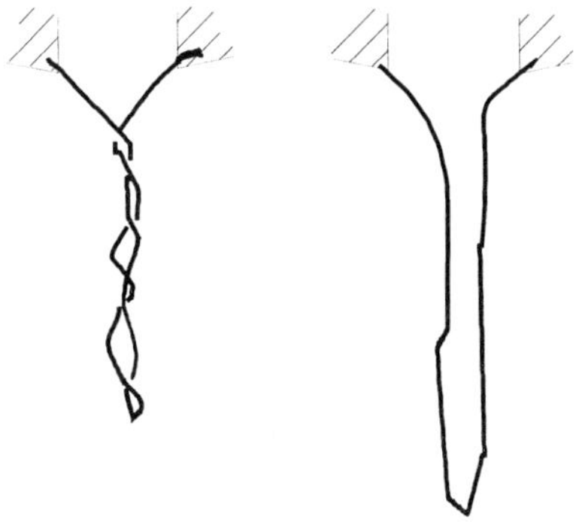

a. Twist lively b. Torque free

Figure 3.88 Twist lively and torque free yarns

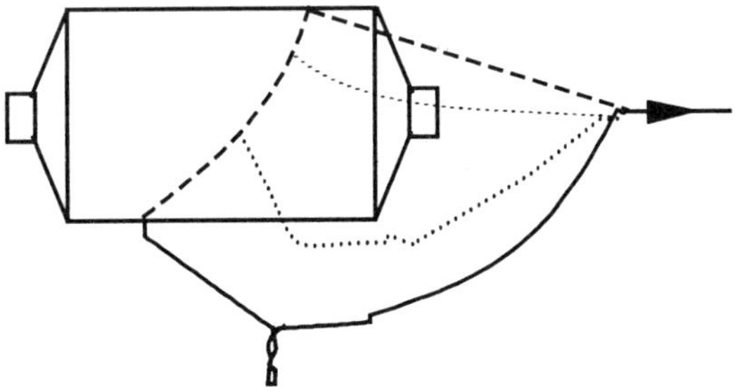

Figure 3.89 Schematic illustration of the snarl formation during yarn wind-off from a bobbin at slow speeds

Means of torque reduction or elimination. To avoid these difficulties arising from the twist liveliness of the false twist textured yarns, the following methods are used throughout the texturing industry:

1. SET texturing. As previously explained, the post-exturing heat treatment, i.e., SET operation reduces the twist iveliness together with the bulk and stretch of the yarn. By increasing the second heater temperatures to the levels that are near, equal to, or more than the first heater temperature, the internal stresses can be almost completely released, and then the internal torque can be eliminated. This relatively harsh setting operation, however, will always accompany the removal of bulk and stretch, which sometimes becomes detrimental to the purpose of texturing. In conclusion, it may be said that the SET operation cannot altogether eliminate the twist liveliness. Therefore, some other means of torque elimination is often required.

2. Doubling - comingling. When the two ends textured by Z and S twists in the texturing zone are brought together before or after the second heater, the opposing torque directions offer the potential to balance each other and create a torque-free yarn. The doubling of the two ends, however, requires a coherence imparting medium between these oppositely torqued filaments. This coherence is usually imparted by a comingling jet.

Because the two ends are involved in producing a single-end yarn, the machine capacity is halved, which renders itself to be a costly operation. Furthermore, to produce a given linear-density single-end torque-free yarn, two ends of finer yarns are required. This also increases the costs of a doubled yarn because the finer yarns are usually more costly than the coarse ones.

3. Double- and single-end twisting. The above-mentioned doubling operation may also be carried out by slight twisting of the two ends on a twisting machine. In this case, the real twist is the coherence-imparting medium. Because the twist level is usually so low, the final yarn is regarded as a torque-free yarn.

It is obvious that such second-step operations are always cost intensive and used for only specific applications. However, a single end can also be twisted in the opposite direction of the texturing operation, and the internal torque could be balanced by the real torque imparted through real twisting. The twist level in such operations is usually as high as 80 twists/m. This real twist naturally reduces some of the bulk and extensibility of the textured yarn.

4. Single-end sizing. The internal torque of the filaments may be balanced by the application of heavy sizes. The sizing material is applied to the extended textured yarn on a separate sizing machinery where the size is later dried. Such sizing material has to be removed by washing after the fabric manufacture. Both the application and removal of the sizing material are mostly expensive and environmentally unfriendly processes. Such a torque balancing method is, therefore, rarely employed by today's modern cost- and environment-conscious textile industry.

5. Steam setting. The twist liveliness may also be reduced by relaxing the yarn in a steam autoclave. The bobbins are thus steam set, and the stress that create twist liveliness is released. This is, again, a slow and costly operation.

6. On-line torque elimination by an air jet. Demir[120] during his development work at Heberlein Maschinenfabrik AG, has developed an air jet that reduces, or completely eliminates, the internal torque of the SET textured yarn during the production. This air jet, named as the DetorqueJet, is placed after the SET heater and provides an on-line torque elimination (Figure 3.90).

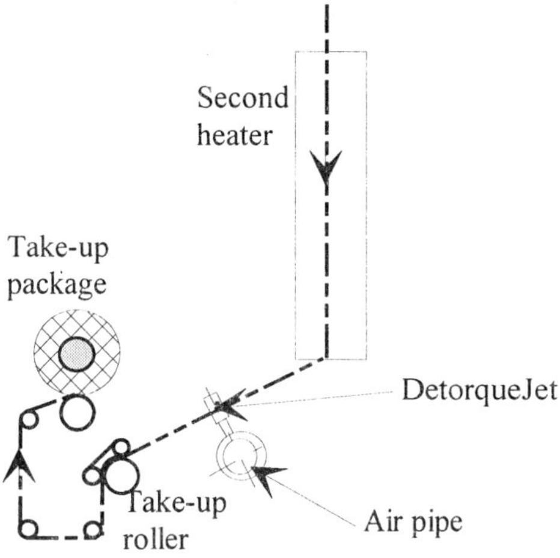

Figure 3.90 Positioning of the DetorqueJet after the second (SET) heater

In principle, the DetorqueJet assists the filaments to come to a torque free configuration in the SET heater and get set there. Such a set filament does not show any apparent torque and twist liveliness, but retains, most of the texture as imparted in the first stage of the texturing operation.

Because the detorqueing by an air jet does not alter the molecular structure but only rearranges the filaments' geometrical configuration to balance the internal torque, the molecules possess the tendency to attain their natural (low energy) state. This tendency naturally results in a partial recovery of some of the torque in time.

3.2.15.4 Intermingling

Compactness of the yarn is the key to trouble-free unwinding. As a rule, intermingling cannot be dispensed with. The arrangement of the jets in the machine, yarn path, and yarn tension in the mingling zone affect the mingling density. This subject will be covered in the section on intermingling/interlacing in further detail.

3.3 STUFFER-BOX TEXTURING

This method was first developed for outerwear yarns under the brand name Ban-Lon. Yarn is propelled by a pair of feed wheels into a tube where it is restrained at the exit and forms a plug of filaments deformed into a crumpled zig-zag configuration. This deformation is set by heat and gives the processed yarn a high bulk and soft handle (Figure 3.91). Its stretch properties are appreciably less than those of false twist textured yarns, and it has no torque[14].

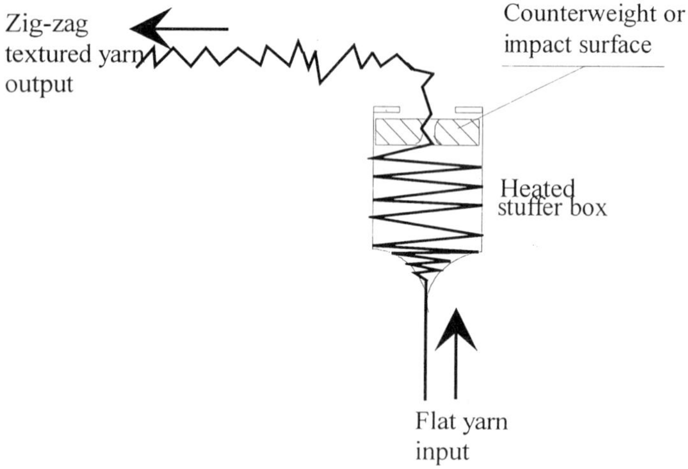

Figure 3.91 Principle of stuffer-box texturing

Lack of uniformity limited its development for a while, but improvements in machine design over the years have led to better uniformity and much higher production rates and renewed interest in this method for the production of textured yarns for apparel use.[14]

An important feature of the bending and stuffer crimping techniques is that in contrast to the twist crimping, the heat setting occurs in the tensionless state or under extremely low tension. The structural changes incurred with the contraction, i.e., the disorientation of the molecules and the increase in crystallinity, result in a pronounced reduction in crimp stability, especially in the case of polyester. This is one of the main reasons for the relatively small-scale application in the textile linear density range.Conversely, the application is concentrated on high-linear-density polyamides and also to some extent polypropylene, where the described effect is of less disadvantage.[15] Stuffer box textured yarns have been used in warp knitted and woven fabrics, but their main outlet has been knitted outerwear. Applications also include knitted shirts, sweaters, men's socks, underwear, and upholstery fabrics.[14]

An important use of stuffer-box system is for the production of the heavy-textured continuous filament yarns that are widely used in tufted carpets and rugs. They have been produced for years by the original mechanical feed system, but increasingly use is being made of jets of hot air or steam to propel the heavy yarn into a stuffer box or against a perforated plate in order to deform and set it. Speeds of up to 2,500 m/min can be used on 3,000 dtex yarns (i.e., production rates of up to 45 kg/hour); in some applications, a single machine will extrude, draw, and texture such yarns in one continuous operation.[14]

3.3.1 Mechanical feed (conventional) stuffer-box texturing

To introduce some feature to the continuous filament yarns of the early 1950s, an already known technique used for crimping of carpeting-wool fibers was used. Figure 3.92 shows this technique in diagrammatic form. A pair of feed rollers, shown at the bottom of the figure, push the supply yarn, which is a thermoplastic type, into a constricted space, an electrically or otherwise heated tube, known as the stuffer box. The individual filaments are sharply bent by the compression of the yarn as it leaves the feed rollers and meets the wad of previously delivered filaments, and the crimp is heat-set while the yarn is in its packed state. The exit end of the tube is unheated; thus, the yarn is cooled before leaving the stuffer box, and the crimp is permanently set. The bulk level is controlled by sensing the height of the yarn column that is allowed to form after the constriction in the stuffer box.

If the stuffer box is cylindrical, the zig-zag crimp could be three-dimensional, but if the box is flat, the crimp would be roughly planar. This affects the behavior of the filament bundle. Furthermore, the friction between the individual filaments, and the stuffer box walls affect the distribution of forces and, thus, crimp amplitude.

Stuffer box technique is generally used for high-linear-density yarns, carpet yarns that are characterized by saw-toothed crimps. Unfortunately, this promotes "snagging," but the yarns have a soft warm handle and give greater covering power than equivalent weights of conventionally spun yarns. Their moisture absorbency is increased well above that of untreated filament yarns.

Although the stuffer box textured yarns do not have intrinsic stretch in the sense that false twist textured yarn has, they can be produced with a voluminous and soft handle. When they are knitted, a desirable combination of comfort and stretch is the end-result.

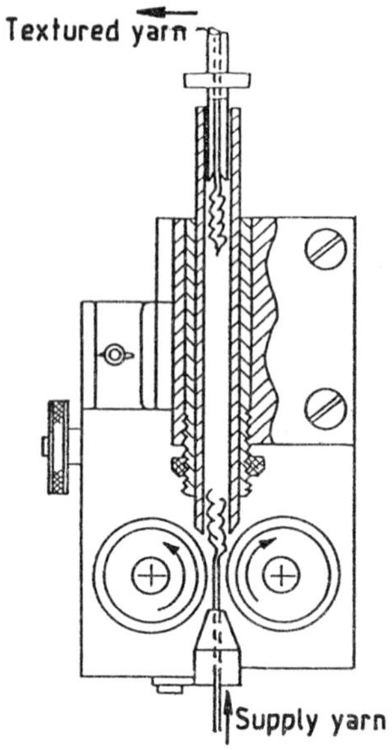

Figure 3.92 Mechanical feed stuffer box texturing

When compared with the false twist textured yarns regarding the bulkiness, the stuffer box textured yarns are superior. Additional advantages are the freedom from torque liveliness and the absence of cross-sectional deformation.

Heavy denier carpet and upholstery yarns are the major domain of the stuffer box textured yarns. However, in recent years there seems to be a renewed interest in the technique and increasingly lower- linear-density yarns for apparel industry are being textured.

The stuffer box, together with the feed rollers, heating element, plug- sensing equipment, and let-off mechanism is obviously the most crucial part of the process. Over the years, many different designs for each of these elements have been developed. Feed rollers, for example, were cylindrical in early days, then some disk type feed in systems were developed. The modern machines, however, favor a fluid feeding system that does heating and plug sensing at the same time. The latter design (Figure 3.93), usually known as hot-fluid texturing, has almost completely replaced the mechanical stuffer box technique in just more than 10 years since the early 1970s. The reasons are much higher production speeds (2,700 m/min typical) and improved bulk resulting from the three-dimensional crimp.

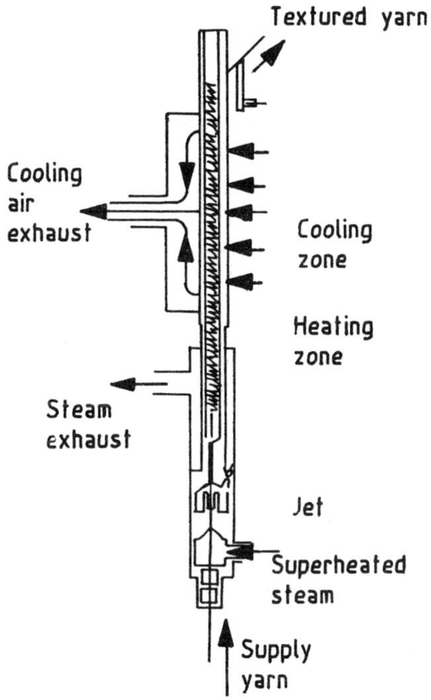

Figure 3.93 Fluid-jet-feed stuffer-box texturing

Wilson and Kollu[111] report that the hot-fluid-texturing process is currently the only known process for producing textured yarns in a single operation, i.e. spin-draw-texturing. So far as the actual stuffer box is concerned, several modifications have been adopted over the years. Different cross-sections – circular, square, rectangular, and slit; different shapes– cylindrical, conical, reverse conical, and J or U shaped; stuffer medium with moving walls; or stuffing between a rotating carding wheel and the hot-fluid jets are all applied design features.

Because supply yarns with thermoplastic properties are prerequisites for the process, the heaters and heating elements possess particular importance. Similar to the heaters used for false-twist texturing, heaters for stuffer-box texturing could be heated electrically or by hot-fluid oil or vapor. Each stuffer box is equipped with its own heater and independent temperature control, which is required to provide ±1°C accuracy.

Secondary heater after texturing operation is not common for the stuffer-box process. But intermingling nozzles are generally deployed to impart some mingling to the textured yarn.

Contrary to many positions on false twist texturing machines, the stuffer box texturing machines possess only a few (usually four) positions. The last ITMA exhibitions have, furthermore, indicated that mechanically driven stuffer-box machines are dying out, but hot-fluid jet texturing machines with speeds reaching up to 6,000 m/min are becoming widely accepted.

Perhaps its special end-use area, i.e., carpeting and upholstery, the stuffer box textured yarns are the longest-lived yarns, and they are expected to keep their moderate market in the future as well.

3.3.2 Fluid Jet Feed Bulked Continuous Filament (BCF) texturing

Hot fluid jet texturing principle appears to have been introduced in early 1970s for the processing of heavy-linear-density yarns. The basic principle is using a jet of hot air or superheated steam to plasticize and blow the feed yarns into a stuffer tube or onto the surface of a rotating card wire covered cylinder (Figure 3.94). This action causes the individual filaments to assume a randomly deformed state in the tube or on the wire points of the cylinder, and they are held in this condition until set.[112]

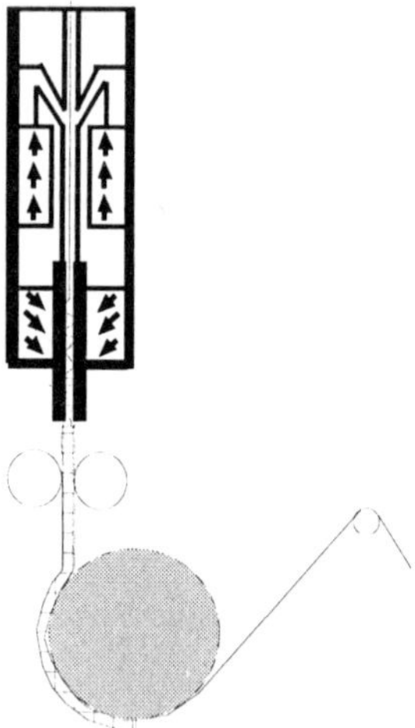

Figure 3.94 Fluid-jet stuffer-box texturing with rotating perforated cooling drum as used in BCF production

A number of fiber producers make their heavy-linear-density BCF yarns by fluid jet method. However, modern texturing machines for BCF yarns virtually all operate with hot air or hot steam jets to texture even low-linear density yarns. This technology has firmly established itself in recent years and may now be regarded as state of the art.[47]

BCF technique permits high production speeds, reaching about 6,000 m/min and possibly higher. Therefore, BCF process is now completely coupled with the filament spinning process and provides spin-draw-texturing operation. In such a BCF texturing machine, the yarns are passed from the spinning unit to two rolls via lubricant applicators and from there to the first pair of heated rolls. From there the yarns go to the second pair of heated rolls, which operates at a much higher temperature. Drawing between the sets of rolls 1 and 2 is generally by a factor of 4.5. From the pair of rolls, the preheated yarns reach the texturing jets where they are transported by means of hot-air stream, additionally heated, and accumulated into plugs in the lower portion of the jets. They are then deposited on a cooling drum. An unheated third pair of rolls draws the yarns from the cooling drum and passed to the automatic winder mostly via an entangling jet in which the textured yarn is given the structure required for tufting or weaving.[47]

In BCF texturing, the crimp is no longer of a saw-tooth shape but has a random three-dimensional curvilinear form.

The hot air or steam jets are the core of the BCF texturing machine. The yarn enters the texturing jet at a temperature near its softening point. There it is heated further by a stream of hot gas and transported into a stuffer box and compacted into a plug that is gradually ejected from the stuffer box.[47]

Figure 3.93 demonstrates a particular type of fluid-jet texturing set up for low-linear density yarn texturing. The jet is placed at the bottom of the system. Yarn is fed in via a labyrinth seal. In this process mainly superheated-steam is used but other fluids as hot air and other gasses can also be used. In the jet area, the steam speeds are sonic. To achieve the desired degree of pressure drop to optimize turbulence patterns, a bleed valve system is adopted. The textured yarn then enters the plug former and is separated from the steam. The steam is led up the outside of the plug chamber and continues to process the yarn and equalize out any temperature-history differences introduced in the jet. Then in the cooling zone, air is sucked through the plug and cools it almost to room temperature under practically tensionless conditions.[113]

At high speeds, the running time of a bobbin could be as short as 15 minutes. This calls for easy threading up of nozzles and automatic doffing systems. The former is accomplished by so-called split nozzles or hinge open nozzles. These also facilitate easy cleaning. For the latter, fully automatic and loss-free changing heads and package conveyers are fitted on the machines.

Current BCF machines as manufactured by such companies as Barmag, Neumag, and Rieter incorporate the necessary drawing mechanisms with so-called duos, an intermediate stuffer box and cooling zone, as well as the high-speed winding machine, fitted with automatic doffers.[50]

Modern texturing machines for the production of BCF yarns are all equipped with hot-air or steam-texturing jets. This technology was developed over many years and is, today, regarded as state of the art.[114]

Two basic process types are currently in use for the production of BCF yarns: draw, and spin-draw texturing. Production speeds for draw and spin-draw texturing are between 1,800 and 6,000 m/min depending on the yarn type and linear-density range. The machinery for the two processes differ in only detail. Therefore, a yarn path description for a spin-draw texturing unit as seen in Figure 3.95 will suffice. The yarn ends lead from the spinning stack over the spin finish application heads to two pretension rollers and from there to the first heated roll duo. A small amount of draw of 1% to 2% is normally applied between the pretension rollers and the first duo to steady the thread run on the first godet. The first, relatively low-temperature heated godet pair, is wrapped up to 10 times, depending on the yarn linear density. From there, the thread passes to the second roll duo, which is heated to a substantially higher t emperature. The draw ratio of up to 4.5 is applied between the first and second godet pairs. From the second duo, which is also wrapped up to 10 times, the preheated thread passes into the texturing nozzles. Here it is conveyed by hot air, heated, and accumulated in the form of a plug in the lower portion of the nozzle. Through the impact of the yarn on the congested plug, the three-dimensional crimp desired in the end product is obtained. The plug is slowly ejected by the overpressure in the nozzle and laid onto the cooling drum where it is cooled by aspiration of the ambient air. From the cooling drum, the thread is drawn off over the unheated duo 3 and then passed via the intermingling jet, in which the textured yarn receives the structure necessary for tufting and weaving, to the automatic winder.[114]

3.3.2.1 The jet

A hot-gas texturing jet works by the preheated yarn coming from the duo 2 entering the jet at the top with a temperature near the softening point of the material. A pressurized hot gas stream entering from the side heats the yarn further up and forwards it into the stuffer section of the jet (Figure 3.96). This chamber is designed to release the gas pressure in such a way that the yarn opens into individual filaments, forms a plug in the lower part of the chamber, and forwards this plug through the chamber by the remaining gas pressure. The intensity of the plug and, therefore, the texturing effect is controlled by the type of the pressure release and the design of the stuffer chamber itself.[114]

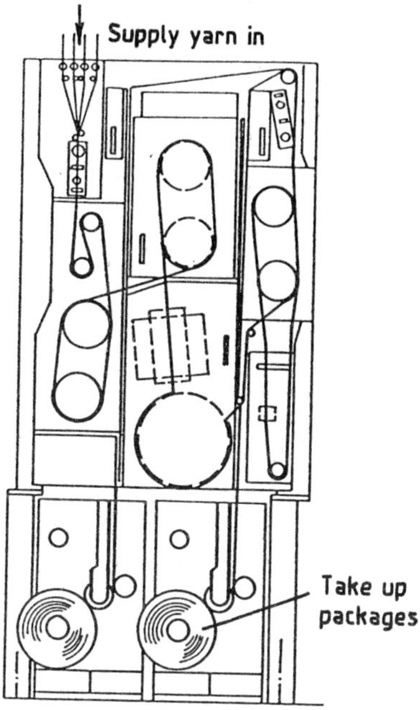

Figure 3.95 Yarn path on a spin-draw-texturing unit

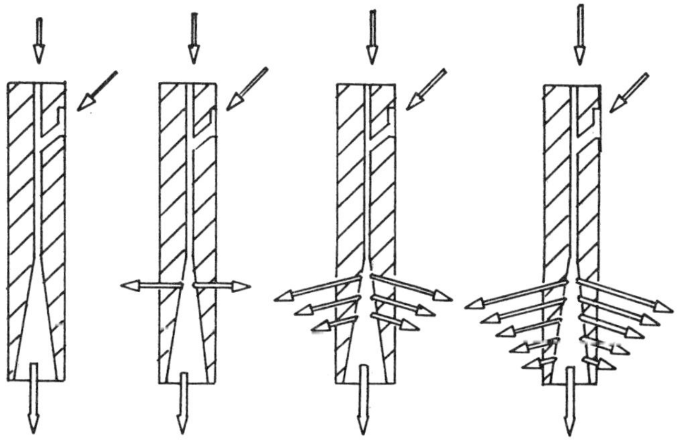

Figure 3.96 Various configurations of hot-fluid-jet texturing jet

The gas-flow conditions that determine plug formation and plug forwarding are as follows:

1. No pressure release in the stuffer chamber. The yarn cannot form a plug; it is only forwarded through the entire jet. The yarn is not textured.
2. The entire pressure is released in the top of the stuffer chamber. The yarn forms a dense plug, and the texturing values reach the highest level under these conditions. However, the plug is not forwarded into the chamber. It will form at the top of the chamber and hamper the pressure release. This causes the process to be unstable and to break out after a short running time.
3. The pressure in the stuffer chamber is partially released through several small bores at the top of the chamber. Only a loose plug is formed under these conditions. The plug is pushed out of the chamber at a relatively high speed. The yarn is only slightly textured.
4. The pressure in the stuffer chamber is released over the entire length of the chamber, i.e., through small slots. An intense plug is formed under these conditions. A certain self-regulating effect is obtained between plug density and plug push-out. The texturing effect under these conditions is satisfactory, and the process is stable.

3.3.2.2 Plug control

The structure and shape of the stuffer chamber plays an important role in the plug formation and plug forwarding. A structural change on the inside wall, for instance, can greatly vary the behavior or otherwise identical chambers. This method of plug control, however, is not recommended because the stuffer part in the jet is greatly influenced by wear caused by the yarn.[114]

Figure 3.97 shows the traditional stuffer-box arrangement. Yarn is stuffed into a heated tube and crimp imparted, and a sort of plug is formed. Under steady-state conditions, if as much yarn is put in as is taken away, the plug should remain constant. However, in practice, workers in the stuffer-box area find that plug height varies, and so they have developed a variety of detection systems sensing the top of the plug and via a control system either speeding up or slowing down the input rollers or, conversely, the wind-up rollers.[113]

Because the mass-balance equation for the stuffer-box operation under the steady-state conditions is

$$V_S L D_S = V_f L D_f$$

where

V_S = supply yarn input speed
LD_s = supply yarn linear density
V_f = textured yarn output speed
LD_f = textured yarn linear density

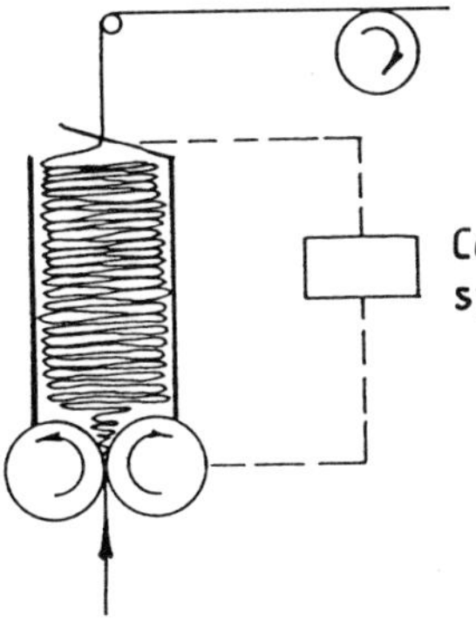

Figure 3.97 Plug-control loop

the only potential variable in this equation would be the linear density of textured yarn LD_f. If the process parameters, such as overfeed and tension are kept constant and only the steam temperature increased, the amount of bulk, i.e., LD_f, is increased. This phenomenon facilitates a feedback control system for fluid-jet stuffer-box texturing operations. If the plug height L increases, it is because LD_f is too low; therefore, the steam temperature should be increased to increase bulk and vice versa. Figure 3.98 shows diagramatically how the previously described plug control approach is implemented on a BCF system. As a sensor for the top of the plug, a linear photo-cell is used. The signal from this cell is fed back through a control system to activate a heating element placed as near to the jet as mechanically possible. This heater quickly increases the already superheated steam temperature; hence, the yarn bulk increases. Such a closed-loop plug control system also compensates for several variables such as the following:

1. Variation in the bulking performance of the supply yarn
2. Variations in the manufacturing tolerances of the jet,
3. Variations in the performance of the jets resulting from deposition of spin finish
4. Changes in steam pressure[113]

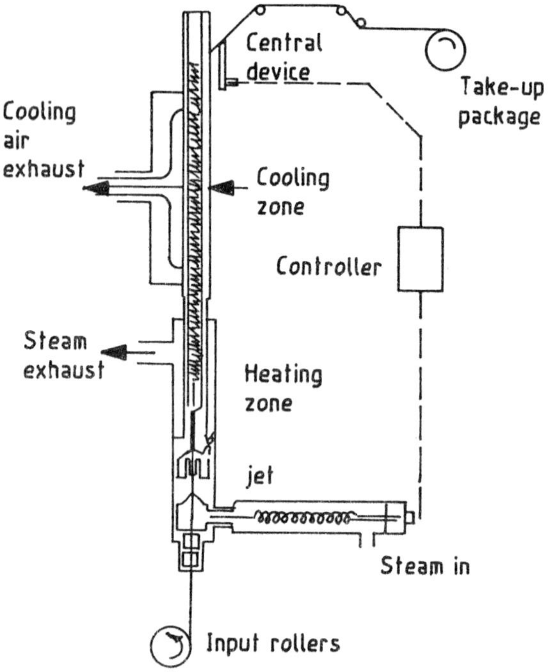

Figure 3.98 Plug-control approach on BCF system

3.3.3 "Moving-Cavity" Texturing

At higher texturing speeds, the stuffer-box method requires higher steam temperatures to achieve good texture levels. The high-temperature steam impacts directly on the plug and then vents through the exhaust ports. This direct impact of high-temperature steam causes yarns to fuse together, and the fused yarn may block the exhaust ports. Also, the longer residence time of the plug exposed to high-temperature steam will severely degrade plug, because it moves at only a fraction of the yarn speed. The following section describes a technique that improves the key features of the stuffer-box texturing method.

3.3.3.1 Basic principles of moving-cavity texturing

The basic principles of the moving-cavity texturing can be illustrated with the diagram shown in Figure 3.99. Steam is directed by a nozzle into an energy tube where the yarn is aspirated and heated to an optimum temperature. The energy tube is inclined at an angle between 15° and 75° with respect to the tangent line drawn along the periphery of the screen surface at the intersection point of the screen and the energy tube. The yarn and steam directly impact on the screen surface where the crimping of the yarn is initiated. The crimped yarn, after randomly deflecting from the screen surface, is further crimped in plug form and heat-set within a cavity.[115]

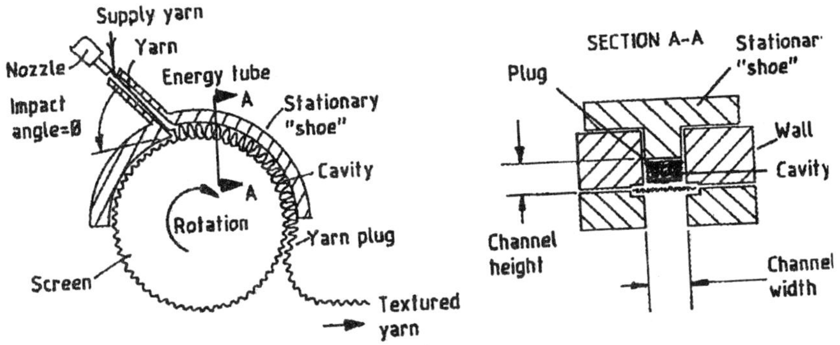

Figure 3.99 Basic principles of moving-cavity texturing

The cavity or chamber space is formed within two parallel walls, a screen that rotates as an integral unit, and the stationary shoe. The length of the cavity can be properly designed to heat-set the yarn plug in the optimum time interval before the yarn is subjected to tension that may lower the texture level. The screen surface carries the plug forward through the confined cavity. In addition to that, the rotation of the screen-cavity serves two important functions:

1. It controls plug stability and density under a broad range of process conditions. For instance, if the steam pressure is doubled, we simply adjust the cavity (or screen) surface speed to maintain the same degree of plug stability.
2. Equally important, a major portion of steam jetted from the nozzle is vented through the screen whereas only the residual portion contacts the slow-moving plug. This minimizes degradation of the plug or textured yarn. It should be stated that the plug travels only at a fraction of the yarn speed.[115]

This moving-cavity technique is capable of texturing 150 dtex 34 filament polyester yarn at a speed of 2,700 m/min.[115] It is also argued that the textured yarn crimp and mechanical properties are comparable with the false twist textured yarns.[115]

3.4 OTHER THERMOMECHANICAL TEXTURING TECHNIQUES

3.4.1 Trapped twist texturing

Trapped twist textured yarns are processed by a different twisting approach (Figure 3.100) in which two ends are twisted together around a common axis rather than each end being twisted on its own axis as in false-twist texturing. The twisted ends are heat-set in the twisted state as they pass over the surface of a heater, then are separated and wound on two separate packages.[116]

The machine used for the production of trapped twist textured yarns was the Turbo Duotwist machine made by Turbo Machine Co. The trapped twist texturing technique is, however, hardly used in today's texturing industry.

A key characteristic of trapped twist textured yarns is their compliant stretch resulting from the relatively large-diameter spring-like coil configuration of the filaments comprising the yarns. This makes for the production of sheer stretch stockings that give perfect fit and smooth comfort without wrinkling, bagging, or localized strain.[116]

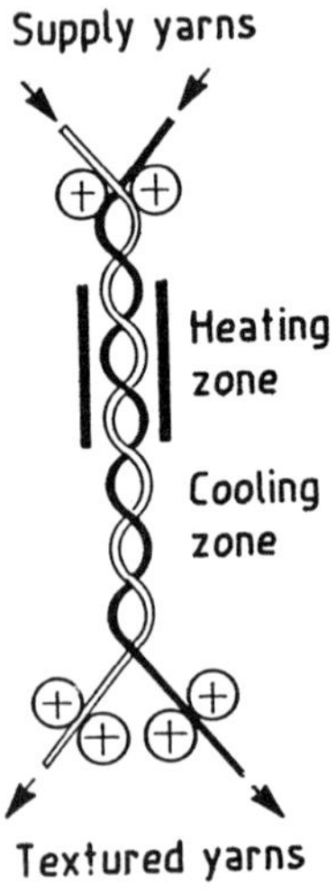

Figure 3.100 Twist trapped between two rollers

3.4.2 Knife edge texturing

The knife-edge texturing process was developed from early attempts to produce a hairy continuous filament yarn by drawing it over a knife edge. In its commercial form as The Agilon process, the method consists of simply passing a yarn, preheated to a plastic condition, over a cold knife edge where the inner part of each filament is compressed and shortened while the outer part is stretched. The yarn cools as it passes over the edge and along the knife blade.[14]

The texturing of yarn in this technique is achieved by the creation of internal stresses in the cross-section of individual filaments. When a filament is pulled over an edge, the molecules in and near the scraped side tend to reorientate. This change in orientation will effectively shorten the fiber on the scraped side and will, therefore, induce a crimp.

The second function of the bending the yarn over an edge is to create extension on one side of the filament and the compression on the other (Figure 3.101). The creation of these extension and compression in the filament cross-section becomes even easier when the thermoplastic filament is heated above a certain temperature.

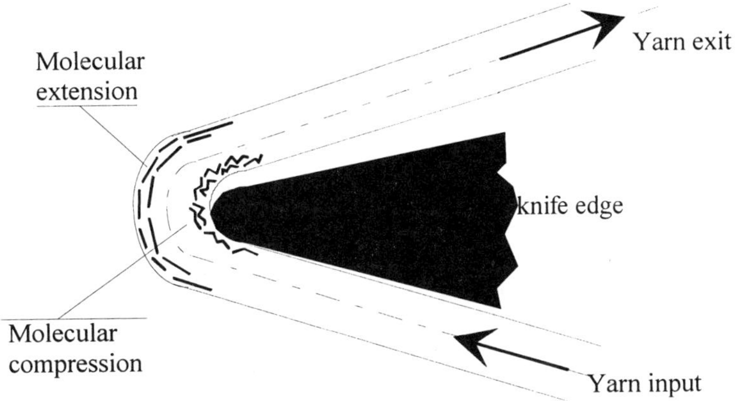

Figure 3.101 Molecular extension and compression in a filament running over a knife edge

Because of the preceding two actions (i.e., orientation and compression/ extension) in the same cross-section, the filament molecular structure reaches a high energy state that should be reduced most probably by the filament's physical deformation. This occurs in the form of consecutive Z and S twists, i.e., a false-twisting action. It is due to this false-twisted crimp structure that the knife edge textured yarns are absolutely torque-free, rendering an easy processing in the further texture processes.

As already noticed, for the texturing to occur, every single filament should get in contact with the sharp edge and get somehow scraped. This naturally will restrain the number of filaments (between one and five), which is few compared with the other texturing processes. Therefore, it has a narrow end use outlet.

A patent[117] claims a method of edge crimping by means of several blades, of which at least one is submerged under hot water. The yarn may contain a lubricant and is claimed to help the crimping process by providing a uniform source of heat, and lubricating the contact between the filaments and the blade. The patent gives a speed of 15 m/min, which is too slow to be commercially attractive.[117]

In the early and mid1960s, when the knife-edge texturing was widely deployed, the knife edge textured yarns were used extensively in the manufacture of ladies seamless and panty hose, to which they import an excellent two-way stretch, resulting from three-dimensional distortion of the knit stitch. A stocking with two-way stretch naturally fits much better and is far more comfortable to wear than a conventional stocking.

It is also reported that a certain amount of the edge textured yarns has been used in carpets and upholstery fabrics.

However, when the edge-textured yarn is stabilized by a heating action and plied so that the yarn linear density is increased. It can then be used for several other purposes, such as in sweaters, dresses, and golf shirts.

The knife edge is the functioning part of the process. Various edge materials are specified, steel, alumina, and silicon carbide. The edge should not be at "utmost sharpness" but should have a radius of 0.0025 to 0.01 mm. As far as the heating methods are concerned, there are many ways. The knife edge could be directly heated, or some other contact-type heater could be placed nearer to the knife edge. The external heater depending on their position heats the yarn either from the extension side or the compression side. However, the heated edge texturing method is more widely used because it is technically more convenient. The heaters are required to apply a constant temperature dependent on the type of yarn used, (e.g., 173°C for nylon 6.6), the control accuracy should be ±1°C as for other texturing techniques.

The famous process of the early 1960s has nearly lost its attractiveness for today's textile industry especially because of the process speeds and the constraint over the number of filaments in a yarn. The Hobourn Machine, probably then the most modern machine was only able to process at about 124 m/min.

3.4.2.1 Rotating knife-edge texturing

Although it is largely regarded as an ancient texturing method and died out completely, a recent work[118] appeared to put some impetus to rejuvenate the process. The report claims that by a rotating-knife-edge crimper (See Figure 3.98), the process speeds of 1,500 m/min can be obtained.

The basic concept of the high-speed crimper is to reduce the relative speed, or to increase the residence time, between the yarn and the rotating knife edge while the heated yarn is sliding over the rotating knife edge under tension in a sharp angular path.[118]

The previously mentioned stationary knife edge texturing process yields hosiery yarns that are well accepted in the trade. The stationary knife-edge process is, however, limited to rather low-exturing speeds of about 50 to 100 m/min. A substantial increase in linear yarn speed was accomplished by designing an apparatus in which the knife edge is caused to move parallel to the direction of the moving yarn at a speed that is lower than that of the moving yarn, the movement of the knife-edge being in the same direction as that of the running yarn.[118]

3.4.2.1.1 Principles of rotating knife-edge texturing

The basic design of this method consists of a knife-edge assembly and a godet, both mounted on the same common axis, as seen schematically in Figure 3.102. Both the godet and the knife edge rotate around this common axis independently and at different rotational speeds. As seen from the Figure 3.102, yarn is supplied from a bobbin. The yarn is fed through a yarn-tensioning device, and then wrapped around the godet roll and the knife edge with as many wraps as required. The yarn tension after sliding over the knife edge is always greater than before the knife-edge. The knife edge is heated, and its temperature is regulated through slip rings.[118]

The godet surface is highly polished and chrome finished. While it rotates, there is no slippage between the yarn and the godet surface. The knife edge revolves at a slower speed around the godet roll surface in a plenary type of motion. The mounting position of the knife edge frame is adjustable to control the angle of the yarn with the knife edge and the tension before or after passing the knife edge.[118]

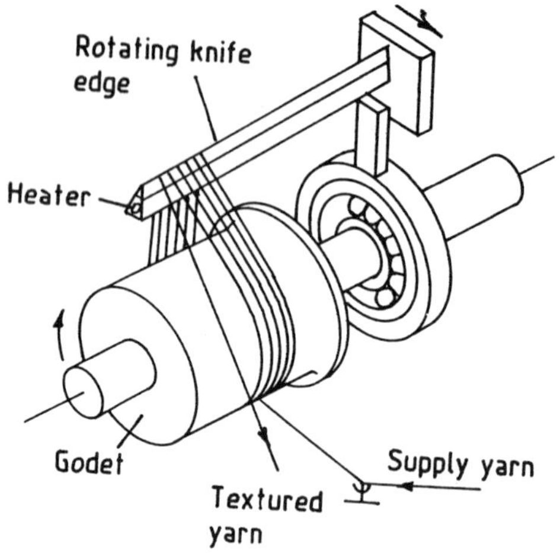

Figure 3.102 Rotating knife edge crimper

The basic difference between the rotating knife edge technology and a stationary knife edge is the number of wraps the yarn makes around the knife edge. There are two basic reasons for the multiple wraps. One reason is that the yarn sections ahead of the knife-edge have to catch up with the knife-edge in order to obtain a yarn textured along its entire length. This leads to a minimum number of wraps requirement. To obtain sufficient texture during crimping with the rotating knife-edge crimper, one uses, in general, a multiple of this minimum number of wraps, which results in each yarn section passing the knife edge at least two times.[118]

The reason for multiple passes is that the total included angle of the yarn over the knife edge is larger in the case of rotating knife edge than for a stationary knife edge. This is a result of the fact that sharper angles of the yarn over the knife-edge would involve greater heights of the knife edge over the godet (or a smaller godet diameter). A greater distance between knife-edge and godet roll surface would result in large fluctuations in the outlet tension of the last wrap at each cycle of rotation. Such large tension fluctuation yields variation in texture level and frequent filament breakage. To avoid such difficulties, the rotating knife edge crimper is designed with a large angle of yarn over knife edge and low differential speeds between yarn and knife edge, which compensates for the angle deficiency.[118]

This technique showed that good and acceptable texture levels can be obtained with monofilament and multifilament nylon yarn at 1,500 m/min.[118] However, the claim does not appear to have materialized yet. No commercial exploitation has been seen so far on the market.

3.4.3 Crinkle Yarn Texturing

When the introduced filament and yarn deformation is approximately planar, the textured yarn is often named as a crinkle yarn. There are two most common ways producing such yarns, being knit-deknit and gear texturing.

Fabrics made with crinkle yarns are characterized by a pronounced sparkling texture, good stretch and recovery from stretch, and greatly enhanced appearance and hand. Unlike false twist textured yarn, the crinkle yarn is a torque-free yarn.

The crinkle textured yarns are used for knitted outerwear (e.g., sweaters, dresses, sport shirts, warp-knit fabrics, such as underwear, fine-gauge seamless hose, and upholstery and carpets with coarse yarns).

3.4.3.1 Knit-deknit texturing

This technique consists of three fundamental operations that can either be separate but consecutive or continuous (Figure 3.103). These operational steps are the following:

1. Knitting continuous multifilament yarn into tube form.
2. Heat-setting the knitted tubes. This could be done in an autoclave if a three-step process is adopted. For a continuous process, however, the heat setting is carried out on the same machine.
3. Unraveling the yarn from the fabric and winding on to a cone or bobbin.

Irrespective of the flow of the preceding steps whether it be continuous or discrete, the end product is of two-dimensional crinkles. Because the whole bundle of filaments are bent and set in an open loopy form, the bulk of the yarn is removed when the yarn is under stretch. The original bulk and texture, however, is required after the load is lifted.

There is a precautionary point when this type of yarn is knitted for the end use. The gauge setting of the knitting machine must always be different from that used for the production of crinkle yarn. When this precaution is omitted, either large holes will appear on the final product, or the texture of the crinkle yarn will be lost rendering a product with little bulk.

Although in the 1960s and early 1970s there has been some interest and development in this type of texturing, no longer is it available nor is it feasible for the modern industry. That is mainly due to the speed limitations and the establishment of superior texturing techniques that will cover more or less the same end uses. However, the popularity of knit-deknit yarns fluctuates depending on the demands of fashion.

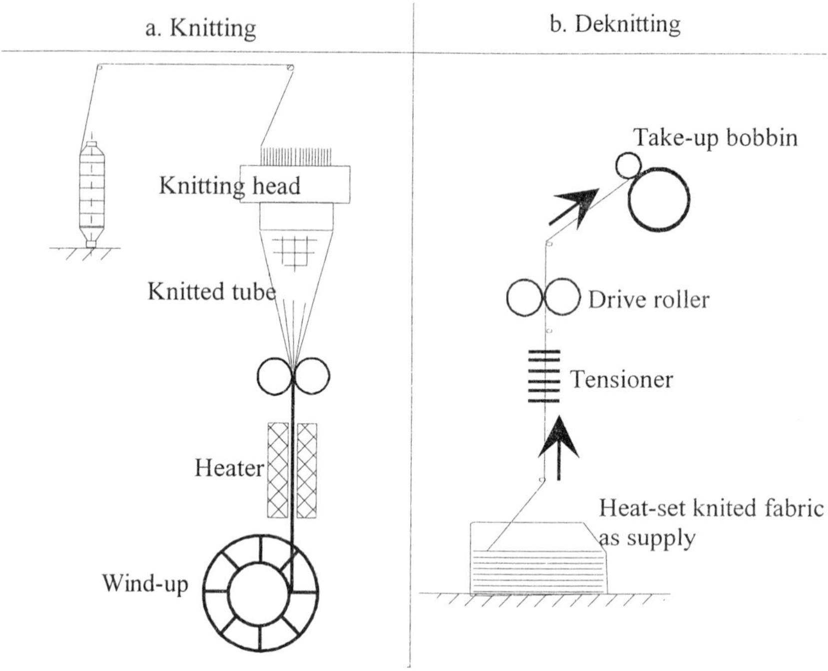

Figure 3.103 Knit-deknit texturing process steps

3.4.3.2 Gear-mesh texturing

The process involves crimp formation as the thermoplastic yarn is trapped between enmeshing sinkers in a rotating head (Figure 3.104). This head is heated so that the crinkle formation is permanently set. The amplitude of the crimp or crinkle, and the number of crimps per unit length can easily be varied by means of varying the pitch and module of the gear-pair over a wide range to suit the desired end-use application. The size of the teeth of the wheels determines the fineness of the crimp. Therefore, these processes require gear wheels of the highest precision, as they have to intermesh in such a way that the yarn is not squeezed off. Furthermore, the gear wheels have to be highly wear resistant[90].

The process is continuous, the yarn being taken from the supply bobbin, textured, and wound onto a package.

Because the texture of the yarn (i.e., crimp height and width) can be positively controlled by means of profiling the gears accordingly, the structure of the textured yarn can be predetermined as having regular crinkles as seen in Figure 3.104. The crinkle is truly planar. Therefore, gear mesh textured yarns impart to fabrics a crisp, dry, silky hand, considerable resistance to creasing, and good recovery characteristics. The inherent stability of the yarns, coupled with its precisely controlled and relatively high-crimp frequency, reportedly results in far better fabrics than its contemporary textured yarns.

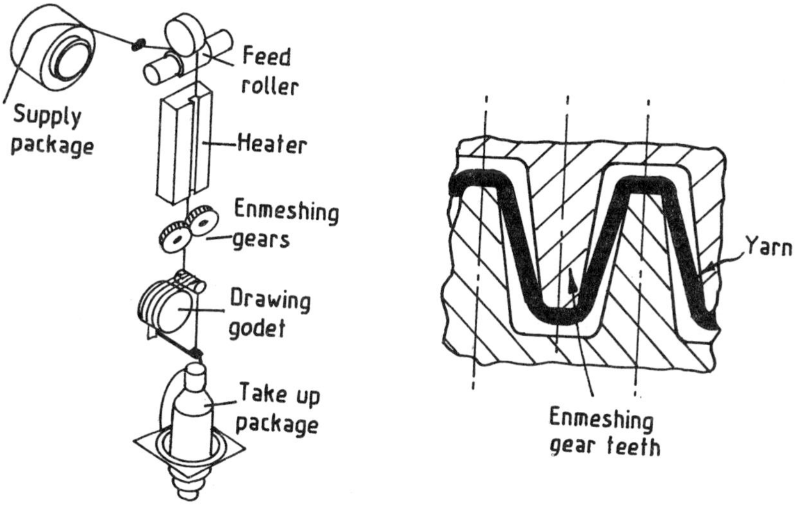

Figure 3.104 Gear mesh texturing process

The most popular uses of the gear mesh textured yarns, when available, were ladies underwear, knitted outerwear, fully fashioned outerwear and children's sweaters. It is also reported that the gear crimped yarns also found end uses in pile fabrics and woven carpets, and because of its versatility to provide optimum crimp frequency to give maximum bulk, the yarns are said to have a considerable covering power in a pile fabric.

The functioning part of the gear mesh texturing process is, of course, the gear pair. The module and pitch of the gears need to be precisely manufactured according to the desired yarn characteristics. The inherent difficulty is the need to change the gear pair when a different type of yarn is required.

It is crucial that the temperature between the texturing heads should be controlled within narrow limits so that the yarns from all positions would be uniform and have even dye uptake and other relevant properties.

Although not available today, the early 1970s have seen the only purpose-built gear mesh texturing machine. This was a 60 position machine equipped with electronic temperature control, stop motion and yarn cutters, and oiling rolls. The machine was capable of texturing supply yarns between 15 to 850 dtex in linear density, which was the usual range for this type of texturing operation. The production speed was limited to maximum 200 m/min.

Later years have seen some patent activity, yet the process is gradually disappearing from the texturing industry.

REFERENCES

1. Piller B., *Bulked Yarns*, The Textile Trade Press, Manchester, 1973.
2. Arthur D. F., and Weller A. F., "The Principles of Friction Twisting," *Journal of Textile Institute*, 1960, Vol. 51, T66.
3. Lord P. R., "Textured Yarns and Fabrics," in *Contemporary Textile Engineering*," Ed. F. Happey, Academic Press, London, 1982, 267-288.
4. Brehm G., "Draw-Texturing Polyamide," *International Textile Bulletins*, 1/78, Spinning, 11-24.
5. Hearle J. W. S., and Yegin A., "The Snarling of Highly Twisted Mono-filaments," Part I: "The Load Elongation Behaviour with Normal Snarling," *Journal of Textile Institute*, 1972, Vol. 63, No. 9, 477-489; Part II: "Cylindrical Snarling," *Journal of Textile Institute*, 1972, Vol. 63, No. 9, 490-501.
6. Denton M. J., "The Development of False-Twist in Bulking," *Journal of Textile Institute*, 1968, Vol. 59, 344-350.
7. Brooks R., "An Experimental Method for Determining the Heat Transfer Coefficient of Polymeric Fibers and Yarns During Rapid Convective Heating," *Journal of Textile Institute*, 1984, Vol. 75, No. 6, 398-404.
8. Morris W. J., "The Use of Heat in the Production of Synthetic Fibers," *Textiles*, 1981, Vol. 10, No. 1, 3-9.
9. Denton M. J., and Morris W. J., "Heat Setting and Yarn Texturing," in *Setting of Fibers and Fabrics*, Ed. J. W. S. Hearle, and L. W. C. Miles, Merrow Publishing Co., Durham, 1980.
10. Backer S., and Yang W. L., "Mechanics of Texturing Thermoplastic Yarns," Part I: Experimental Observations of Steady State Texturing, *Textile Research Journal*, August 1976, 599-610.
11. Riding G., *Journal of Textile Institute*, 1959, Vol. 50, T425.
12. Treolar L. R. G., "A Migrating-Filament Theory of Yarn Properties,"

Journal of Textile Institute, 1965, Vol. 56, T359.

13. Lord P. R., *The Economics, Science and Technology of Yarn Production*, School of Textiles, North Carolina State University, Raleigh, 1981.
14. Morris, W. J., "Yarn Texturing," Part I: The False-Twist Method; Part II: "Texturing Methods Other Than False-Twist, " *Textiles*, 1973, Vol. 8, 46-55.
15. Lünenschloss J., and Fischer K., "Texturierverfahrenihre Entwichlung und Bedeutung," *Chemiefasern/Textilindustrie*, 1978, 38/80, 536-540, 619-621, 701-704.
16. Greenwood K., "The Present Situation and Future Prospects of the False-Twist Process," *Journal of Textile Institute*, 1975, Vol. 66, No. 12, 420-425.
17. Smith P. A., "Texturizing Machinery," *Textile Institute and Industry*, February 1976, 53-60.
18. *The Yarn Revolution*, Conference Proceedings of the 60th Annual Conference of The Textile Institute, 1976.
19. Iyer C., "ITMA 87-Texturizing," *Melliand Textilberichle*, Part I: 1987, Vol. 68, 897, Trans. 1/88 E7; Part II: 1988, Vol. 69, 91-33, Trans. 3/88, E86-87.
20. Rieter-Scragg's Centenary Announcement, *Textile Month,* February 1989, 24-25.
21. Ernest Scragg & Sons Ltd., BP 1,085,052 (30 April 1963).
22. Kang J. T., and El-Shiekh A., "Mechanics of High Speed Texturing," Part I: "Theoretical Analysis of Belt Twisting," *Textile Research Journal*, January 1983, 108, Part II: "Theoretical Analysis of Ring Twisting," *Textile Research Journal*, November 1988, 653-662; Part III: "Experimental Studies of Belt and Ring Twisting," *Textile Research Journal*, December 1988, 719-725.
23. Denton M. J., "Texturing Machines and Test Equipment, *Textile Asia*, December 1982, 67-77.
24. Lünenschloss J., and Fischer K., *International Textile Bulletin: Weaving*, 1978, 143.
25. Thwaites J. J., "The Mechanics of Friction Twisting," *Journal of Textile Institute*, 1970, Vol. 61, 116-138.
26. Weinsdorfer H., "Aspects of Friction Texturing," *Chemifasern/ Textilindustrie*, June 1978, E99-E101, 540-550.
27. Morris W. J., and Denton M. J., "An Improved Method of Friction-Twisting in the False-Twist Texturing Process," Part I: The Development of an Improved Method, *Journal of Textile Institute*, 1975, Vol. 66, No. 3, 116-122; Part II: Theoretical Relations between Yarn and Processing Parameters in the Improved Friction Spindle, *Journal of Textile Institute*, 1975, Vol. 66, No. 3, 123-128.
28. Thwaites J. J., "The Mechanics of Friction-Twisting Reassessed," Part II: Tension and Torque Generation in the Disc Spindle, *Journal of Textile Institute*, 1985, Vol. 76, No. 3, 157-170.

29. Hearle J.W.S., and Morton W. E., *Journal of Textile Institute*, 1957, Vol. 48, T159.
30. Thwaites J. J., "The Mechanics of Friction-Twisting Reassessed," Part I: "The Yarn Path in a Disc Spindle," *Journal of Textile Institute*, 1984, Vol. 75, No. 4, 285-297.
31. Greenwood K., "An Apparent Friction Paradox," *Journal of Physics, D: Applied Physics*, 1977, Vol. L53-L58.
32. Anonym, "Innovation is the Signpost for Throwster's Road to Survival," *Textile Month*, January 1975, 41-44.
33. Anonym, *Draw-Textured Yarn Technology*, Monsanto.
34. Farbir K., and Bock G., "Texturing and Twisting: Further Performance Boosts and Detail Improvements," *International Textile Bulletin: Yarn Forming*, 1/84, 4-19.
35. Denton M. J., "Developments in Yarn Texturing Machinery," *Textile Institute and Industry*, November 1971, 307-312.
36. Arthur D. F., and Jones C. R., "Heat Transfer in Highly Twisted Nylon Yarns," *Journal of Textile Institute*, 1962, Vol. 53, T217-T229.
37. Egambaram T., Afify E. M., and El-Shiekh A., "Heat Transfer in False-Twist Texturing," *Textile Research Journal*, October 1974, 803-812.
38. Coll-Tortosa L., "Beitrag zur Technologieder Torsionstexturierung," Dissertation 1975, RWTH Aachen.
39. Moussa N. A., "Heat Transfer to a Moving Threadline," Conference given at MIT, July 1977.
40. Wang S., "Effects of Laser Heating on the Structure and Properties of False- Twist Textured Yarn," *Melliand Textilberichte* (Eng. Edition), 4/1986, E106-108 + 227-231.
41. Dupeuble J. C., and Fleurier J., "The Advantages of Solid Ceramics Compared with Ceramic Coatings in Friction Texturing," *Textile Institute and Industry*, July/August 1976, 227-229.
42. Atkinson C., "High-Speed Friction Texturing," *Inner Company Report*, Rieter Scragg Ltd.
43. FAG Kugelfischer Georg Schaefer KGaA, "Stacked-disc Friction Texturing Unit," *International Textile Bulletin*, 4/84, 79-80.
44. Whittlestone D. A., "Oiling Problems in High Speed Texturing," *Textile Institute and Industry*, September 1978, 271-275.
45. Twigge R. E., "Innovation in Draw-Texturing," *Inner Company Report*, Rieter-Scragg Ltd.
46. "Revolution in Draw-Texturing," *Unpublished Inner Company Report*, Rieter-Scragg Ltd.
47. Abstracts of Papers, Third Reutlingen Texturing Colloquia, 10-11 October 1984.
48. Schubert G., "Situation Report on High-speed Spinning at the Start of the Eighties," *International Textile Bulletin: Spinning*, 3/80, 229-258.
49. Gerhartz S., "Automation in Continuous Filament Spinning and Texturizing," *Melliand Textilberichte* (Eng. Edition), July 1985, 528-532.

50. Anonym, "High-speed Texturing: A Measure of Progress?," *International Textile Bulletin: Spinning*, 4/79, 525-528.
51. Denton M. J., "Evolution in the Design of False-Twist Texturing Machines," *Textile Month*, August 1976, 49-55.
52. Denton M. J., "The Structural Geometry and Mechanics of False-Twist-Textured Yarns," *Journal of Textile Institute*, 1975, Vol. 66, No. 2, 80-86.
53. Arthur D. F., in "*Modern Yarn Production from Man-made Fibers*," Ed. G. R. Wray, Columbia Press, Manchester and London, 1960.
54. Love A. E. H., "A Treatise on the Mathematical Theory of Elasticity", Cambridge University Press, Cambridge, 1920.
55. Gupta D. K., and El-Shiekh A., "Load Deformation Behaviour of Three-Dimensional Crimped Filaments: Theoretical Analysis and Experimental Verification," *Textile Research Journal*, May 1983, 313-327.
56. Cooke W. D., "Knitted Fabrics from Textured Yarns," *Textile Institute and Industry*, March 1977, 92-95.
57. Derder G., "Textured Polyester Filament Yarns in Spun Yarn Weaving," *Melliand Textilberichte* (Eng. Edition), September 1985, 699-713.
58. Dipak D. E., Mukhopadhya, S K and Majumdar, A, "Effect of Heater Temperature and Contact Time on the Tensile Properties of Friction-Twisted Texturised Yarns," *Textile Research Journal*, February 1974, 91-95.
59. Gupta V. B., Majumdar, A., and Seth, K. K., "Structural Changes in Nylon 6 Yarn on Heat Setting and Friction Twisted Texturing," *Textile Research Journal*, July 1974, 539-544.
60. Gupta V. B., and Kumar M., "Changes in the Structure of Polyethylene Terephthalate Yarn on Texturing," *Textile Research Journal*, May 1975, 382-388.
61. Gupta V. B., Kumar M., and Gulrajani M. L., "Dyeability Characteristics of Textured Polyethelene Terephthalate Yarn," *Textile Research Journal*, June 1975, 463-467.
62. Weinsdorfer H., and Egberg G., "The Draw Texturing Process," *Textile Research Journal*, September 1975, 654-666.
63. Hardegree G. H., and Burchanan D. R., "Physical Properites of Textured Polyester Yarns: The Influence of Texturing Temperature," *Textile Research Journal*, November 1976, 828-833.
64. Warwicker J. O., "Structure, Properties and Performance of False-Twist Textured Yarns," Shirley Institute Pamphlet S25, October 1976, 49-66.
65. Anonym, *Yarn Texturing in the 80's*, Shirley Institute S39, Conference Proceedings, 30 April 1980.
66. Gupta V. B., and Amirtharaj J., "The Effect of Tension and Twist on the Structure and Dyeability of Textured Polyethlene Terephalate Yarn," *Textile Research Journal*, November 1976, 785-790.

67. Smith R. L., Pieters R., and Morrison M. E., "Fundamentals of False-Twist Texturing of Continuous Filament Yarns," *Transactions of Society of Rheology*, 1972, Vol. 16, No. 3, 227-576.
68. Sengupta A. K., Sen K., and Mukhopadyay A., "False-Twist Texturization of Polypropylene Multifilament Yarns," Part I: "Time-Temperature Effects," *Textile Research Journal*, May 1986, 281-286; Part II: "Reducing Feeder Yarn Crystallinity Through Tension Annealing," *Textile Research Journal*, June 1986, 389-392; Part III: "Reducing Feeder Yarn Crystallinity through Melt Blending with Small Amounts of Polyester and Polystyrene," *Textile Research Journal*, July 1986, 425-428; Part IV: "Structural Influences on Dye Uptakes," *Textile Research Journal*, August 1986, 511-515.
69. Sen A., Gupta V. B., and Sengupta A. K., "False-Twist Textured Poly(ethylene Terephthalate) Slit Fiber Yarn," *Textile Research Journal*, November 1982, 710-713.
70. Earnshaw J., Greaves D., and Duncan D. G., "Fiber Finishes for False-Twist Textured Yarns," *Textile Month*, July 1974, 44-48.
71. Anonym, "Progress in Yarn Texturing," *Textile Month*, August 1975, 55-58.
72. Shealy D. L., and Kitson R. E., "An Age-Stable Feed yarn for Draw Texturing by the False-Twisting Process," *Textile Research Journal*, February 1975, 112-117.
73. Peacock G. B., and Jacques D., "Microprocessor Application to False-Twist Texturing Machinery," *Textile Institute and Industry*, March 1980, 63-65.
74. Lünenschloss J., and Farber K., "Study of Relationships between Yarn Tension, Weight and Twist Density Variations and the Texturing Faults in Polyester Filament Yarns During Simultaneous Draw-Texturing," *Melliand Textilberichte* (Eng. Edition), 3/1986, E100.
75. Hanisch M., "Yarn Tension Measuring for On-Line Quality Monitoring in False-Twist Texturing Machines," *Chemiefasern/Textilindustrie*, November 1985, Vol. 35/87, E86, 779-780.
76. Wulfhorst B., and Bruske J., "On-line Measurement of Yarn Twist in False-Twist Texturing," *Melliand Textilberichte* (Eng. Edition), 9/1989, E310, 620-621.
77. Textechno Herbert Stein, Textile Testing Instruments, "Yarn Tester, ELKOMETER II," *Trade Catalogue*.
78. Vidurashwatha K. L., and Mishra S., "Texturizing: Today and Tomorrow in India," *Manmade Textiles in India*, August 1987, 379-387.
79. Fitton S., and Stacey P., "Apparatus for the Measurement of the Crimp Rigidity of Bulked and Stretch Yarns," *Journal of Textile Institute*, 1960, Vol. 51, T435-T437.
80. Gupta V. B., Ailawadi A., and Ganguly P. K., "Crimp Retraction of Textured Poly(ehylene Terephthalate) Yarn," *Textile Research Journal*, September 1983, 520-523.

81. Textechno Herbert Stein, Textile Testing Instruments, "Dynamic Thermal Stress Tester, DYNAFIL M," *Trade Catalog*.
82. DIN 53840, "Determination of Parameters for the Crimp of Textured Filament Yarns," November 1983 (in two parts).
83. Barnes D. S., and Morris W. J., "Rates of Setting in False-Twist Draw-Texturing," Part I: The Effects of Heating Time and Length in the Processing of Polyester-Fibre Yarn, *Journal of Textile Institute*, 1980, Vol. 71, No. 6, 291-298.
84. Seidel L. E., "Texturing's Ongoing Challenges," *Textile Industries*, April 1983, 22-25.
85. Hinchliffe M. G., Eaton D. C., and Lawrence C. J., "Meaningful Package Performance Analysis," *Inter Company Report*, Rieter Scragg Ltd.
86. Brookstein D., and Backer S., "Mechanics of Texturing Thermoplastic Yarns" Part VI: Transient Mechanics of Draw-Texturing, *Textile Research Journal*, April 1978, 198-218.
87. Thwaites J. J., Brookstein D. S., and Backer S., "Deduction about the False-Twist Process from Observations of the Variation of Torque on Detwisting a Twisted Heat-Set Yarn," *Journal of Textile Institute*, 1976, Vol. 67, No. 6, 183-186.
88. Backer S., and Thwaites J. J., "Transient Phenomena During False-Twist Texturing," *Journal of Textile Institute*, 1976, Vol. 67, No. 6, 218-219.
89. Denton M. J., "Transients and Tight Spots in False-Twist Texturing," *Journal of Textile Institute*, 1976, Vol. 67, No. 7/8, 276-279.
90. Klein W., and Trummer A., "Draw-Texturising Techniques with Special Reference to the Friction False-Twisting Process," *Textile Manufacturer*, 1973, Vol. 100, No. 1, 16-23.
91. Du G. W., and Hearle J. W. S., "Threadline Instability in the False-Twist Texturing Process, Part I: "Experimental Observation of Tension Surging"; Part II: "The Influence of Process Parameters on Threadline Instability," *Journal of Textile Institute*, 1989, Vol. 80, No. 2, 203-224.
92. Lünenschloss J., Faber K., and Gökpekin S., "Instability Behaviour during the Simultaneous Draw-Texturing of Nylon 6.6 Filament Yarns Spun at High Speeds," *Chemiefasern/Textilindustrie*, 34/86, April 1984, E33-E35, 264-272.
93. Familant H. M., "Theory on Heater Cross-Over in the False-Twist Texturing Operation," *Textile Research Journal*, June 1980, 336-340.
94. Knowlton R. C., Pierce N. C., and Popper P., "Snow Deposits in Friction Twisting," *Fiber Producer*, December 1978, 18-30.

95. Denton M. J., Eaton D. C., and Baker A. H., "Noise in False-Twist Texturing Plant," in *Texturing Today*, Shirley Institute Publication S.46, Manchester, 1983, 65-107.
96. Erbach W., "Error Sources in Texturizing and Uptwisting," *Melliand Textilberichte*, December 1978, 953-956, (English Translation 895-900).
97. Denton M. J., "The Future of Textured Yarns: Draw-Texturing and Bicomponents," *Textile Institute and Industry*, January 1974, 5-9.
98. Thwaites J. J., "Linear Density Variations in Draw-Textured Yarns," *Journal of Textile Institute*, 1974, Vol. 65, 353-359.
99. Nigam J. K., "The Simultaneous Draw-Texturing of Undrawn Feed Yarns," *Journal of Textile Institute*, 1976, Vol. 67, No. 11, 406-409.
100. Ghosh S., and Wolhar J., "The Influence of Some Machine Setting Variables on the Properties of Yarns in the Friction-Twist Draw-Texturing Process," *Textile Research Journal*, June 1981, 373-383.
101. Denton M. J., "The Conversion of Conventional Texturing Machines to Draw-Texturing," *Journal of Textile Institute*, 1977, Vol. 68, No. 3, 118-120.
102. Gay T. H., "The Simultaneous Draw-Texturing of Undrawn Feed Yarn," *Journal of Textile Institute*, 1977, Vol. 68, No. 3, 121.
103. Yang W. L., "The Effect of Pre-Heating in the False-Twist Draw-Texturing Process," *Journal of Textile Institute*, 1983, Vol. 74, No. 4, 221-223.
104. Cannon C. G., "Filament Deformations in Simultaneous Draw Texturing by the False-Twist Method," *Journal of Textile Institute*, 1979, Vol. 70, No. 6, 243- 252.
105. Thwaites J. J., and Hooper C. W., "The Spectra of Linear-Density Variation in Draw-Textured Filaments," *Journal of Textile Institute*, 1975, Vol. 66, No. 9, 313-317.
106. Familant H. M., "Single-Stage Simultaneous-Draw-Texturing Heat-Transfer Studies," *Journal of Textile Institute*, 1978, Vol. 69, No. 5, 157-159.
107. Jones C. R., and Mason T., "Heat Transfer in a False-Twist-Bulking Process," *Journal of Textile Institute*, 1971, Vol. 62, 147-165.
108. Hsu P., and Backer S., "Mechanics of Texturing Thermoplastic Yarns," Part X: "Effect of Frictional Snubbing on Pseudo False Twisting during Set Texturing," *Textile Research Journal*, September 1983, 530-539.
109. Lüneschloss J., Fischer K., and Murata T., "Effect of Heat Setting Temperature and Overfeed in the Setting Zone and in Winding on the Properties of Textured Polyester Yarns," *Chemiefasern/Textilindustrie*, 24/76, October 1974, 825-829.
110. Smith S. C., and Backer S., "Mechanics of Texturing Thermoplastic Yarns," Part IX: "Yarn Behaviour in the Second Heater," *Textile Research Journal*, December 1982, 761-775.

111. Wilson D. K., and Kollu T., "The Production of Textured Yarns by Methods Other Than the False-Twist Technique," *Textile Progress*, 1987, Vol. 16, No. 3.
112. Anonym, "Yarn Texturing: Slow-Down in Developments," *International Textile Machinery*, 1970, 59-62.
113. Foster P. W., "Fiber M: Non-Isothermal Texturing at High Speeds," *Textile Institute and Industry*, June 1975, 172-174.
114. Schellenberg H., "New Developments in BCF Texturing," *Textile Month*, May 1989, 44-46.
115. Li H. L., "Moving-Cavity Texturing Technology and High Speed Texturing of Poly(ethylene-Terephthalate) Yarns," *Textile Research Journal*, December 1981, 745-754.
116. Wilkinson G. D., "A Knitter's Guide to Texturing Processes," *Knitted Outerwear Times*, June 1970, 57-65.
117. Asahi Kasei Kogyo K. K, BP1.084.616, (Japan, 12 October 1964)
118. Smith P. A., "Yarn Production and Properties," *Textile Progress*, 1969, Vol. 1, No. 2.
119. Li H. L., Oswald H. J., Prevorsek D. C., and Lamb G. E. R., "Rotating Knife-Edge Crimping," *Textile Research Journal*, October 1978, 591-598.
120. Demir, A., "Drallfreie FZ-Texturgarn-Produktion," *Chemiefasern/ Textilindustrie,* 42/94, January 1992, 36-40, E3-E5.

CHAPTER 4

MECHANICAL TEXTURING TECHNIQUES

4.1 INTRODUCTION

Parallel, and, therefore, lustruous arrangement of most continuous filament yarns can be changed to create less lustrous, bulkier, and more attractive human appeal by purely mechanical means. Because no heat treatment is essential for this type of texturing process, thermoplastic and nonthermoplastic filament yarns, such as glass, carbon, and metal, are equally manageable, thus opening a wide spectrum of yarn processing possibilities for the human imagination.

Air-jet (cold fluid) texturing process is the major thechnique utilizing purely mechanical means to impart texture to the filament yarns. A similar use of cold air-jets is seen on mingling process, but, in general, it is used to give some extra cohesion to already textured (mostly by thermomechanical texturing techniques) yarns. However, the latest trend in this field is to blend continuous filament yarns with staple fiber yarns (natural and/or synthetic) by the use of intermingling jets. This process, therefore, could be regarded as a mechanical texturing technique. In the following sections, these two mechanical texturing techniques will be studied in further detail.

4.2 AIR-JET TEXTURING

The air-jet texturing process converts flat, continuous synthetic yarns into entangled, convoluted, bulky, spun-like structured yarns (Figure 4.1).

The air-jet texturing process is by far the most versatile of all the yarn texturing methods in that it can "blend" filaments together during processing, and it, therefore, simulates a desirable attribute of spun fiber yarns. This greater versatility offers the texturizer far greater scope than other types of textured yarns.

Moreover, the feed yarns need not be restricted to the synthetic filament yarns with their good thermoplastic properties. Although polyester and polyamide multifilament types have so far been mainly used for the air-jet yarn products to date, other filament types, such as polypropylene, glass, and viscose and acetate rayons, are also being used for certain speciality end uses.

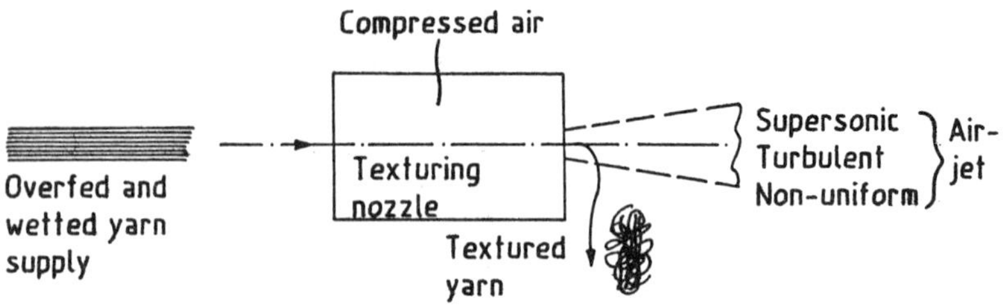

Figure 4.1 Air-jet texturing process

4.2.1 Historical Development of the Process

From the announcement of the first nozzle patent in 1952 to the present day, the air-jet texturing process has experienced many developments and improvements. Although texturing speeds have been increased from 10 m/min up to 600 m/min and the compressed air consumption has been reduced from about 22 m^3/hour to about 6 m^3/hour per nozzle, while the yarn properties have been considerably improved, the process can still hardly be regarded as fully established. High-energy demand resulting from high compressed air consumption has been one of the initial reasons for the slow development of the air-jet texturing process. The lack of a thorough understanding of the basics of the process has also handicapped improvements, such as improved nozzle design and systematic provisions of current process parameters.

Today's air-jet texturing technology requires no pretwisted supply yarns to produce a wide range of textured yarns ranging from fine polyester and polyamide yarns for apparel fabrics to heavy carbon filaments for aerospace technology and glass fibers. In contrast with early texturing machines equipped with only one type of nozzle, (i.e.,Taslan IX), several texturing nozzles are available for modern-purpose-built texturing machines. Such machines also facilitate many other new features, such as drawing POY, to provide process economics, and heat setting after texturing, to satisfy growing market demands.

4.2.1.1 Economic aspects of the process

In the early days of the strictly licensed, and, therefore, highly secret, Taslan-texturing process, the conversion costs were thought to be so unfavorable that no attempts at comparison with the other yarn texturing techniques appeared in the literature. The license fee was, of course, additional to those costs. As air-jet processing gradually improved, with the introduction of more efficient nozzles, the conversion costs decreased to such a level that some comparisons with both false twist textured and spun-staple yarns started to be made. By comparing the costs of per kilo of 100% wool, 40/60 polyester/wool, 50/50 polyester/wool, and 100% polyester worsted yarns with false twist textured yarns of approximately the same linear density, Brehm,[1] in 1979, concluded that the energy consumption of air-jet texturing needs to be reduced for the air-jet textured yarns to compete with false-twist textured yarns. Because air-jet textured yarns closely resemble spun staple yarns and are of completely different characteristics than false twist textured yarns, a competition between the two texturing processes is reported to be unjustifiable.[2]

The energy requirements of the air-jet texturing process have been compared with other conventional spinning processes by many researchers.[3-8] A report by Wilson[8] has illustrated that the energy requirement for 167 dtex ringspun 67/33 cotton/Polyester yarn is 4.41 kWh/kg, whereas the same linear density POY polyester yarn can be air-jet textured at 600 m/min with an energy consumption of only 3.1 kWh/kg. The article also reported that this amount could even be reduced to as little as 1.8 kWh/kg if a two fold 167 dtex yarn is used.

With such economic advantages, the air-jet textured yarns are predicted to have an opportunity of competing for 54% of home furnishings, 15% of industrial fabrics, and 53% of apparel fabrics in the Unitd States.[9] Similarly, the air-jet textured yarns are also occupying a considerable slice of textile yarn application in Western Europe. Despite the estimated total of 24,500 tons of air-jet textured yarns in 1984,[10] the approximate production, however, was reported[11] to be 63% more at 40,000 tons.

4.2.2 Principles of Air-Jet Texturing

Figure 4.2 schematically illustrates the basic requirements of the air-jet texturing process. The process involves the "overfeed" principle where the multifilament supply yarn, as supplied 'over-end' from a creel, is fed into the nozzle at a greater rate than it is taken away. To achieve this degree of overfeed, the yarn passes through feed roller systems FR1 and/or FR2 faster than it does through the delivery roller systems DR2. The overfed filaments enter the nozzle and are blown out from the exit end. They are formed into textured yarn under the turbulent effects of the supersonic air stream provided from a suitable compressed air supply.

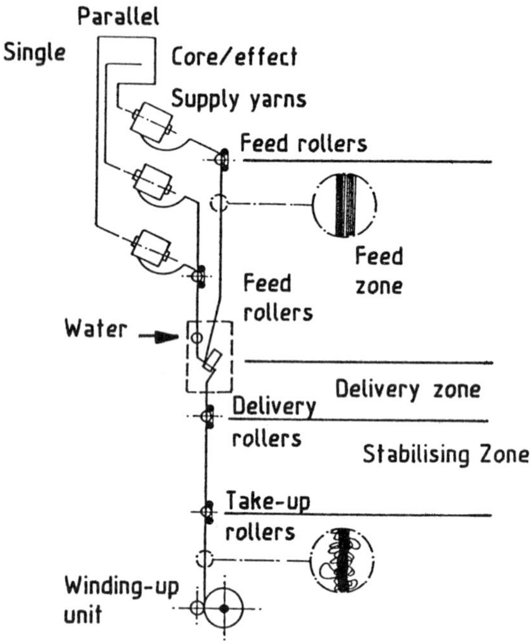

Figure 4.2 Basic requirements of the airjet texturing process

The zone between the feed rollers and the nozzle is termed the *feed zone*, and that between the nozzle and the delivery rollers is termed the *delivery zone*. The supply yarn is normally wetted just before it is fed into the texturing nozzle by passing it through a water bath or through a wetting unit that can either be separate from or integrated with the nozzle assembly. Wet texturing apparently improves the quality of the yarn produced. It is also believed that an impact element, such as a small sphere at the nozzle exit, enhances the quality of the yarn. Texturing nozzles are usually enclosed in a chamber, not only to reduce the noise created by the air jet, but also to collect the used water and the spin finish that is washed away from the filaments during the process.

The following three main types of operations are established to produce a range of yarns:

1. Single-end texturing
2. Parallel texturing
3. Core-and-effect texturing

Single-end texturing denotes the situation whereby a single end is overfed to the texturing nozzle, and the resultant textured yarn is withdrawn. In the case of parallel texturing, two or more ends are fed at the same overfeed into the nozzle so as to facilitate a blend of different types of supply material, or of the same material but with different filament linear densities or number of filaments, or different cross-sectional shapes. Yarns produced by the parallel texturing have a reasonably uniform bulk as the overfeeds of all the component yarns are equal. The versatility of the air-jet texturing process is best observed by the core-and-effect operation in which one or more ends of yarns are fed into the nozzle at a low overfeed to form the core, whereas a second group of yarns is simultaneously fed to the same nozzle at a higher overfeed to create the desired bulk. By changing the overfeeds, a wide range of such yarns can be produced. Any of these three main yarn types can be modified by intermittently varying the overfeed conditions so as to cause periodic longitudinal variations in the bulking effect to create fancy effect yarns.

4.2.3 Air-Jet Textured Yarns

Yarns textured by the methods based on heat setting during mechanical distortion (Figure 4.3) have a common characteristic of high extensibility under low loads. This extreme extensibility arises from the open structures that result in such "stretch" yarns. Conversely, yarns produced by the air-jet technique are of totally different structures in that they much more closely simulate spun yarn structures. The bulkiness of the stretch yarns decreases with the degree of tension imposed on them, whereas the form of air-jet textured yarns can be made to remain virtually unchanged at loads corresponding to those normally imposed in fabric production and during wear. This is due to the "locked-in" entangled loop structure attributed to air-jet textured yarns. Air-jet textured yarns again more closely resemble conventionally spun yarns in that the yarn surface is covered with fixed resilient loops, and these serve the same purpose as the protruding hairs in spun yarns by forming an insulating layer of entrapped still air between neighboring garments.

The combined effects of surface loops with a well-entangled core enable the air-jet textured yarns to simulate the appearance and feel of spun-staple yarns. They are certainly more spun-like than the other types of textured yarns and, consequently, find similar application fields to spun yarns, ranging from woven apparel fabrics to heavy "technical" yarns.

When air-jet textured yarns are woven or knitted into fabrics, they offer an excellent performance with air permeability approaching that of worsted fabrics, better covering power, and no snagging because of the locked-in yarn structure. Because the texturing process does not affect the individual filament cross-section, the fabrics do not glitter in the brightest sunshine. Furthermore, air-jet textured yarns offer superior abrasion resistance, dimensional stability, and crease resistance than worsted fabrics.

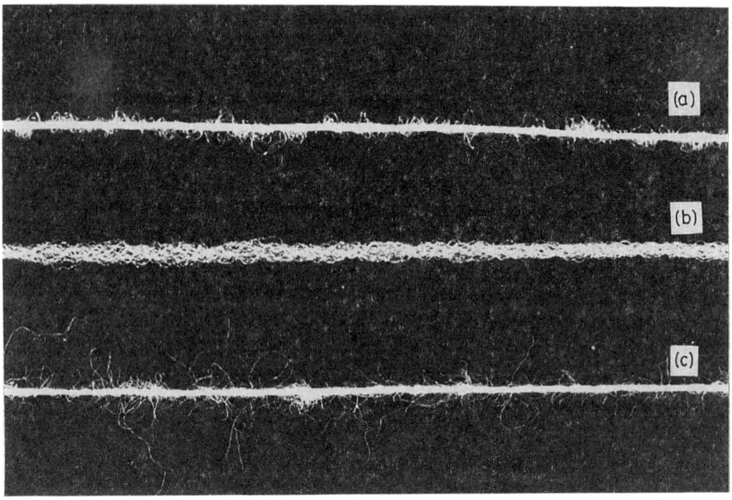

Figure 4.3 Visual comparison of textured yarns

4.2.3.1 Studies on the properties of air-jet textured yarns

The air-jet textured yarns are of different characteristics to their parent yarns, in that they are of higher bulk, exhibit increased covering power, have a more subdued luster, and are warmer in hand. Therefore, they possess some unique properties that require investigation.

The work on the properties of the air-jet textured yarns starts with Wray's[12] research. He defined some of these unique properties such as increase in linear density resulting from texturing, loop size and frequency, physical bulk, and instability, and the suggested improvements in their measurement techniques.

The Du Pont Company has suggested a microscopical test method[13] by which the overall diameter, core diameter, and loop size and frequency of a Taslan air-jet textured yarn can be assessed. Wray,[14] criticizing this method as being tedious and inaccurate, devised a quicker and yet more accurate graphical method[15] for measuring the yarn diameter and average loop size. Ten years later, Fischer[16] reported a more elaborate system that uses photodiode cells to detect the loops and arcs around the core and sorts out the loop size and frequency of the yarn. Bock[17] made use of such a system for his investigations. However, Wilson[5] reported that a fray counter developed by Toray Co. offered possibilities of counting the loops and evaluating their sizes. Also Kollu[18] described another system that makes use of a microdensiometer and ancilliary equipment. Using such a system in a graphical way, Kollu obtained an estimate of the loop size and number. Acar et al[19] developed a highly sophisticated device with a line of 256-monolithic charge coupled devices as the image recognition element. They claim that this device, together with a microcomputer and suitable software, will enable the user to measure the mean yarn diameter, loop size, and frequency at test speeds up to 10 m/min.

Depending on the amount of overfeed and the effectiveness of the texturing conditions (viz. nozzle type, air pressure, and texturing speed), the physical bulk of the textured yarn is increased. To quantify physical bulk, Du Pont[13] recommended a test that compared the package densities of the yarn before and after texturing. In addition to this, Þen and Wray[20] utilized another method in which he compared the specific volumes of fabrics woven from both textured and supply yarns. A specially designed device was used by Rozmarynowska and Godek[21] in 1966 to measure the apparent volume of the air-jet textured yarns. This bulkiness test consisted in gauging a skein of yarn subjected to a known initial tension. Þen and Wray[20] have refined the Du Pont's physical bulk test by weighing and accurately measuring the outside diameters of the two packages of supply and textured yarns. They also devised a water absorption technique to evaluate the bulkiness of the textured yarn, and concluded that the water absorption test was much quicker and simpler to use than the package density tests.

The stability of an air-jet textured yarn is its resistance to having the loops pulled out by tension. This unique property is tested by means of instability measurements. Both Du Pont[13] and the Heberlein Yarn Technical Center[22] have suggested their own instability measuring techniques.Conversely, three more instability measuring techniques have also been proposed by Wray,[23] and Acar et al.[24] Therefore, it is the common concern of many[22,25,26] that a standard, objective instability measuring technique is later defined by Demir.[90]

In an attempt to establish a quality control criterion, Piller[26] defined an SVS (Stabilitat, Voluminositat, Schligenhaufigkeit) value that is an aggregate total of instability, bulkiness, and loop frequency, but this value does not seem to have attracted any interest to date. Piller and Lesykova,[6] using idealized models, also attempted to build up a mathematical model for the structure of the air-jet textured yarns, but they came to the conclusion that the structure of an actual yarn differs considerably from that of an idealized model, thereby confirming the view that determining the characteristics of the yarn seems to be complex if not impossible. Another mathematical approach to the mechanics of the air-jet textured yarns was also attempted by Kollu.[27]

The overall quality assessment of air-jet textured yarns is made to limited extent by means of some of the previously discussed test techniques together with simple visual inspection and tensile tests. Most of these tests leave much to be desired in that these are laborious, time-consuming, and largely subjective. In contrast, the modern textile industry requires fast and objective on-line quality control[28,29] whenever possible.

4.2.4 End Uses of Air-Jet Textured Yarns

The application and end uses for the various types of air-jet textured yarns are normally determined by their properties. However, because of the previously mentioned versatility of the process and yarns, air-jet textured yarns find application in almost every corner of the textile industry, such as shoe inseaming thread, men's shirting, shoe laces, sheeting, tent fabrics, upholstery, drapery, neckties, women's sweaters,[30] men's jacket, women's blouses, dresses, sewing threads, automobile upholstery, towels, blankets, swimsuits,[31] tapes, embroidery articles, certain medical fabrics,[26] anoraks, structural and coated applications,[2] chaffer fabrics, luggage, rainwear, sportswear,[32] canvas-type fabrics for luggage, glass fibers in printed circuit boards, and automobile body parts.[15]

4.2.5 Air-Jet Texturing Nozzles

Being at the heart of the entire air-jet texturing process, the various industrially used texturing nozzles have been developed through practical experience rather than by informed theoretical considerations. Their continuous evolution may be divided into the following four major groupings:[33]

1. Early stages and "false start" with the Taslan IX nozzle
2. Upsurge of improved nozzles (i.e.,Taslan X and XI)
3. Further improvements to the Taslan nozzles (Taslan XIV and XV) with impact elements and interest from textile companies besides Du Pont
4. Contemporary stage with the Taslan XX and cylindrical-type nozzles (e.g., HemaJets)

At the first stage of the development, nozzles[34] utilized the compressed air so inefficiently and unreliably[35] that they not only created a negative public image of the process by being energy intensive, but they also diverted the attention of nozzle designers to inadequate designs that occupied the first 10 years of development. Therefore, the hesitant beginnings of the process was perhaps rightly termed as a "false start" by Phillips[32] and Isaacs.[36] However, when the first patented nozzle,[34] shown in Figure 4.4, is studied, it is interesting to note that this apparently little-used nozzle design covered almost all the features of the modern Taslan nozzles. It had an axial yarn feed, whereas the compressed air was admitted to the nozzle body through a side entry. It also had an impact element of a flat plate type. Nevertheless, the lack of aerodynamic design of the inner configuration caused tremendous losses to the air flow resulting in inefficient texturing. Another early design reported by Piller[37] also suffered from similar deficiencies.

Following the failure of this nozzle, the nozzle designers were forced to change the basic configuration and introduced a completely different design.[38] The Taslan IX nozzle shown in Figure 4.5 had an axial air flow that was accelerated by well-established converging-diverging geometry, while the highly pretwisted overfed yarn was admitted to the main channel through a stepped hollow needle inclined at 45° to the nozzle axis. This nozzle had serious wear problems and was difficult for an inexperienced operator to set and operate[33] because of the crucial importance of the needle setting. Nevertheless, it was used by licencees of the Taslan process for well over a decade.

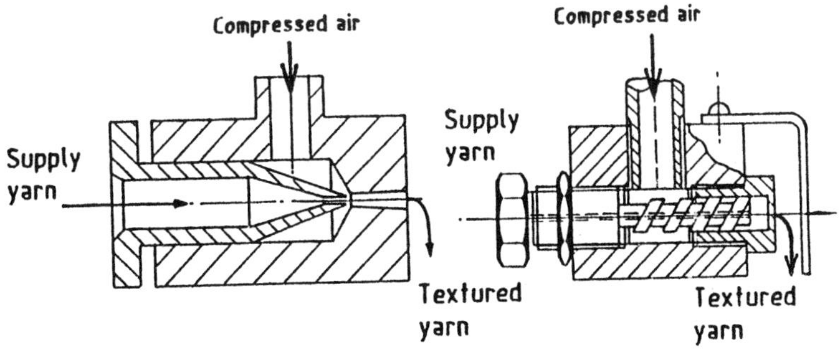

Figure 4.4 Early nozzle designs

Du Pont's nozzle designers, dissatisfied with the poor performance of the Taslan IX nozzle, had to resort to the basic features of their earlier design (See Figure 4.4). With improved aerodynamic design of the inner surfaces of the nozzle as well as the implimentation of the converging-diverging nozzle geometry, a long-lasting breakthrough began and the Taslan X nozzle[39] (Figure 4.6) gradually emerged, from behind all the closed doors of the Du Pont laboratories, as the most acceptable nozzle for the Taslan licensees. During this period, Wray and Entwistle[44] had thoroughly investigated the Type IX nozzle, as revealed in the patent literature, and had publicly revealed many drawbacks of its design. They, furthermore, suggested a modification that improved the performance by reducing the air consumption up to 65%. This modification was also reported to be eliminating the sensitive needle setting of the original Taslan IX nozzle while having little negative effect on the yarn quality. The Taslan XI[41] and other nozzles[42] (Figure 4.7) then followed with improving performance by creating asymmetrical flows in the nozzle venturis. The asymmetry in the flow is created by one-sided air entry to the so-called turbulence chamber in the Taslan XI and the eccentric fitting of the needle in the Taslan XII. Through these developments in the nozzle design, the texturing speeds have been increased in the order of 10 times (i.e., to 600 m/min from the earlier 50 m/min).[33]

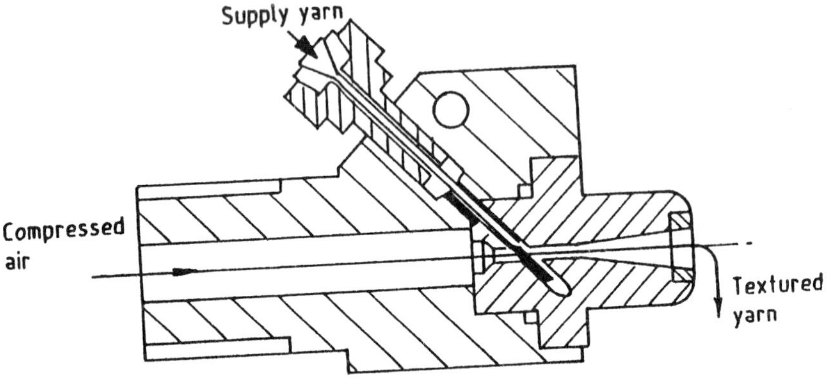

Figure 4.5 Taslan Type IX

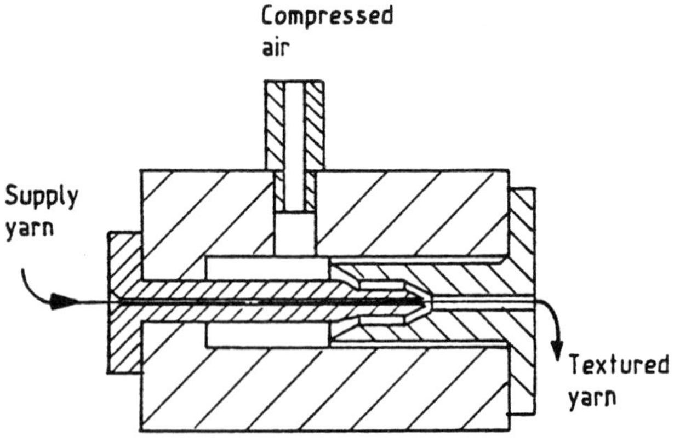

Figure 4.6 Taslan Type X

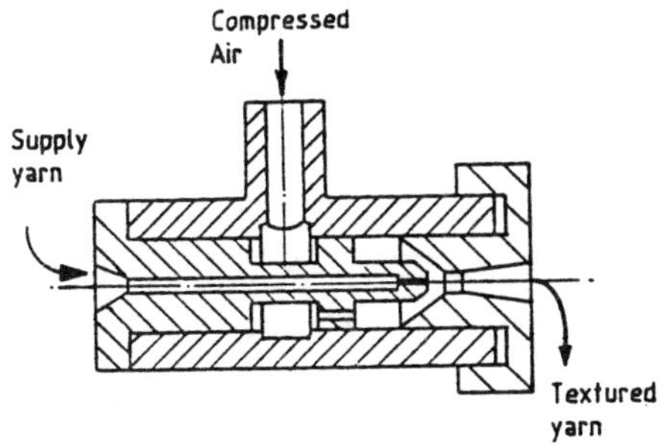

Figure 4.7 Taslan Type XI

The introduction of external impact elements, such as plates, bars, and spheres, immediately after the nozzle exit was another development stage. The Taslan XIV[43] and XV[44] nozzles (Figure 4.8) with impact plate and bar, respectively, have been introduced, which have allowed the use of twist-free supply yarns. In 1985, Du Pont patented[45] its Taslan XX nozzle, which possesses an easy string-up facility with no needle or venturi adjustment (Figure 4.9).

Despite the continuous development and improvement attempts, certain drawbacks, such as high compressed air consumption, nozzle-to-nozzle inconsistency resulting from the need for precise adjustment of both the needle and the venturi, and the frequent contamination of the venturi with spin-finish material washed away from the yarn surface, thereby requiring regular cleaning and servicing,[33] still needed to be solved. Such factors have induced some other

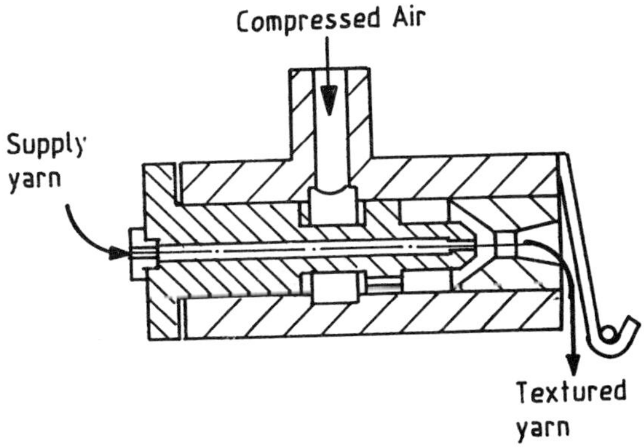

Figure 4.8 Taslan Type XIV

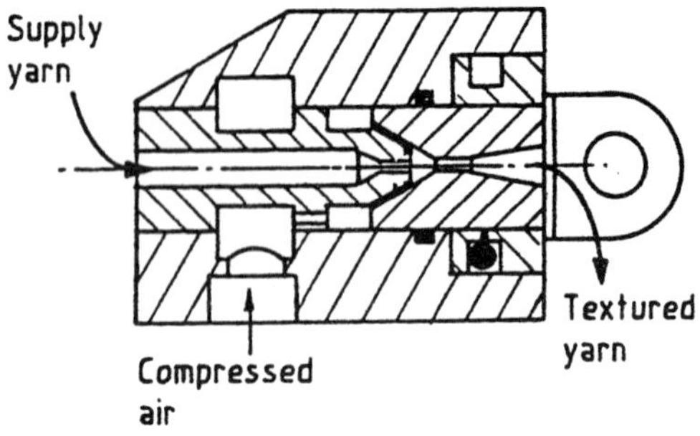

Figure 4.9 Taslan Type XX

manufacturers to take an interest, particularly after the expiration of the main Du Pont patents, which formed the basis of the Taslan process. They are too numerous to mention them all, but two, namely Enterprise Machine and Development Co. (EMAD) of the United States and Heberlein Maschinenfabrik A.G. of Switzerland stand out. The latter has recently made even more contributions to the air-jet texturing process than Du Pont. Although the nozzles[33] developed by EMAD (Figure 4.10) have also followed the similar configuration to the Taslan nozzles, having converging-diverging geometry, the Heberlein nozzles are of completely different geometry.

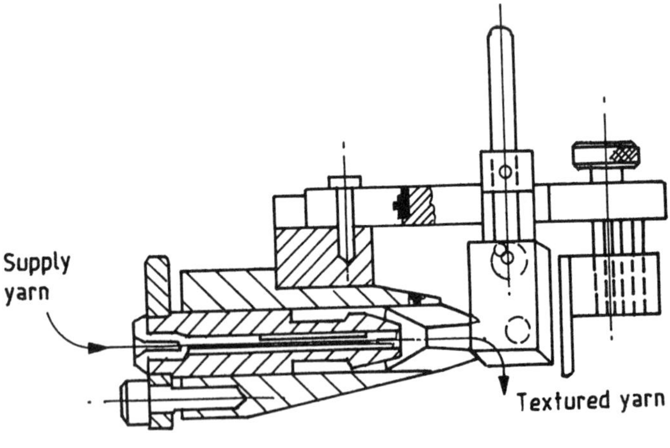

Figure 4.10 EMAD jet

The most significant developments in the nozzle design have, therefore, taken place in the last decade. Heberlein took over the air-jet technology developed by Berliner Maschinenbau A.G. of Switzerland in 1977 and introduced its first HemaJet nozzle called the Standard-Core HemaJet shown in Figure 4.11 in 1978.[46] This nozzle is reported to be similar to an earlier design, (i.e., the Mirlan nozzle from Czechoslovakia)[37] (Figure 4.12). In this design, the air is fed into the main duct of the nozzle through three small inlet holes, where it impinges on the overfed supply yarn, which follows a straight axial path in the nozzle. Because the Standard-Core HemaJet had no adjustable parts, the need for a skilled operator for manning the Taslan nozzles has been eliminated. Better nozzle-to-nozzle consistency is obtained by keeping the machining tolerances tight. The design has a self-cleaning facility, as spin-finish constituents are washed away from the yarn surface because of the straight geometry of the main duct which is constantly in contact with the moving yarn.[7] Heberlein later introduced other nozzles designated as T100, T311, T341, T110, T140, and T350 (Figure 4.13) with the first digit indicating the provision of either one or three inlet holes. Some of these nozzles for the apparel range have reduced the air consumption drastically.

It appears that most of the nozzles to date have been developed by the industry through practical day-to-day experience. Acar,[47] and Demir,[48] using the results of their theoretical and experimental investigation with the HemaJet nozzles designed novel nozzles based on these more fundamental studies. The prototypes of these nozzles were privately manufactured by Heberlein for the research purposes and were reported to yield satisfactory texturing and provided further substantial compressed air savings.

4.2.6 Air Flow in the Nozzles

Because the actual texturing process (i.e.,loop and entanglement of the overfed supply yarn), is carried out by the air flow in the nozzle and the air jet created by this flow, a detailed study of the aerodynamics of the nozzles is of paramount significance for the improvement of the nozzle performance. Therefore, in parallel to the nozzle development the flow properties have been studied. While Wray and Entwistle[49] studied the air flow in the Taslan IX nozzle, and Þen[50] extended his research to investigate the air-jet created by that nozzle. Sivakumar[51] later applied the theory for converging-diverging nozzles to both the Taslan IX and X nozzles. However, the more recent Taslan nozzles (Taslan XIV and XV) have been examined by other researchers.[52-55] Acar et al[56] have, conversely, studied the air flow through the Standard-Core HemaJet and also built a mathematical model for the flow.[57] Finally, Demir[58] and Versteeg et al[59] carried out an extensive study on cylindrical texturing nozzles mostly by experimental means.

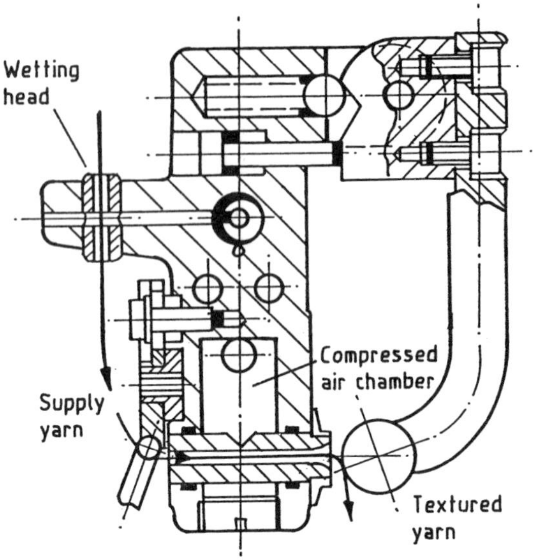

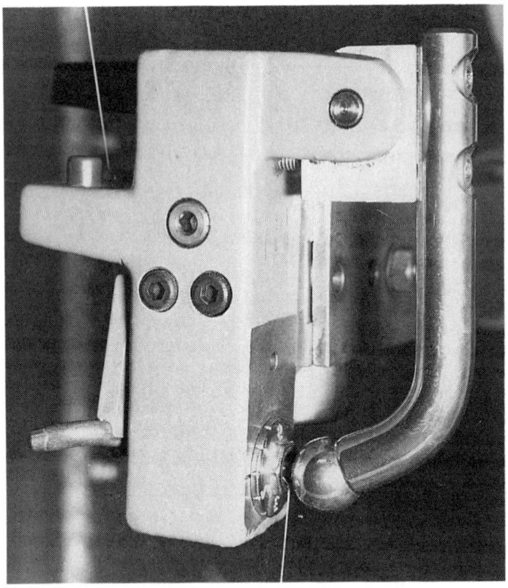

Figure 4.11 Standard-core HemaJet

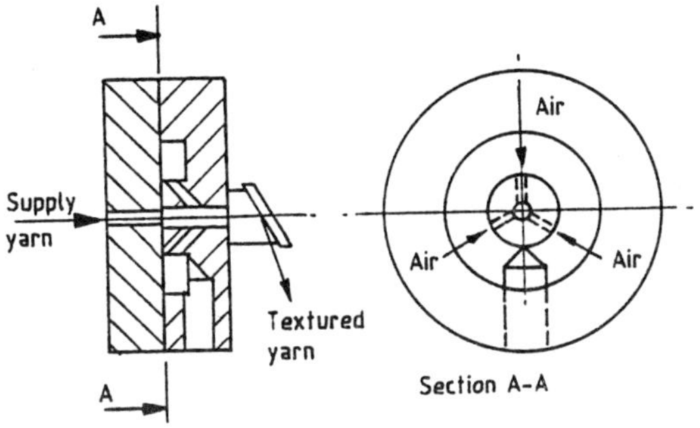

Figure 4.12 Mirlan jet

In the 1960s, Wray,[12] and Wray and Entwistle,[49] concentrated their flow studies on the wake flow created by the tilted needle of the Taslan IX. By extending the turbulent flow theory of flow passing over a cylinder, they envisaged the vortices created in the wake of the needle. They also concluded that the overfed and pretwisted yarn is untwisted by the effect of such rotational flow, resulting in a loop and entanglement formation. Such an explanation of the flow led Wray and Entwistle[40] to modify the Taslan IX nozzle and achieve up to 65% compressed air savings. Þen[50] also proved this theory to be incorrect. He used scaled-up transparent versions of the Taslan IX nozzle to verify that the actual texturing occurs outside the nozzle and that the Reynolds number, based on the nozzle throat diameter, is well aboýve the range at which such vortices at the wake of a cylinder could exist. Þen, furthermore, visualized the air jet by means of the Achlieren technique and showed the formation of oblique shock waves. Þen's flow velocity and turbulence measurements by means of Hot Wire Anemometry also showed that the air jet created by the Taslan IX nozzle was supersonic and highly turbulent. The asymmetry in the velocity profile was attributed to the presence and setting of the feed needle.

Sivakumar[51], conversely, studied the air flow through the Taslan IX nozzle without the needle by theoretical means and thereby predicted the formation of shock waves at the throat. By making a nozzle similar to the Taslan X patents, with his theoretically predicted dimensions, Sivakumar carried out some experimental research and concluded that the loop formation and the formation of the shock waves at the exit of the nozzle are closely related. He also surmized that pretwisted overfed yarn collapses when it reaches the shock wave (i.e., a pressure barrier) and that this creates the entanglement.

	Standard Jet Cores				Special Jet Cores		
	● T100	● T311	T321	● T341	T110	T140	● T351
Pressure range (bar)	4-14	4-14	4-14	4-14	4-14	4-14	4-14
Production speed, (m/min)	50-650	50-650	50-650	50-500	50-750	30-500	30-500
Total feeder range * (dtex)	50-700	30-300	150-800	250-2,000	150-1,000	500-5,000	500-5,000
Single-filament count (dtex)	1.5-2.5	1.0-2.5	1.5-4.0	2.5-6.0	1.5-4.0	up to 22	up to 22
Overfeed + before jet (%)	30	60	60	80	50	35	80
Air consumption (m3/hour at 9 bar)	6.1	5.4	8.1	14.6	8.1	20.0	20.5

* Guide values: dependent on the different feeder yarn properties and the requirements for the air-jet textured yarn

+ Dependent on feeder yarn, machine setting, kind of operation, yarn path, and frictional conditions

Figure 4.13 Technical data of HemaJets

Bock and Lünenschloss[52,53] reconstructed the flow path in the Taslan XIV nozzle by observations of the surface structure of the deposits within the nozzle venturi resulting from the spin finish washed away from the yarn inside the turbulence chamber. Using this technique, they studied the effects of supply pressure and nozzle geometry, and concluded that the air flow is affected by both.

When the feed needle channel is blocked, Taslan nozzles behave closely similar to the classic converging-diverging nozzle for which the flow of compressible fluids, such as air and steam, is well established by experimental and theoretical means. Therefore, Bock and Lünenschloss made use of such established theory for their study and extended their work into the air jet created by the Taslan XIV nozzle; using the Schlieren technique, like Þen earlier, they visualized the jet flow and measured the total pressure distribution in the jet. Together with the theoretical study, they constructed a picture of the jet consisting of high and low pressure zones.

A later work of Bock and Lünenschloss[54,55] examined a flat, transparent version of the Taslan XIV nozzle. The Schlieren technique enabled them to observe the turbulence chamber and the formation of oblique shock waves in and out of this nozzle. The turbulent nature of the air-jet was also verified by both the instantaneous flow visualisation and pressure measurements with a high-frequency (100 kHz) transducer.

A scaled-up model of the Standard-core HemaJet was used by Acar et al.[56] Through velocity measurements at the exit plane of the nozzle, it was concluded that the Standard-Core HemaJet nozzle also creates supersonic, turbulent, slightly asymmetric jets with a nonuniform velocity profile. The result is apparently in agreement with previous researchers conclusions. The air jet was also visualized by means of interferometric and shadow graphical techniques. Acar et al, measuring the static pressures in the main channel of the scaled-up nozzle through wall static pressure tappings, reached the conclusion that the flow in the nozzle near the air-inlet bores is complex and contributes to the asymmetry and nonuniformity of the jet.

To predict the flow properties created by a converging-diverging type nozzle, the steady one-dimensional isentropic flow equations as described in standard texts can be utilized. Nevertheless, the cylindrical geometry of all HemaJets precludes the use of such equations. Therefore, Acar et al[57] attempted to develop a mathematical model for such nozzles and claimed that the developed theoretical model provided reasonably close predictions of the flow properties at the exit plane of the Standard-Core HemaJet.

4.2.6.1 Cylindrical texturing nozzles

The commercial texturing nozzles in this category consist of a cylindrical main duct into which one inlet hole or three inlet holes of smaller diameter are obliquely opened. The compressed air is admitted into the main duct through this inlet hole(s), while the yarn follows a straight path in the nozzle as shown in Figure 4.14. The yarn inlet section of the cylindrical duct is called the secondary flow section and the other end of the duct is the primary flow section usually with a trumpet-shaped divergent exit. In the early types of the HemaJets (e.g., the obsolete "Standard-Core"), this divergent exit had a smaller radius (small trumpet). However, all the recent HemaJets have trumpet shaped exits with larger radii. A comparison of exit shapes of early and recent HemaJets is also seen in Figure 4.14.

The major difference between various different HemaJets is the number of inlet holes and, with the exception of the Standard-Core, these are designated according to the number of inlet holes they possess. The three inlet holes of the Standard-Core HemaJet are not only radially equispaced but are also longitudinally slightly staggered. This configuration is said to impart a swirl to the air flow[60]. However, the recent HemaJets with three inlet holes (e.g., T341) appear to have inlet holes on the same circumferential line (i.e., the staggered configuration has been eliminated).

The two most common features of all cylindrical nozzles are that the ratio of the sum of inlet hole areas to the main duct area is always less than unity and the inclination of the inlet hole(s) to the main duct is at the same angle. These two features of HemaJet nozzles, together with other geometrical parameters, are broadly studied by Demir.[58] Through this investigation their optimum values (from the air flow standpoint) are sought.

4.2.6.1.1 Air flow through cylindrical nozzles

Figure 4.15 shows a HemaJet nozzle mounted in a nozzle housing. It can be said that the compressed air that flows through the inlet holes is divided into two streams in the main duct. The *primary* flow, which has the largest momentum resulting from the inclination of the inlet hole; flows vigorously; and creates a supersonic, asymmetrical, nonuniform, and turbulent air jet, as claimed by Acar et al.[56] The *secondary* flow having smaller momentum, flows in the opposite direction to the primary flow and is always subsonic and uniform. This description of the flow is insufficient, however correct, for the objectives of this study. Therefore, the flow field will be divided into subsections and investigated in detail.

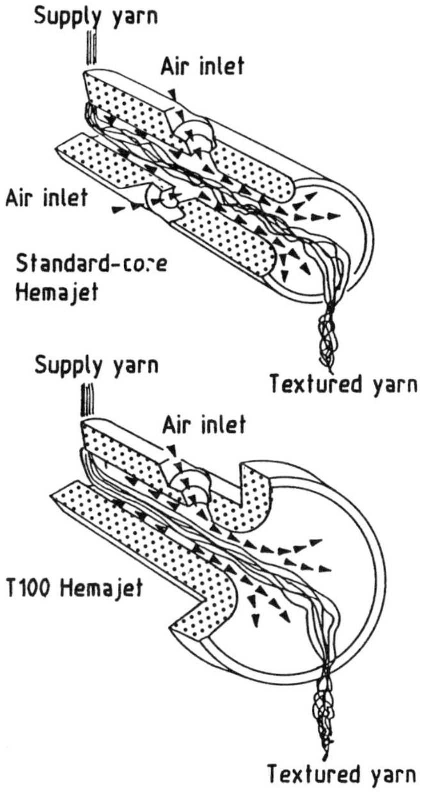

Figure 4.14 Air flow and yarns path in cylindrical nozzles

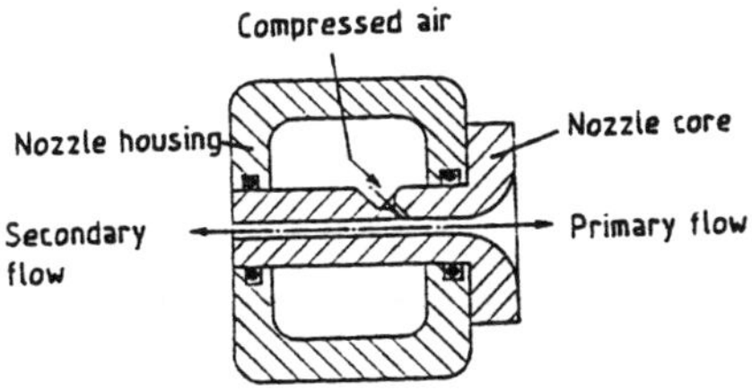

Figure 4.15 Cylindrical type (HemaJet) nozzle mounted in a nozzle housing

This subdivision, as dictated by the geometry of the nozzle, can be defined as follows:

1. Inlet hole
   a. Convergent section,
   b. Sonic section,
   c. Abrupt enlargement and backward deflection
2. Main duct
   a. Oblique impingement and division of the flow
   b. Primary flow
   c. Secondary flow
3. Jets
   a. Primary jet (supersonic jet)
   b. Secondary jet (subsonic jet)

Inlet hole. The experimental investigation of the air flow through the inlet hole has been done on a simplified scaled-up model as shown in Figure 4.16. The first simplification is that the jet created by this model is assumed to expand into an infinitely large area at atmospheric conditions, although the cross-sectional area of the main duct of the texturing nozzles is finite and pressure builds up because of the static pressure recovery in the main duct. As the second simplification, the originally circular main duct is assumed to be rectangular. This facilitates visualisation of the flow at the very intersection of the inlet hole with the main duct. Furthermore, to facilitate the explanations that follows, the inlet hole is divided into three subsections as seen in Figure 4.16.

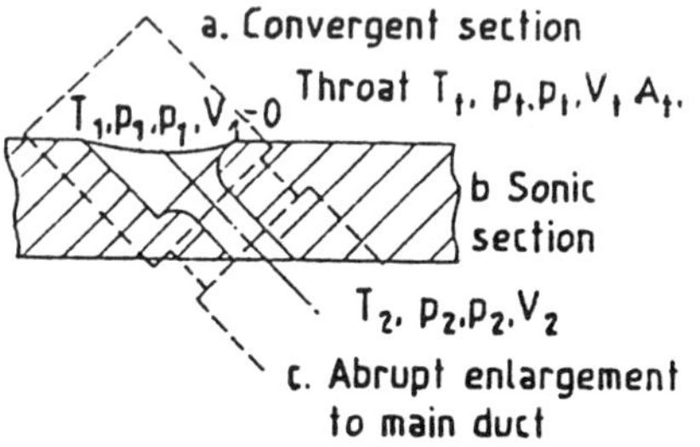

Figure 4.16 Three subsections of an air inlet

Convergent sectio. The theory for compressed air flow in converging channels can be applied to the converging section of the inlet hole. The theory shows that if the reservoir pressure is above the critical value, the flow chokes at the throat of the converging section and delivers only the critical mass flow rate, the flow velocity being sonic. Because the working pressures of the texturing process are well above the critical value, texturing nozzles operate under choking conditions. Hence, at a given pressure, a nozzle delivers only the critical mass flow rate.

Sonic section. The sonic section (Figure 4.16) is a cylindrical channel that continues from the throat of the convergent section. Because the length of this channel is short and no area change is observed, the flow conditions are assumed to remain constant throughout this section.

Abrupt enlargement and backward flow deflection. The total area of the inlet holes is always less than the cross-sectional area of the main duct. An abrupt enlargement of the sonic flow is, therefore, inevitable. Pressure losses occur there because of the abrupt enlargement of the flow; therefore, pressure energy is not fully converted into kinetic energy. Flow velocity reaches supersonic levels, but some energy is wasted because of the pressure loss.

When a sonic flow passes over a sharp convex corner, the flow accelerates and turns through large angles without separating; such deflection of a flow is termed "Prandtl-Meyer deflection" and is described in most textbooks.[61-63] A Prandtl-Meyer deflection occurs at the intersection of the inlet hole and the main duct of the nozzle.

Shadow graphs of the jets created by the simplified and scaled-up model of the inlet hole with the oblique opening have revealed that the total "backward" deflection of the jet is a function of the driving pressure. This backward deflection increases the angle of impingement of the incoming jet with the opposite wall. The secondary flow velocities, therefore, increase. Furthermore, these higher angles of impingement cause more flow losses and, in turn, create primary flow velocities.

Main duct. Being a straight cylindrical channel, the main duct is of no aerodynamic significance. However, the impingement and division of the incoming jets occurs in this channel; moreover, the primary jet, which carries out the actual texturing, is formed at the end of the primary flow section of the channel. Although the secondary flow is always seen as a source of energy loss, this flow does contribute to the texturing by blowing off the applied water and the spin-finish constituents from the surfaces of the filaments. Therefore, the main duct will be investigated by dividing it into three subsections.

Oblique impingement and division of the flow. In the case of cylindrical type jets with one air inlet hole, the incoming jet carries the whole mass flow of the air into the main duct. The impingement of this jet on the opposite wall of the main duct, and its division, have been experimentally investigated by Demir.[48] This investigation illustrated that the incoming jet creates an oblique shock wave detached from the opposite wall. The mass of the incoming flow was diverted to the primary flow section of the nozzle after this shock wave. It was also noted that the expansion waves crossed the whole secondary flow section, whereas in the primary flow section they only crossed less than one-half of the channel. In both directions, after the expansion waves, the pressure became less than the upstream pressure. In the secondary flow section, this corresponds to a pressure equal to or less than the atmospheric pressures.

The entire flow divides into the primary and secondary flows in such a way that the secondary flow momentum flux (depending on the working pressure), only varies between 15% and 30% of the primary flow momentum flux.

Primary flow. **Because of** the inclination of the inlet holes, a greater portion of the incoming flow, despite the backward deflection of the incoming jets, goes to the primary flow section of the nozzle. The speed of the flow is supersonic, provided that the reservoir pressure is well above the critical value.

An acceleration or deceleration of a supersonic flow requires an area change. Being a cylindrical channel, the main duct of the nozzle fails to fulfil this requirement. Consequently, no velocity change is observed by the flow, until the divergent section of the duct is reached. As soon as the flow area increases, the speed of the flow increases at the expense of pressure, in other words the flow expands and the flow pressure decreases. The rate of the divergence is so large that the flow cannot follow the full contour of the surface and separates as is seen in Figure 4.17 which shows the shadow graph of the primary flow from the transparent walled square nozzle.

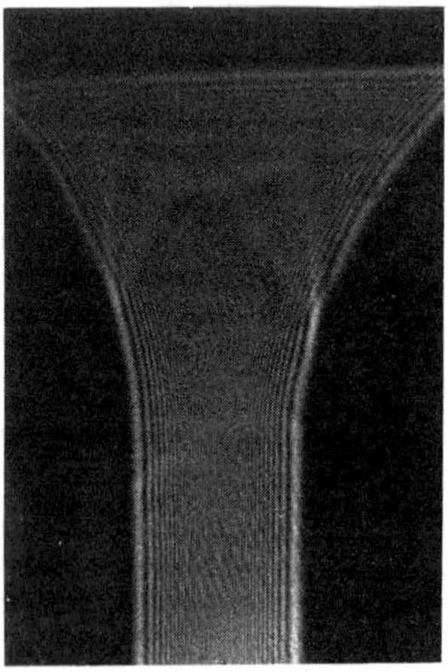

Figure 4.17 Shadowgraph of the primary flow from the transparent walled square nozzle

Secondary flow. The static pressure measurements obtained by Acar et al,[56] using wall tappings, and the static pressure measurements which have been carried out by Demir,[58] using a centre line static pressure tube, both indicate that the secondary flow pressure is atmospheric. Therefore, the secondary flow is almost exclusively subsonic for all the working pressures and appears to have no significance.

Jets. Both the primary and the secondary flows create free jets when they leave the nozzle walls. Being supersonic, the primary jet possesses many characteristics, whereas the secondary jet is fully subsonic and uniform.

Primary jet (supersonic jet). Essential characteristics of the supersonic jet created by the air-jet texturing nozzles may be outlined as follows.

Center-line velocity fluctuations. To elaborate on the behavior of the supersonic jet emerging from an underexpanded nozzle, the center-line total pressure distribution is superimposed on Figure 4.18a and b. It is seen from this presentation that the flow decelerates up to the point where the intersection of the expansion waves occurs. Having passed this intersection point, the flow recovers rapidly, then slows down up to the point where the shock waves have intersected. Further downstream from the jet, these rises and falls of the center line total pressure are repeated and they finally fade out because of the viscous and turbulent mixing of the jet into the surrounding atmosphere.

Nonuniform axial velocity distribution. The axial total pressure distribution illustrated in Figure 4.19 is not of the same pattern at various distances from the exit plane. It is fairly uniform just at the exit plane, while a central deficit is observed in the downstream section of the jet. However, further downstream where the viscous mixing through a thick boundary layer dissipates the flow, the traverse velocity distribution across the jet becomes bell shaped.

Axial velocity distributions are observed at different distances from the exit plane. The level of the central deficit varies from one point to another because of the formation of expansion or compression waves that cause acceleration or deceleration of the flow as seen in Figure 4.18c. Further downstream, the central deficit dissappears because of the dissipation of the density discontinuities (i.e., shock or expansion waves). Thus the velocity profile becomes bell shaped. For this reason, the axial velocities measured at the exit plane of the T100 nozzle (which is located at 4.5D, with D being the nozzle diameter further downstream from the beginning of the diverging section where the flow separates) do not show any central deficit. Neither do the center-line velocities fluctuate as is seen in Figure 4.20a and b.

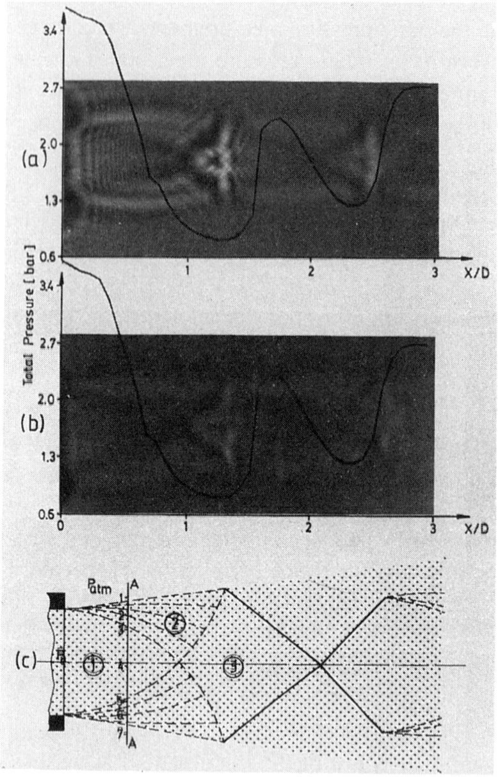

Figure 4.18 Shadow graph of a supersonic jet with graphically superimposed center-line total pressure distribution

Secondary jet (subsonic jet). When the secondary flow leaves the nozzle, it creates a secondary jet that is subsonic. The pressure in the jet is atmospheric and no density discontinuities are present. Figure 4.21a shows the secondary jet axial velocities at various working pressures for the T100 HemaJet nozzle. These are all subsonic and of uniform distribution. In addition to this, the center-line velocities for this nozzle at various pressures are shown in Figure 4.21b. The figure shows that the flow velocity remains constant indicating a potential core.

4.2.6.1.2 Description of air flow through cylindrical nozzles

The air flow through HemaJets, the cylindrical type nozzles, may be described by the following points:

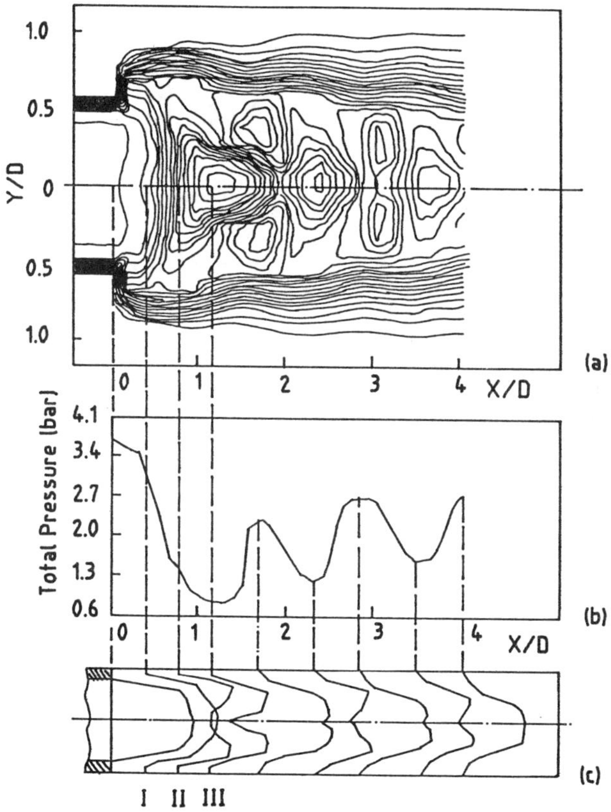

Figure 4.19 Topographical distribution and center-line fluctuations of static pressure of free jet

1. The flow is choked at the throat of the inlet hole and delivers only the critical mass flow determined by the reservoir pressure.
2. Flow in the inlet hole is only sonic and possesses excess pressure again determined by the resevoir pressure.
3. Because of the sudden opening of the inlet hole into the main duct, abrupt enlargement losses that create energy loss are inevitable.
4. The oblique opening of the inlet hole deflects the incoming jet backward and consequently increases the angle of collision to the opposite wall.

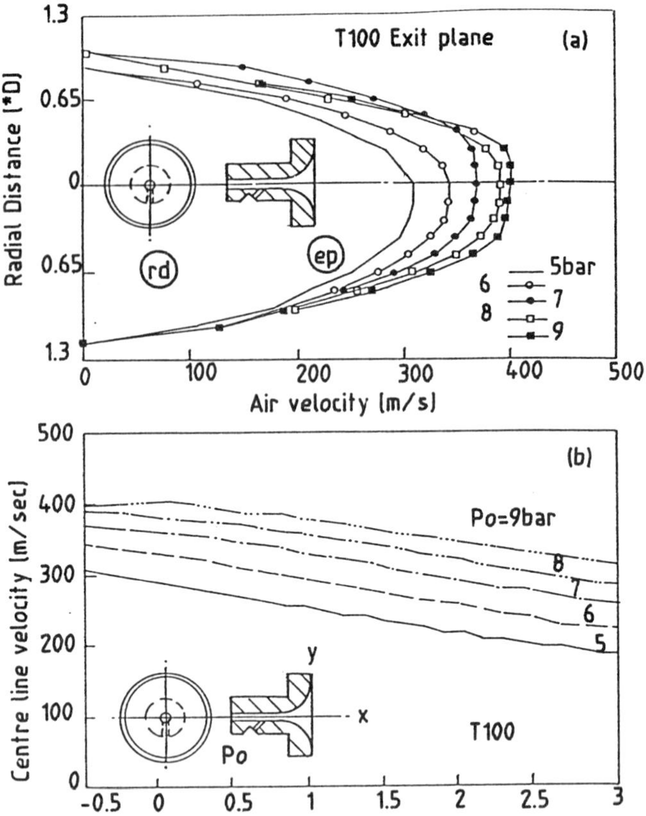

Figure 4.20 Axial velocities distribution and the centdine velocities measured at exit plane of the T100 nozzle

5. The oblique collision of the incoming jet is a paramount source of energy loss.
6. The primary flow is supersonic in the main duct, whereas the secondary flow remains subsonic both in the main duct and in the jet form.
7. Further acceleration of the primary flow occurs at a place where the divergence of the main duct starts. However, the flow soon separates from the wall of the nozzle because of the large divergence of the trumpet-shaped exit.
8. Because the supersonic jet has pressures above atmospheric, it creates an underexpanded supersonic jet. The properties of this jet has been investigated by means of another underexpanded jet created by a straight pipe. The results are applicable to the jets created by all the texturing nozzles under consideration.

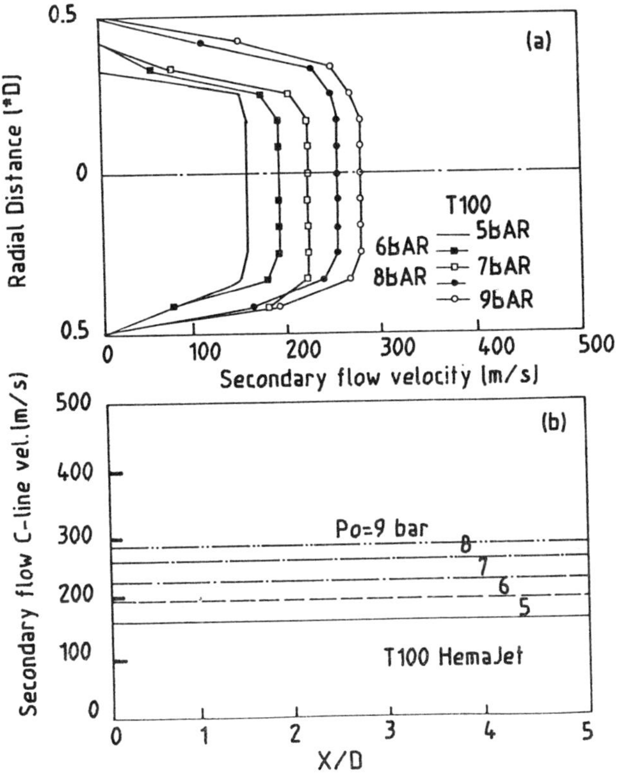

Figure 4.21 Secondary jet axial and centerline velocities at various working pressures

9. The axial velocities from a supersonic jet fluctuate on the center-line because of the semiperiodic expansion and compression of the flow.
10. The nonuniform distribution of the flow velocities is due to the formation of pressure waves, and the viscous and turbulent mixing of the flow on the boundary layer.
11. Being a subsonic jet, the secondary jet has the characteristics of all well-defined subsonic jets.

4.2.6.2 Mathematical models for texturing nozzles

Designs of the texturing nozzles vary in detail but remain similar in underlying principles because of the common need to create a supersonic, turbulent, asymmetrical, and nonuniform air flow. The known texturing nozzles, according to their construction, can be categorized in two main groups as follows:

1. Converging-diverging (de Laval) type nozzles (i.e., converging-diverging nozzles) are attached to the yarn exit end of the nozzle assembly (e.g., Du Pont's Taslan Type 14 nozzle)
2. Cylindrical nozzles (i.e., one or more air inlets opening with an angle to a cylindrical straight uniform main flow duct of the nozzle, e.g., Heberlein's HemaJet).

4.2.6.2.1 Model for converging-diverging nozzles

Properties of the flow in this group of nozzles are determined by the converging-diverging shaped nozzle attachment at the yarn exit end, providing that the air passages and gaps are larger than the throat area. At typical texturing pressures, the air flow is "choked" at the throat of the nozzle, because the critical condition of unity Mach number is reached. It then continues to accelerate in the diverging part making the flow supersonic. Properties of the flow are determined by the well-established theory of compressible flow in a converging-diverging nozzle, and these can be calculated for a frictionless nozzle by applying the principal equations for the steady one-dimensional isentropic flow.

4.2.6.2.2 Model for cylindrical nozzles

The flow in this group of nozzles is more complex. The main flow duct is a straight, uniform, and cylindrical where several, usually convergent air inlet bores enter radially at an angle. Flows from these inlet bores are intermixed at their meeting point in the main flow duct and are divided into the following two axially opposing flows:

1. Primary flow; in the direction of the yarn exit
2. Secondary flow; in the direction of the yarn inlet

This division into primary and secondary flows is determined by the geometry of the nozzle. The total mass flow rate through the cylindrical-texturing nozzles is usually determined by the throat areas of the air inlet bores, because the flow is choked when conventional texturing pressures are used.

No readily available mathematical model can be adopted to these types of nozzles, and, therefore, Acar[47] developed a new model. Acar made the following simplifying assumptions:

1. Flow is steady, one-dimensional, and isentropic throughout (as is assumed in the case of the converging-diverging nozzle).

2. There are no frictional losses.
3. Summation of the air inlet bore throat areas is smaller than the cross-sectional area of the main duct.
4. Flow from air inlet bores meet at one single point on the center line of the main duct.
5. Static pressures at the exits of both the primary and secondary flows are atmospheric.
6. Velocity profiles of the primary and secondary flows are uniform.
7. Air is assumed to be free of any water or other particles.

Because the structure of the cylindrical nozzles and, hence, the flow in these nozzles is highly complicated, the laws of fluid dynamics governing compressible flows cannot be directly applied to them. Therefore, for mathematical treatment, Acar[47] conveniently divided the nozzle into three smaller sections or control volumes as shown in Figure 4.22, and accordingly apply the principles of gas dynamics and thermodynamics to each of these. The properties of the outgoing flow from one control volume are fed into the succeeding control volume as the properties of incoming flows; hence, the final properties of the primary and secondary flows at the exit planes are calculated.

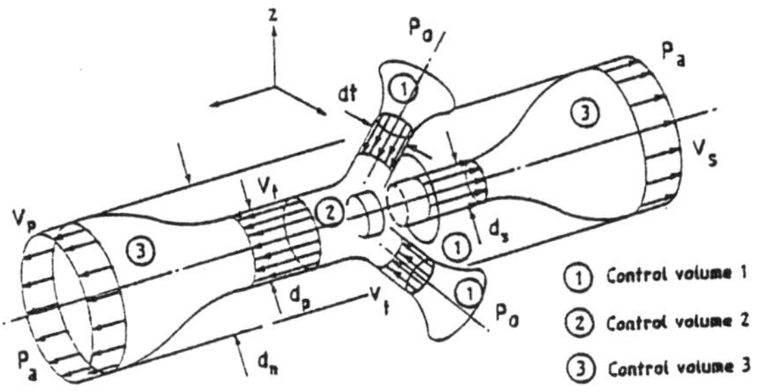

Figure 4.22 Three smaller subsections or control volumes of air flow[47]

4.2.6.3 Features of most recent texturing nozzles

4.2.6.3.1 Nozzles

Four industrial designs, namely, the Standard-Core, T100 and T341 HemaJets, and the Taslan XIV nozzle are investigated by Demir.[48] Figure 4.23 illustrate the general features of the three HemaJets and the Taslan XIV nozzles, respectively. The figures also show the technical drawings of these nozzles.

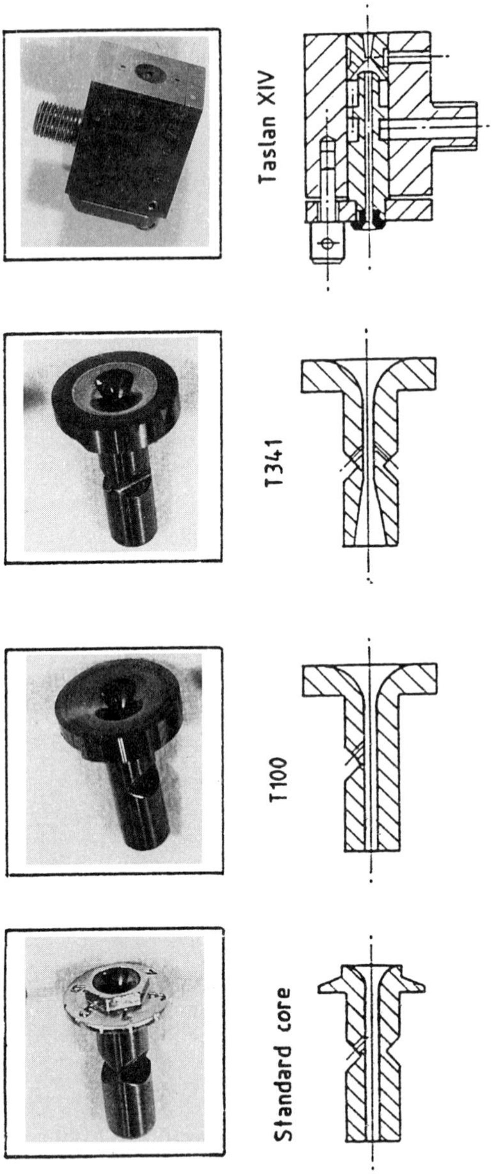

Figure 4.23 General features of three HemaJets and Taslan XIV nozzles

In a Taslan XIV nozzle, air is passed through a one-sided hole by which an asymmetry as well as a swirl is said to be imparted to the flow.[17] The flow then is admitted through an annular gap between the needle and the so-called venturi to the converging section of the nozzle. The width of this gap is adjusted by moving the converging-diverging nozzle part backward and forward via a thumb screw. The annular gap and the throat area are determining factors for the air consumption of such Taslan nozzles. The optimum working condition is, therefore, adjusted by this gap during the texturing trials. Under the working conditions, the air flow is choked at the throat of the venturi (reaching the sonic velocity) and is then accelerated in the diverging section which extends up to the exit plane of the nozzle without an exit shape.

The amount of secondary flow is determined by the air pressure, the size of the needle hole, and the annular gap. The needle hole of the Taslan XIV nozzle may be restricted by a sapphire insert. The size of this restriction varies according to the supply yarn linear density; therefore, several interchangeable sizes of needles can be used with a single nozzle body[44] for use with different yarns.

The air consumption levels of these nozzles at 7 bar are given in Figure 4.24. The figure shows that the T341 HemaJet consumes the highest amount of air compared with the other nozzles, and the T100 HemaJet, with only one inlet hole, appears to be the most economical nozzle.

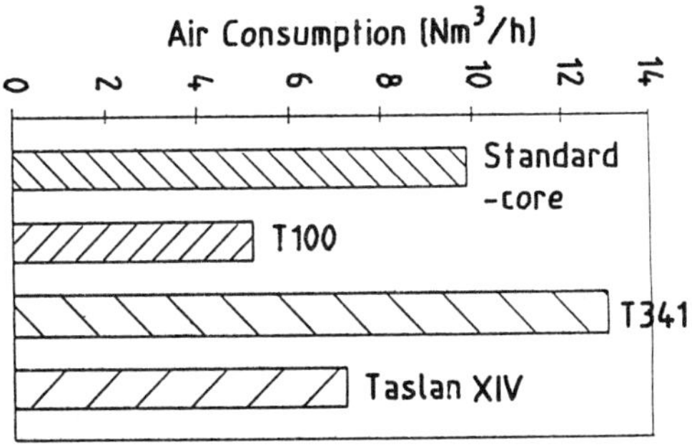

Figure 4.24 Air consumption levels of texturing nozzles at 7 bar

4.2.6.3.2 Air flow properties

Axial velocity profiles: At varying air pressures, the axial velocity distributions at the exit plane of the Standard-Core, T100, T341, and Taslan XIV nozzles are given in Figures 4.25a to d respectively. These velocity distributions are also presented in three-dimensional and topographical form in Figures 4.26a to d for a typical 7 bar supply pressure.

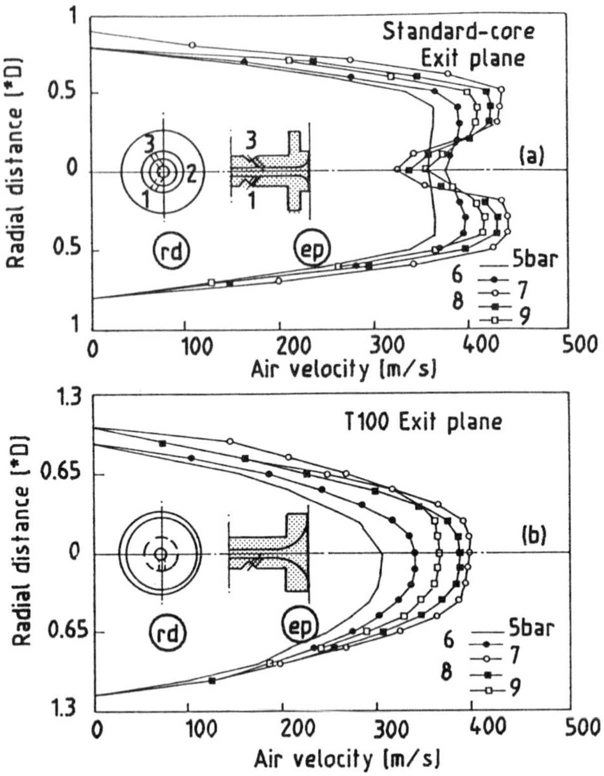

Figure 4.25a, b Axial velocity distributions at the exit plane of Standard Core and T100 nozzles at varying air pressures

From these figures, the supersonic and nonuniform nature of the flow can be seen. The asymmetrical nature of the flow is obvious in the case of the T100 jet (Figure 4.25b), whereas the other nozzles more approximate to symmetrical velocity distributions. Some dissimilarities in the axial velocity profile from these three nozzles were observed.

Shock waves and center-line velocities. Flows from all these nozzles at different pressures were also studied by using a shadowgraph system to obtain a better understanding of them. From the shadow graphs and the center-line velocity fluctuations (which are also indicators of the occurrence of shock waves) as shown in Figures 4.27a to d, it can be concluded that the strengths of the shock waves produced by different nozzles are varying. Because the shock waves are known to affect the velocity distribution, it is not surprising to see different velocity profiles from different nozzles with varying degrees of shock strengths.

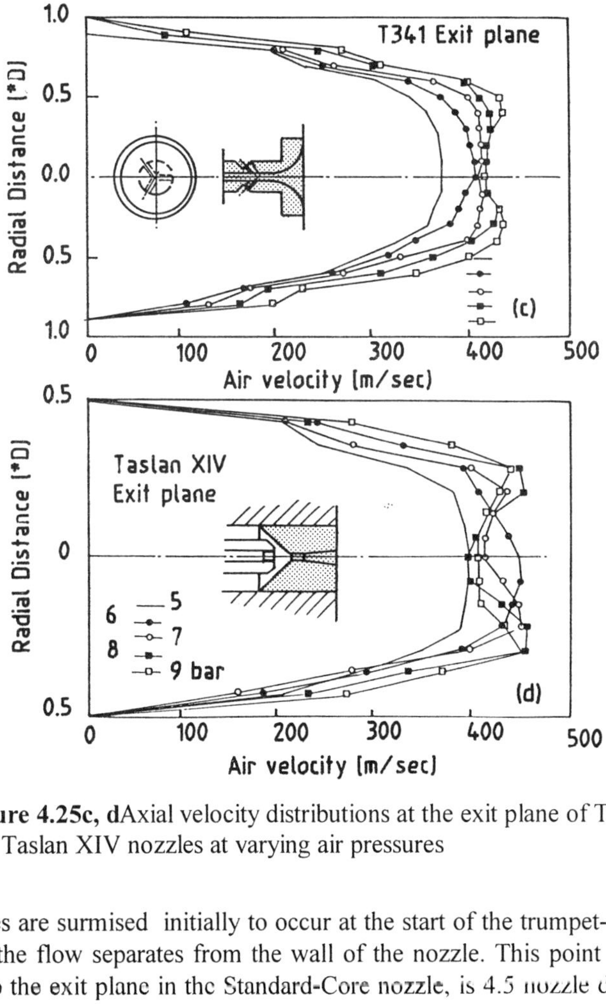

Figure 4.25c, d Axial velocity distributions at the exit plane of T341 and Taslan XIV nozzles at varying air pressures

Shock waves are surmised initially to occur at the start of the trumpet-shaped exit part where the flow separates from the wall of the nozzle. This point is 2 nozzle diameters to the exit plane in the Standard-Core nozzle, is 4.5 nozzle diameters in the T100 nozzle, and is 3 nozzle diameters in the T341 nozzle. However, it corresponds to the exit plane in the case of the Taslan XIV nozzle. Because the repetitive shock waves will gradually diminish and expand to the ambient atmospheric conditions, waves of relatively lesser strengths are highly unlikely to be observed in the emerging flows from different nozzles with various diverging shape and sizes. For instance, the T341 nozzle, even at high pressures, produces no clearly visible shock waves outside

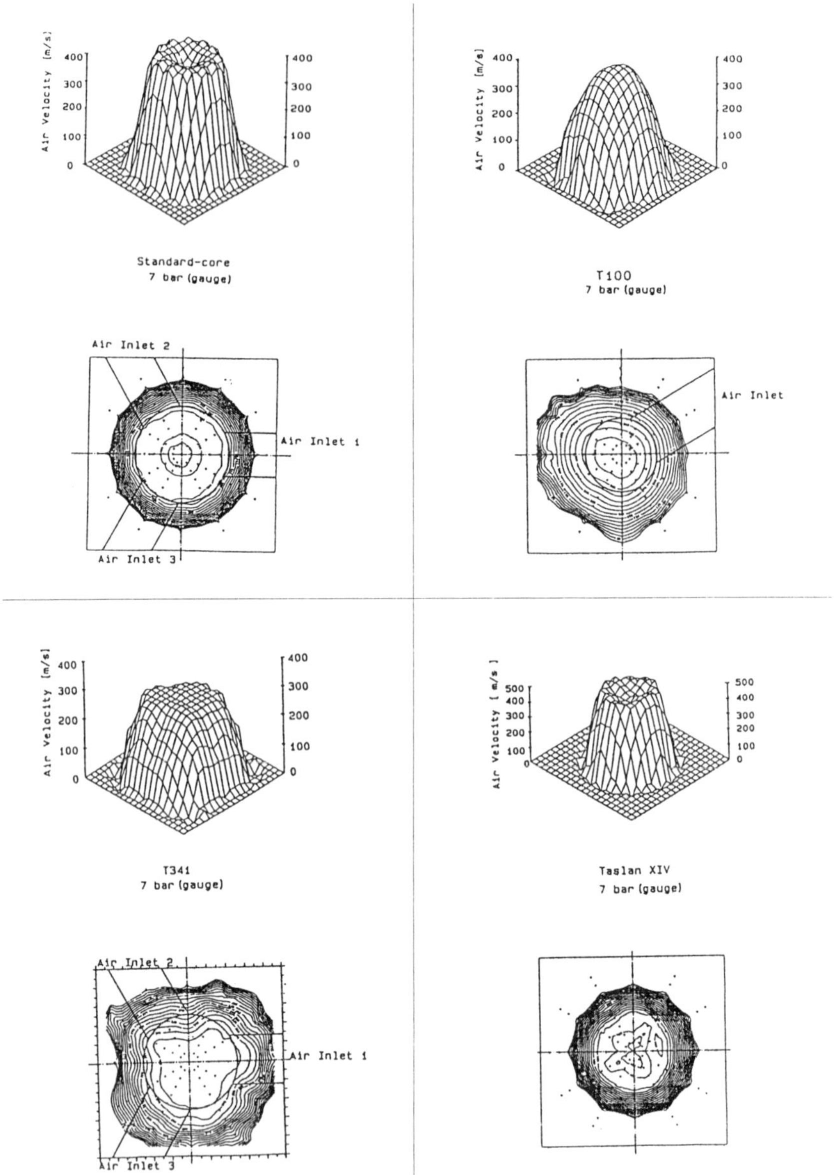

Figure 4.26 Three-dimensional and topographical form of air -low velocities at a typical 7 bar supply pressure

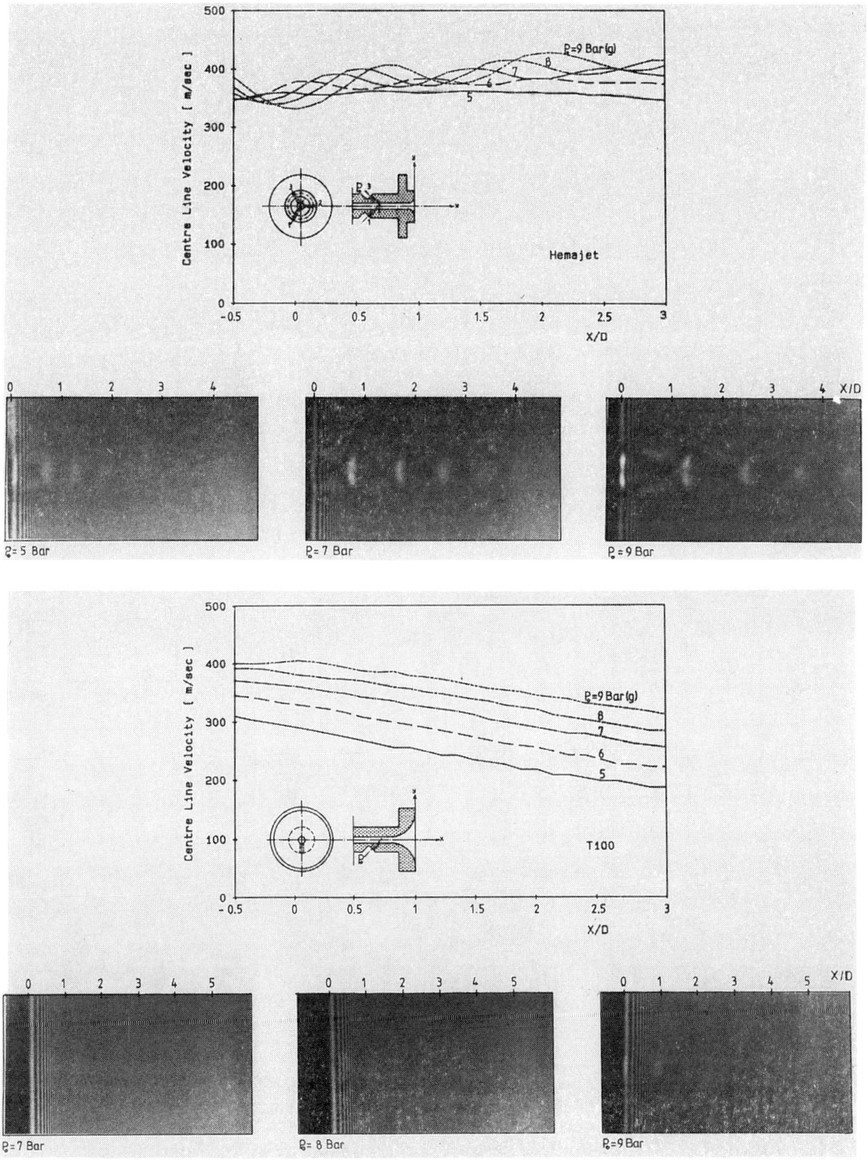

Figure 4.27a, b Shadow graphs and centerline velocity fluctuations of Standard-Core and T100

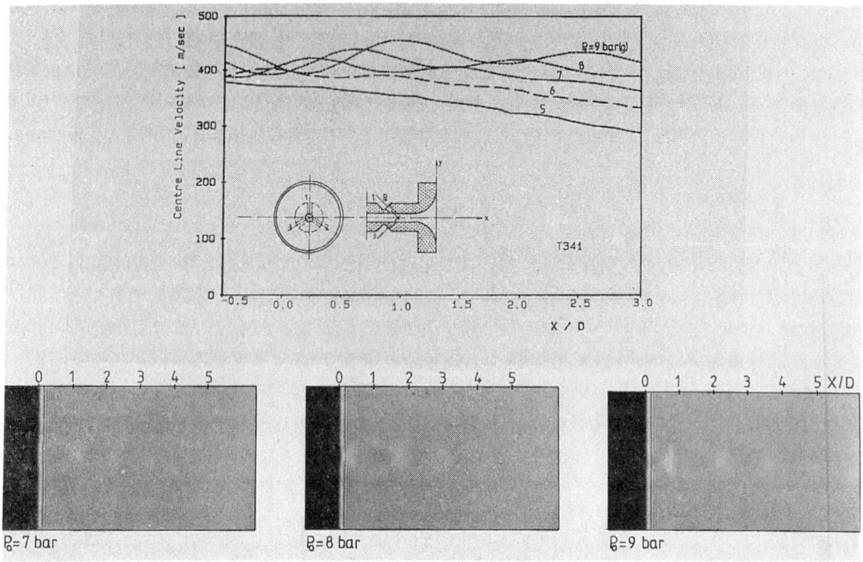

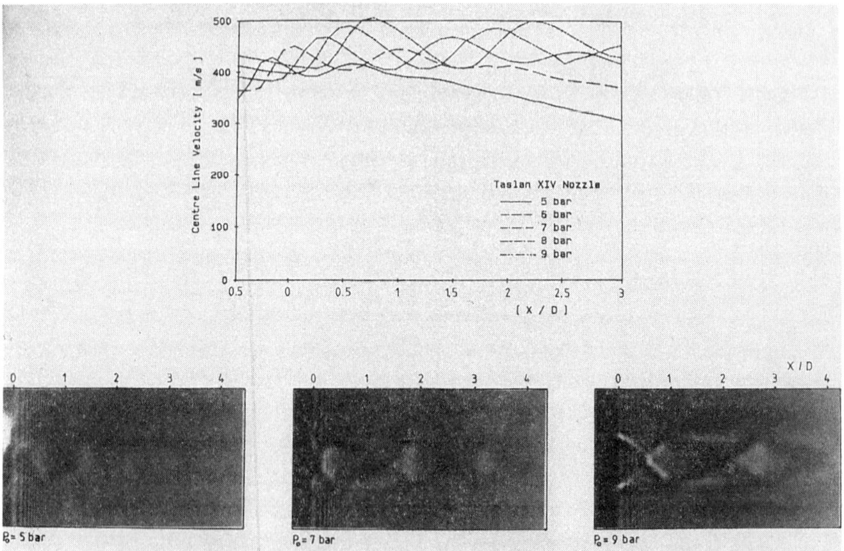

Figure 4.27c, b Shadow graphs and centerline velocity fluctuations of T341 and Taslan XIV

the nozzle, whereas the Taslan XIV and Standard-Core nozzles produce relatively strong shock waves (Figure 4.27). Because the strong shock waves do not dissipate in short distances and repeat themselves periodically in the further downstream of the flow as seen in Figure 4.27d, they are expected to be visible (by a flow visualization technique) outside the nozzle. Therefore, it can be argued that different nozzles produce shock waves of different strengths.

Turbulence. Wray and Entwistle,[49] in their explanation of the mechanism of texturing, refered to turbulence in the flow in Taslan Type IX nozzle, which was later experimentally verified by Þen.[44] Bock and Lünenschloss,[46-55] although they did not give experimental evidence of the turbulence, referred to its role in the texturing process with a Taslan Type XIV nozzle.

In such high velocities provided by cylindrical nozzles, intense turbulence is expected. An attempt was made to measure it, but because of the difficulties encountered in employing hot-wire anemometry, no satisfactory results were obtained. However, these preliminary trials indicated a typical turbulence level in the region of 25% to 30%. Observations of the filament behavior, also showed that intense turbulence occurs in such supersonic flows from a texturing nozzle. During texturing this turbulence will contribute to the movement of the filaments across the nozzle, which will, in turn, cause further disturbances in the flow and, hence, increase the intensity of the turbulence. Therefore, it is believed that the turbulence and movements of the filaments could mutually contribute to each other's activity.

The turbulent nature of the jets created by several nozzles was demonstrated by means of a high-speed shadow graphical technique. Figure 4.28 shows such photographs obtained from jets created by (a) the Standard-Core HemaJet and (b) the Taslan XIV nozzle at 9 bar supply pressure. It is seen that the instantaneous shadowgraph of the jets from the Standard-Core HemaJet and Taslan XIV nozzles illustrate the local density variations caused by the highly turbulent nature of the flow.

Furthermore, by comparing these shadowgraphs, it could be argued that the Standard-core HemaJet creates jets which are more turbulent than the Taslan nozzles. It could be suggested that this higher turbulence is created by the oblique impingement of the incoming jets of the HemaJet nozzles.

Interactions between the air jet and filaments

High-speed photography. The behavior of the filaments in the HemaJet texturing nozzle, under various texturing conditions, was studied by Demir[48] using high-speed still photography, taken with 400 ns exposure time, and cine photographs taken at 20,000 frames per second.

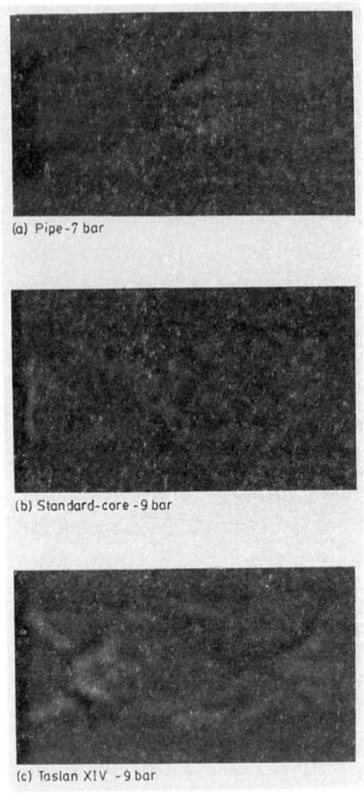

Figure 4.28 Instantaneous shadow photographs of jets

Still photograph.: Figure 4.29a shows a high-speed still photograph of a yarn being textured under wet conditions. This illustrates that the loops are being formed as the filaments emerge from the nozzle and that these occupy the lower half of the nozzle outlet. This also indicates that the tension in the textured yarn, as a result of shortening the overall length of the yarn because of the effective loop formation, is sufficiently high to pull the yarn close to the nozzle exit in a straightened form.

When the loop formation is not so effective (e.g., as in dry conditions), the overfed lengths of the filaments are not completely taken up by loops formed, and the tension in the yarn becomes so low that the textured yarn slackens and is convoluted as shown in Figure 4.29b. Occasionally, the whole bundle of filaments is blown straight away from the nozzle as a consequence of low tension in the yarn, with this being caused by poor loop formation as shown in Figure 4.29c, where the filament separation is seen to be poor. The filament bundle emerges at about the center line of the nozzle in a parallel and compact form where the velocity profile of the air flow is reasonably uniform. Therefore, the forces acting on the individual filaments are approximately equal; hence, each filament emerges from the nozzle at about the same speed causing scarcely any longitudinal displacement relative to each other; this hardly leads to any loop formation.

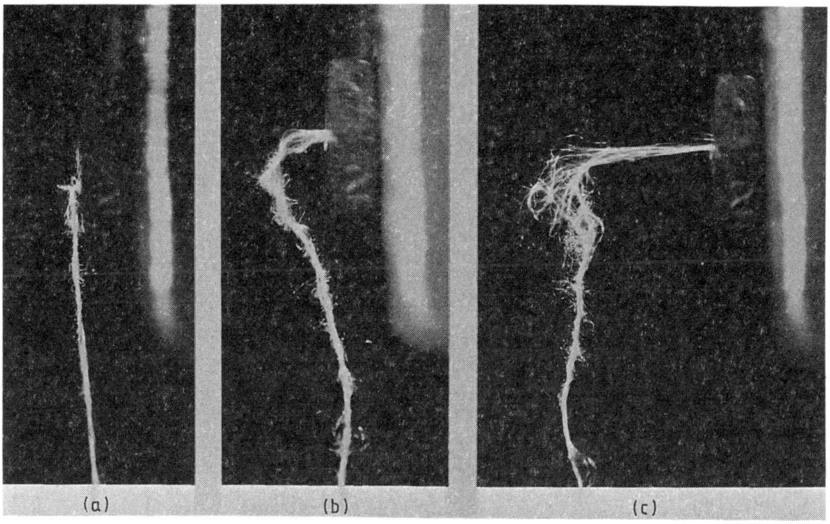

Figure 4.29 High-speed still photograph ofyarn being textured under wet conditions

Cine photography. All the preceding observations were amply confirmed by the high-speed cine photography but these results are naturally difficult to produce in a printed paper. They provided even better evidence that the dry-texturing process becomes unstable; straight compactly arranged filaments being frequently blown out from the nozzle, whereas wet texturing usually displayed stable process conditions. Figure 4.30 shows prints from such high-speed films showing (1) stable process conditions where the yarn is tensioned and kept close to the nozzle exit by effective texturing and (2) unstable process conditions where the yarn is blown out from the nozzle because of less effective texturing.

The emerging filaments. To give more insight into the loop formation mechanism, 100 still photographs of each dry and wet texturing process were taken and individually analysed by Acar.[47] As shown in Figure 4.31, the point most remote from the nozzle exit plane at which loops were instantaneously being formed, was measured by the horizontal coordinate, *x*, from this plane and by the vertical coordinate, *y*, from the nozzle axis. The vertical distance, *d*, from the nozzle axis to the uppermost filament in the emerging bundle, for a downward take-up, was also measured. The results, summarized in Figures. 4.32a and b, show that the filaments occupy the lower one-half of the nozzle exit area in every case when wet textured and in most cases when dry textured. They also show that the filaments are pulled further down and closer to the nozzle exit in wet texturing conditions than in dry texturing. Shortening the overall length of the textured yarn is more effective in wet texturing, because of better loop formation achieved, consequently causing a rise in the yarn tension, and, in turn, pulling the emerging, loop-forming filaments down and against the nozzle, for a downward yarn delivery.

Figure 4.30 Prints of high-speed cine films

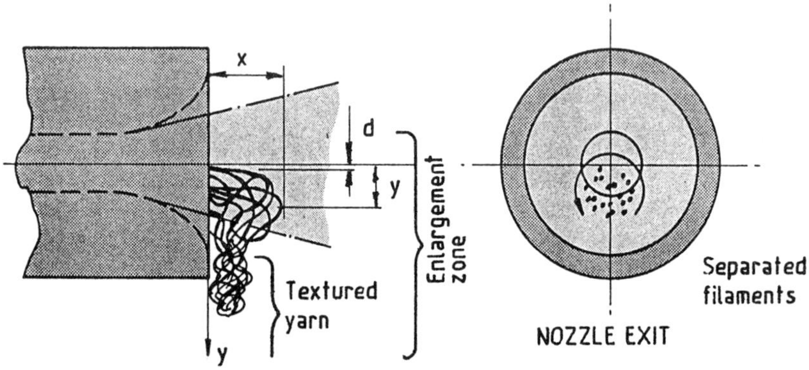

Figure 4.31 Hundred still photographs of each dry and wet texturing process analyzed by Acar[47]

Average filament speed. Because the filaments are overfed into the nozzle, some excess lengths of filaments are free to travel at a speed induced by the supersonic, turbulent, and nonuniform air flow. Therefore, the speeds of the filaments during texturing are similar to those of free filaments in an air stream; these would aproach the air velocities when the frictional losses are small. Nonuniform velocity distribution of the air flow, turbulence, swirl imparted to the filaments, friction between the filaments themselves, and friction between the filaments and the contacting sufaces will cause variations and fluctuations in the individual filament speeds across the nozzle, and these, in turn, will cause longitudinal displacements of the filaments relative to each other as evidenced by Wray.[23]

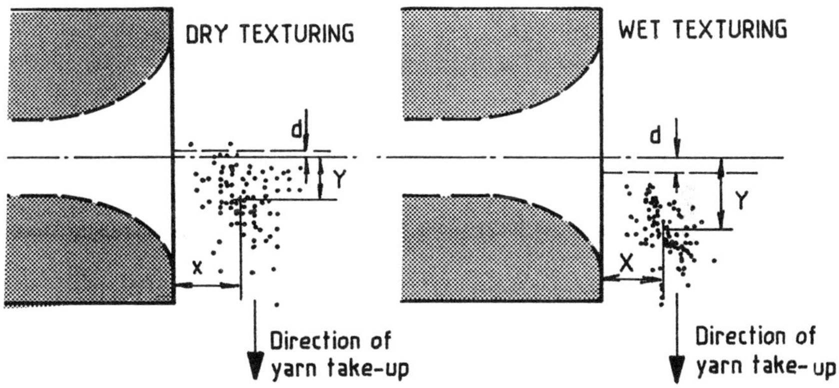

Figure 4.32 Results of analysis of hundred still photographs

Yarn tension during texturing. An examination of the structure of the textured yarns shows that they consist of randomly distributed sections with loops interspersed with sections of virtually straight parallel filaments with very few loops if at all. The sections with loops occur more frequently in a yarn textured under wet conditions (effective texturing) than those textured under dry conditions (poor texturing). This suggests that loop formation is an instantaneous process, which occurs intermittently, as discussed in previous sections, and, thence, causes the yarn tension to increase as the loops are formed. Therefore, such fluctuations in the yarn tension in the texturing zone could provide evidence regarding the frequency of loop formation.

The average value of the tension in the yarn is another useful property that could provide information regarding the effectiveness of the texturing and the stability of the loops produced. Because the tension in the yarn between the nozzle and the delivery rollers is a result of loop formation itself, it gives a measure of the effectiveness of the texturing. Also because less stable loops are removed under applied loads, whereas more stable loops offer a resistance that gives rise to increased tension in the yarn, an indication of the stability of the loops formed during the process can be obtained by measuring the tension in the stabilizing zone.

The magnitude of the tension varies according to the stabilizing ratio applied to the yarn; therefore, comparative results can be obtained for different process conditions by keeping the stabilizing ratio constant at a preset value. This tension could be related to the eventual yarn instability.

4.2.7 Loop Formation Mechanism

Flow from the texturing nozzles, at the usual texturing pressures, is supersonic, turbulent, slightly asymmetrical and of a nonuniform profile. The mean velocity of the flow and the degree of its nonuniformity increases with increasing texturing pressures. Asymmetry of the flow is most pronounced in planes close to the air inlets but diminishes with distance toward the nozzle exit.

Shock waves occur in the flow, the strength of which varies depending on the type of the nozzle. These shocks are at least partially destroyed by the presence of the filaments in the nozzle during the texturing process; nozzles providing varying degrees of shock strength are equally effective in the texturing process. Therefore, the effect of shock waves on the filament motion is negligible.

Intensive turbulence has the effect of separating and scattering the filaments across the nozzle cross-section. During texturing, the filaments are pulled down to the lower part of the nozzle (for downward delivery) because of the tension created in the yarn as a result of texturing itself. Shock waves appear to play no role in changing the direction of the filaments when they are left free to be dragged along with the air stream.

Some of the filament segments within the nozzle travel at much faster speeds than the typical yarn texturing speeds. Because the flow-velocity distribution is nonuniform and the fluid forces acting on the filaments are a function of the local air velocity, filaments that are separated and scattered across the nozzle are under the action of different driving forces and, therefore, move at different speeds. This, in turn, causes longitudinal displacements of the filaments relative to each other. Friction between the filaments and the solid surfaces plays a significant role in affecting the resultant force acting on the filaments and counteracts the effects of the fluid forces.

Wetting of the filaments, which is known to improve the quality of the textured yarns, has a lubricating effect on the filaments; this, in turn, reduces the friction between the filaments and the solid surfaces, thereby resulting in increased forces acting on the filaments. Moreover, wetting also reduces the friction between the filaments themselves and, hence, enhances their longitudinal displacements and assists entanglement as they emerge from the nozzle. Consequently, all these factors improve the texturing process.

The amount of water required to create the desired effects of wetting is only a small fraction of the total amount used in the process, because most of it is blown away by the secondary flow allowing only a small fraction to be entrained into the nozzle. Therefore, the possible effects of water particles on the primary flow are insignificant; nevertheless, these effects act so as to cause a reduction in flow velocity. This small yet adverse effect of water particles on the air flow rules out the claims that state that water alters the flow properties such as to improve the texturing conditions and the resultant yarn quality.

Filaments with different cross-sectional areas and different cross-sectional shapes behave differently in the air stream. For a circular cross-section, while the drag forces that act on the filament increase in proportional to the filament diameter, the mass and inertial resistance of the filament to these fluid forces will increase as a function of the square of the diameter. This will cause coarser filaments to travel more slowly in given flow conditions and, therefore, will result in inferior conditions. Filaments with equal fineness but with smaller area and polar moments of inertia and with increased surface areas, such as elliptical cross-sections, are more suitable for air-jet texturing than are the solid circular filaments.

4.2.7.1 Mechanism of Loop Formation

Usually there are many filaments in a supply yarn, but to explain the loop formation mechanism, Acar[47] considers only a few filaments emerging from the nozzle as shown in Figure 4.33. At any instant some of these filaments will be moving at faster speeds than others because of the relatively greater fluid forces acting on them. The free excess lengths provided by overfeeding the filaments enable the faster moving filaments to slip and be displaced longitudinally with respect to the relatively slower moving filaments. The amount of these longitudinal displacements is affected by local drag and frictional forces instantaneously acting on the filaments, and also by the overfeed ratio.

The textured yarn is delivered from the nozzle at right angles to the nozzle axis and travels at the yarn texturing speed. Because the filaments are converted into loops, the resultant textured yarn is shortened, and a tension is created of a magnitude determined by the effectiveness of the texturing. Thus, on the one hand the emerging filaments are blown out of the nozzle along the direction of the air stream at much faster speeds than the yarn-texturing speed; conversely the tension created in the yarn, as a result of texturing, pulls the "leading ends" of the emerging filaments in the direction of the yarn delivery (i.e., at right-angles to the nozzle axis). The "trailing ends" of the filaments in the nozzle are blown out at very high speeds, whereas the "leading ends" of the filaments are held within the core of the much slower moving textured yarn, and these are pulled downward and kept close to the nozzle exit; therefore, the filaments are forcibly bent into bows and arcs by the fluid forces acting on them. These are than entangled with other emerging filaments, which are formed into fixed stable loops with in the textured yarn.

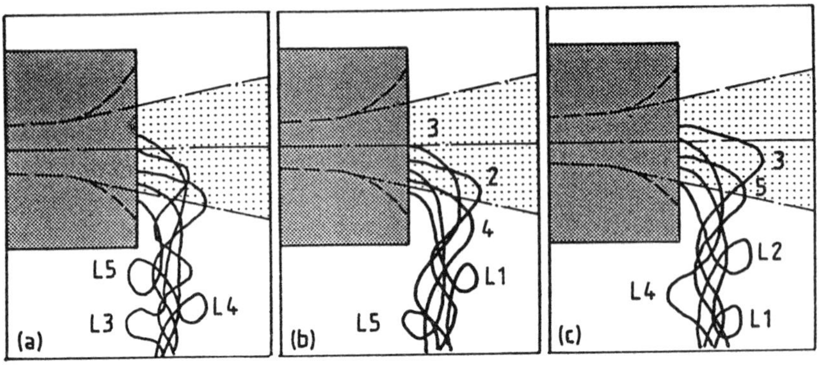

Figure 4.33 Loop formation mechanism

Lengths of such filaments, as they form loops, are fixed in the core of the textured yarn, become shorter; consequently, the tension in them increases. The increased tension resulting from loop formation and subsequent entanglement causes loop-forming filaments to migrate toward the lower part of the nozzle because they will assume the shortest possible path between the "trailing" and the "leading ends." An instant later the tension in the segments following these entangled filaments may be relieved because of the overfeed ratio, and these filaments may be blown out by the air stream and its associated turbulence to form new loops. Therefore, every filament goes through this process at different instants and the cycle repeats itself randomly.

This can be illustrated in Figure 4.33, which is a simplified schematic diagram with only a few filaments representing the behavior of a more complex multifilament yarn. In Figure 4.33a, filament 1 is the fastest-moving filament having the greatest longitudinal displacement relative to the others and is blown furthest out of the nozzle to form a loose bow or arc. An instant later, in Figure 4.33b, it is formed into a fixed loop L1 within the textured yarn as a result of mutual entanglement of the filaments under the action of the air stream. This newly formed fixed loop L1 increases the tension in filament 1, thereby causing a change in its position and also contributing to the total yarn tension, which is pulling the yarn down closely to the nozzle. Meanwhile, filament 2 comes under the action of a greater drag force as a result of changes in the positions of the filaments across the nozzle due to the turbulence and varying tension in the individual filaments, and this now becomes a faster moving filament so causing it to be blown out and displaced longitudinally relative to the others to form a loose bow or arc. Immediately afterward (Figure 4.33c), while filament 2 is being similarly entangled into a fixed loop L2, a further filament 3 commences a similar loop formation process.

Because there are many filaments in the actual supply yarn rather than the five illustrated in Figure 4.33, several loops are formed at any particular instant, and these help each other to be fixed and locked within the yarn structure by mutual entanglement. Other filaments, although they do not form visible loops on the surface of the yarn, have their originally closely packed parallel structure changed into a tangled, voluminous structure with increased bulk and linear density.

This description of the mechanism of loop formation is valid for all types of texturing nozzles, despite detailed differences in their design, because the underlying requirement to create a supersonic, turbulent, asymmetrical, and nonuniform flow is common to all satisfactory texturing.

4.2.7.2 Factors affecting loop formation

To be able to analyze the effects of various parameters on the loop formation mechanism, Acar[47] considers a filament emerging from the texturing nozzle as shown in Figure 4.34a. Point A on the filament is the "trailing end" of the filament, which makes a bow on emerging from the nozzle and moves at a speed dictated by the air flow (i.e., filament speed), V_f. Point B is the leading end of the filament and moves downward at the texturing speed (i.e., at the speed of the textured yarn) V_y. The ratio of these two speeds is estimated to be of the order of 10 (i.e., $V_f/V_y > 10$). In unit time, points A and B will move to positions A' and B', respectively, as illustrated in Figure 4.34b. The length of filament L, which is blown out in this unit time is determined by the filament speed, V_f, and by the excess free length of filament as provided by the overfeed ratio, whereas the length BB' is determined by the texturing speed, V_y. The filament length L_o between point, A and B will become $(L_o + L)$.

The effects of any changes in the texturing conditions on this filament will be analyzed in the light of the described mechanism of loop formation, and by reference to illustrations Figures 4.34a to d.

4.2.7.2.1 Effects of air pressure

When the air pressure is increased, the speed of a filament will also increase as a result of increased air velocity, and, therefore, point A will move more rapidly,

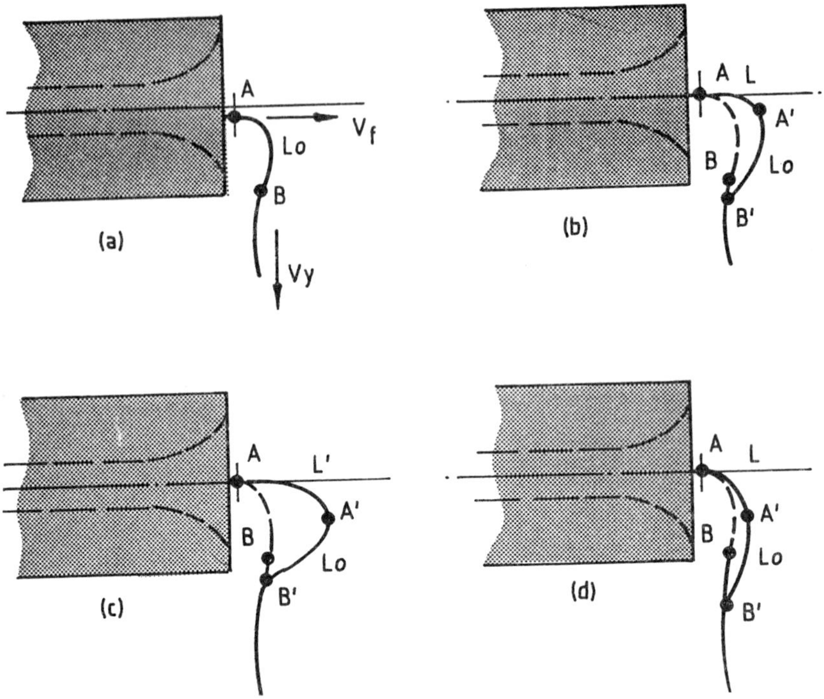

Figure 4.34 Filament emerging from texturing nozzle

and the time taken to blow out the free length of the filament will become shorter. High air pressure will also enhance the differences between the forces acting on the filaments and cause greater relative longitudinal displacements. Because the loop formation is an intermittent process, the increased filament speed could increase the frequency of the loop formation and the entanglement process could then become more effective in flows with high velocities and intensified turbulence. Therefore, more effective texturing is likely to be achieved by increasing the air pressure.

4.2.7.2.2 Effects of overfeed ratio

When the overfeed ratio is increased a relatively longer filament will be blown out in unit time (i.e., $L' > L$), as illustrated in Figure 4.34c. Because the yarn texturing speed is not changed point B will not move very far and the longer filament between A and B' will form a larger loop or several smaller loops in this unit time.

Therefore, increasing the overfeed ratio will result in an increase in loop size and loop frequency, which, in turn, causes an increase in the linear density and a decrease in the yarn stability, because the higher the number of loops the higher becomes the chance of these loops being pulled out under applied loads.

When the overfeed ratio is excessively high, the bow formed by the filament becomes large, and will consequently reduce the tension in the yarn, causing the filament and the textured yarn to be blown away from the nozzle. This will result in a failure in the loop-formation process.

4.2.7.2.3 Effects of texturing speed

Increasing the texturing speed will cause point B to move faster and in turn cause the distance BB' to increase in a given unit time (Figure 4.34d). Because the total length of the filament L that can be blown out in this unit time is not changed, increased yarn speed will cause the filament between points A and B' to form a flatter bow. This bow is unlikely to form into a stable loop and will be removed under applied loads. Therefore, the number of stable loops formed will be reduced, the tension in the yarn will drop, and the filaments will be blown away from the nozzle, causing failures in the continuity of the texturing process.

To counteract, at least partially, the adverse effects of increased texturing speeds, the flow velocity could be increased, which may cause the process to become more effective. This requires high operating pressures, which makes the process uneconomical. Therefore, the scope of increasing the productivity by increasing the production speed is limited.

4.2.7.2.4 Effects of other parameters

Wetting will reduce the friction between the filaments themselves, and between them and other solid surfaces. This will cause an increase in the resultant fluid forces acting on them and the texturing process will become more effective; hence, more stable loops will be formed.

The impact element is unlikely to have a major effect on the flow inside the nozzle, because the element is usually situated at a distance of about one nozzle diameter from the nozzle exit, and is, thus, remote from the immediate nozzle exit region, where the loop formation process occurs. One minor influence concerns the role of the impact element as a physical barrier to those filaments, which are blown away from the nozzle.

4.2.7.2.5 Effects of the supply yarn properties

Increased numbers of filaments increases the likelihood of mutual entanglement, and, thus, may be expected to improve the effectiveness of the texturing process and the resultant yarn quality. The linear density of the individual filaments is also a significant factor, as a consequence of reduced inertial resistance opposing fluid forces acting on the finer filaments. Finer filaments have reduced stiffness, which makes bending and other forms of deformation of the individual filaments much easier during the loop formation process, which, in turn, causes better textured yarns to be produced. The cross-sectional shape of the filaments may also have considerable effect on the loop-formation process as, for instance, a thin elliptical section will have a preferred bending axis, which leads to reduced stiffness of the filament, while the extended surface and projected areas will tend to increase the fluid forces.

4.2.8 Wetting of the supply yarn

Wetting or moistening of the supply yarn before its entry to the nozzle is an integral and indispensable part of the air-jet texturing process today, because it highly improves the texturing conditions, thereby resulting in a more uniform and stable yarn.

In modern industry, dry air-jet texturing (i.e., texturing without the application of water) is rarely used because it does not produce successfully textured yarns with the desired qualities. The entanglement and surface loops of such yarns are easily removed by the effects of even slight tensions.

Several methods for water application have been developed alongside other developments in air-jet texturing. In the early days of the process, the conventional way of wetting the yarn was by threading it through a water bath as shown in Figure 4.35a. Modern machines, however, make use of a simpler technique of wetting in the form of a wetting unit either integrated with or separate from the nozzle housing (Figures 4.35b and c). Both of these methods have their advantages and disadvantages. It could be argued that more uniform wetting would be achieved by means of a water bath because the yarn is immersed into the water for a longer period. Conversely, the water in the tub is contaminated by the spin-finish washed away from the yarn surface; consequently, this contamination reduces the wetting properties of the water and results in a less effective application. However, the main superiority of the wetting unit over the wetting bath is that the former facilitates the precise control of the water applied and always provides fresh clean water. "Spray Wet"[64] (See Figure 4.35) is a newly developed spray wetting system. Under high pressure, jets spray water into ceramic yarn guides. The spray mist wets the front of the yarn and then penetrates into the filament bundle. The water draining from the yarn guide wets the back of the yarn end. This ensures complete cover and penetration of the filament bundle with water.

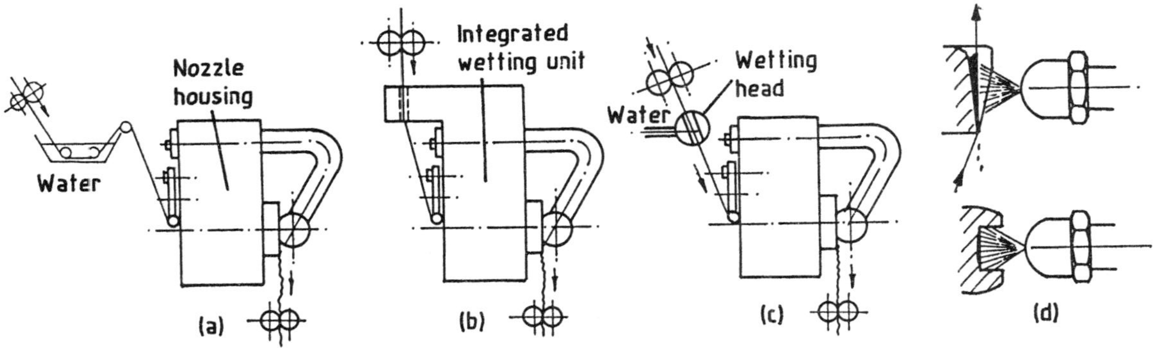

Figure 4.35 Water application techniques

Wet texturing, although it produces superior yarns, has its own drawbacks. First of all, because most of the spin-finish on the filaments is washed away by the water, as illustrated in Figure 4.36, which shows the spin finish content of the yarn before and after both wet and dry texturing; the wet-textured yarn requires the application of oil in order to gain improved antistatic and reduced frictional properties. Second, the effluent water contaminated by the spin finish materials needs to be cleaned if it is to be recycled and the texturing nozzles should also be cleaned frequently.[65] Idle machine time during this cleaning procedure obviously increases the cost of the textured yarn.

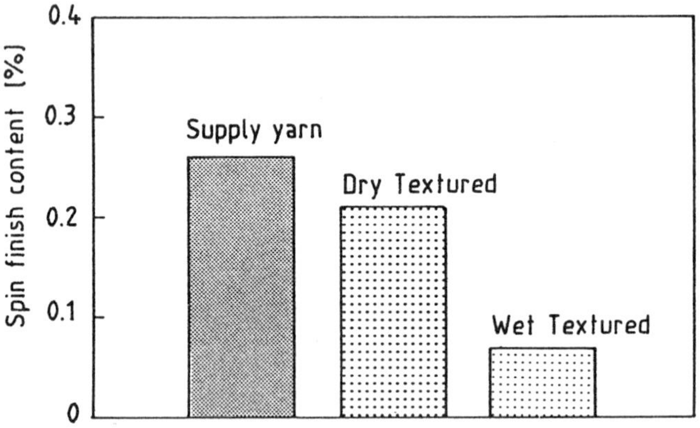

Figure 4.36 Spin-finish remaining on yarn before and after both wet and dry texturing

The research by Demir[48] showed that the amount of water which is necessary for an effective texturing is as little as 0.06 liter/hour which is even less than those claimed by both Bock and Lünenschloss,[53] and Acar et al.[66] This minimum amount of required water varies with the secondary flow momentum flux as determined by the air pressure.

4.2.8.1 Frictional effects

According to Olsen,[67] the temperature and the viscosity of a lubricant are two important factors that affect the frictional behavior of textile yarns and fibers in the hydrodynamic region; other factors are the yarn speed, lubricant viscosity, yarn or filament linear density, pretension, roughness of the guides or yarn surfaces, and guide diameter.

It is known that, for some textile applications, water has good lubricating qualities. The lubricational characteristics of both water alone and some water-wetting agent mixtures are discussed in this section, and the effect of increasing the water temperature is also reported.

In the Figures 4.37 and 4.38, the coefficient of friction is plotted against yarn speed. Figure 4.37 clearly indicates that water reduces the filament/solid surface friction a great deal, whereas only slight variations in the friction have been observed by adding wetting agents into the water. Conversely, Figure 4.38 shows that the lubricating capability of water is slightly enhanced as its temperature is increased.

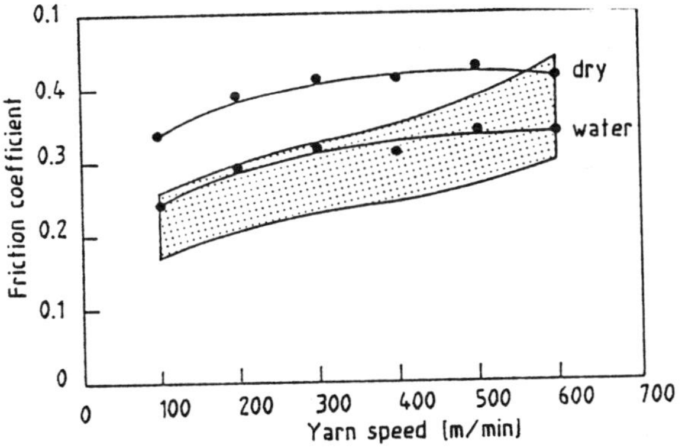

Figure 4.37 Coefficient of friction against yarn speed

4.2.8.1 Water temperature

It is known in practice that the texturing process is adversly affected when water at low temperatures (near to the freezing point) is used. Therefore, Demir[48] thought that by changing the temperature of the water used in the texturing process, the frictional properties of the yarn could be affected, because water temperature affects both the yarn temperature and the guide temperature. The viscosity of the water, which is said to play a significant role in frictional properties in the hydrodynamic region,[68] is also affected by the temperature. Nevertheless, Demir's investigations revealed that although the frictional properties of both the filament/solid surface and filament/filament interaction are altered by the temperature of the lubricant, the texturing could not be significantly improved.

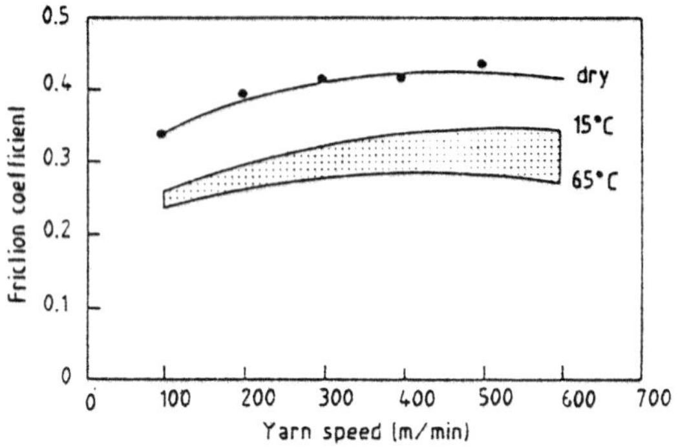

Figure 4.38 Lubricating capability of water is enhanced with increasing temperature

4.2.8.2 Wetting agents

By reducing the surface tension of the water, its wetting properties could be improved so as to penetrate the individual filament surfaces more effectively and thereby further reduce the friction. Several wetting agents were added by Demir[48] to the water and their effects on the texturing were ascertained by producing yarns and testing them. They indicate that any improvement in the process accomplished by adding the wetting agents into the water is too insubstantial for any positive conclusion to be drawn.

For the purpose of further investigation, therefore, two extreme frictional conditions are made use of and ascertain their possible consequences. One of these extreme conditions was to eliminate the filament/solid surface friction completely from the process. The other is to increase the filament/solid surface friction just before the nozzle by means of a rough surfaced yarn guide.

4.2.8.3 Elimination of filament/solid friction before the nozzle

By realigning the yarn path in a straight line prior to the nozzle, Acar et al[66] virtually eliminated the yarn-to-yarn guide friction. This alternative yarn path, however, without wetting the yarn made no improvement to the process and produced similar yarns to the conventional yarn path arrangement. Therefore, no further attempt was made to test the yarns textured with a realigned yarn path under dry conditions, whereas the yarns textured under wet conditions with the realigned yarn path arrangement were tested, and found that little improvement in the process is achieved, although no friction between the yarn and the guide occurs. Therefore, it could be concluded that the yarn/guide friction seems to play an insignificant role in texturing. Two further friction sources, however, still exist. These are the friction between the yarn and the nozzle walls, and interfilament friction.

4.2.8.4 Increasing the filament/solid friction

Had there been a substantial effect of friction occuring between the supply yarn and the yarn guide before its entrance to the nozzle, it could have been expected that the high friction between these surfaces would have worsened the texturing. To prove this hypothetical prediction, the surface roughness of the yarn guide was increased by coating it with a highly rough-surfaced sheet with that of a conventional yarn guide. In conclusion, it is found that neither the complete elimination nor an increase of the friction occurring between the yarn and the yarn guide just before the nozzle could eliminate the necessity for water application during the air-jet texturing.

In conclusion, it could be said that the amount of water that is typically been used in practice is much beyond the required amount. Because the excess water creates additional problems, the amount of water should, therefore, be reduced. Even further improvements in yarn wetting could be achieved by injecting water into the nozzle housing, provided that the cost of supplying high pressure water is feasible and also that the wearing problem of nozzles is solved. In the case of water injection, the process becomes cleaner because all the water injected mixes with the air forming only a mist, and less spin-finish material is surmised to be washed away from the yarn surface.

The experimental work has shown that no additives (i.e., wetting agents) make any significant improvement either in the lubricating properties of water or in the texturing process. Some slight improvement in the process was achieved when the water is heated up to 65°C.

It can be concluded that the filament/solid surface friction occurring between the yarn and the yarn guide just before the nozzle (for the conventional yarn path arrangement) is of no significance.

Although there is evidence that water enters the primary flow, no condensation has been observed because of the fact that the water particles are in the form of water droplets at both the entrance and exit of the nozzle. Consequently, no phase change occurs in the nozzle or in the downstream section of the air flow. Therefore, the mechanisms for water effects on the texturing, which are based on the formation of condensation shock waves, are disapproved.[48]

The most important effects of water droplets are seen in the primary flow section of the nozzle. Having ruled out the possibilities for the formation of condensation shock waves, the only effect of water droplets would seem to reduce the interfilament friction, and the filament-to-nozzle wall friction. It could be suggested that water injection into the air reduces the friction further, and, in turn, this could result in better texturing by facilitating the easy relative movements of the filaments.

4.2.9 Machines

In parallel with developments in the process, several texturing machines have also been developed by various textile machinery companies. Most of the machines used in the early days were either spinning or winding machines retrofitted with the texturing nozzles or in-house built machines. In more recent years, false twist texturing machines were retrofitted for air-jet texturing. Today, most air-jet texturing machines are specifically designed and built for the purpose and, therefore, provide optimum working conditions for yarn processing.

In Europe, the first purpose built machine was introduced in early 1970s by the Eltex Company of West Germany. In 1975 and 1978, two other companies, namely Berliner Machinenbau A.G.[69] of Switzerland and Barmag Company[65] of West Germany, respectively, introduced their machines. Although these existing companies were improving their machines, several new ones also appeared in the market and Rieter-Scragg, Enterprise, Murata, Toray, Theiler, ICBT, RPR, Giudici, and Aiki are a few worthy of mention in this respect.

With these modern types of machines, mechanial speeds up to 1,000 m/min can be reached. Most of them also facilitate the drawing of POY yarns and provide heaters for heat setting after texturing. The choice of the nozzle type is generally left to the operator.

The air-jet texturing machines may be classified into two main types:

1. Mass production oriented line shaft machines,
2. Special yarn production oriented individual position machines.

Most of the false twist texturing machine manufacturers have readily modified their existing machinery design to suit for the air-jet texturing process. In addition to basic process requirements, these machines are also equipped with all possibilities, such as sequential drawing for POY supply yarn, post-texturing stabilizing extension, post- texturing heat setting, free fiber ends and slub-forming devices, on-line quality monitoring, as well as robot doffing. Nevertheless, this approach had limited success because of the fact that the air-jet textured yarns are mostly superior in speciality yarn applications. Therefore the machine manufactuers designed machines specifically for the air-jet texturing machines. These machines had either individual position drive or a maximum of four positions are driven independently. This has given a high flexibility to both optimize the production and to change the produced items quickly. Most of these machines offer the following process possibilities (Figure 4.39).

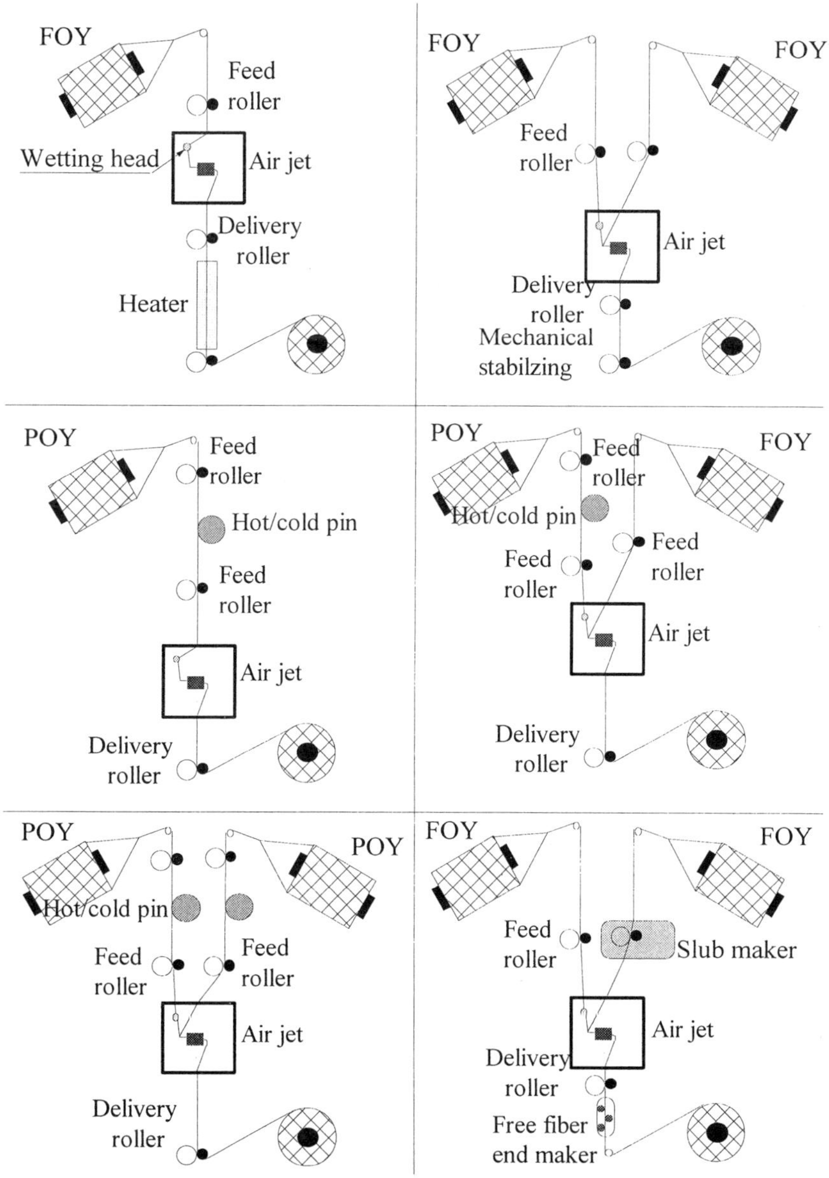

Figure 4.39 Proces possibilities with modern air-jet texturing machines

4.2.10 Properties of air-jet textured yarns

4.2.10.1 Instability

A close examination of a typical sample of air-jet textured yarn reveals that some of the loops that characterize such yarns can be pulled out by applying tension. The number of loops remaining intact depends on the "stability" of the loops. This is determined not only by the supply yarn properties, such as material, linear density per filament, number of filaments, and shape of the filament cross-section, but also by the process parameters, such as type of nozzle, air pressure, texturing speed, overfeed, and whether wet or dry processing is involved.

The easy pulling out of loops under working tensions would be a disadvantage in fabric-forming processes, because the bulk of the yarn would be reduced and the possibility of fabric irregularity increased. Therefore, a quality control test is required to determine the stability of the yarn.

4.2.10.2 Increase in linear density

Air-jet texturing causes longer lengths of synthetic filaments to be compacted into shorter lengths with an entangled structure. The resultant yarn is stable, does not stretch, and does not lose its interlaced and intermingled structure even under relatively higher loads. Therefore, a considerable linear-density increase is expected from an air-jet textured yarn. An increase in the yarn linear density at varying process conditions but at a given overfeed can be interpreted as an increase in the effectiveness of the texturing process.

Theoretically, it is expected that the increase in linear density will be equal to the amount of overfeed irrespective of the process and the supply yarn parameters. Nevertheless, this theoretical expectation is never met in practice because of both the on-line stabilizing process and the winding up tension. Because a well-textured yarn will resist the stabilizing and winding-up tensions, the loops will stay intact within the well-entangled core; consequently, its linear density will be higher than that of a poorly textured yarn. This, therefore, implies that the amount of increase in linear density is influenced both by the process itself and those supply yarn parameters that affect the degree of entanglement and loop formation.

4.2.10.3 Strength

Rozmarynowska and Godek[21] and Kollu[70] have disclosed their research findings on the effects of the air-jet texturing process on the individual filaments. The former researchers used physicomechanical tests such as Scanning Electron Microscopy (SEM) scanning, strength measurement, birefringence and criystallite orientation tests. The latter has only deployed SEM scanning and strength measurement on individual filaments before and after texturing. However, both have come to the same conclusion that the air-jet texturing process, however severe it may be (up to 5 bar air pressures in the former and up to 11 bar pressures in the latter), caused no damage on the individual filaments in the form of crack, local cross-sectional shrink (thinning), or even chip formation on the filament surface. Consequently, it is concluded that any decrease in strength after texturing results solely from the entanglement of the originally parallel arrangement of the constituent filaments of the supply yarn.

Comparisons of the load elongation curves (Figure 4.40) of untextured and textured yarns reveal striking differences. All of the filaments in an untextured yarn simultaneously share the applied load to the yarn and firstly deform elastically. When the load is increased beyond the elastic limit, all of the filaments are then plastically deformed, the filaments exhibiting an increasing strength under this increasing load (i.e., hardening) up to the yield point where they start to elongate rapidly. When the stress in an individual filament exceeds its breaking stress, it will break irrespective of the other filaments' conditions. Therefore, in a supply yarn the filaments break singly at different times, probably due to slight variations in their diameter, and some of the filaments appear to elongate more than others. Nevertheless, for the purpose of the strength tests, the load that causes most of the filaments to break is taken as the breaking load for the yarn and the corresponding elongation is regarded as the breaking elongation.

Textured yarns, however, exhibit totally different load elongation characteristics. Because the filaments are randomly entangled, and some of these local entanglements and loops are removed under the applied load, the deformation of a textured yarn starts with permanent elongation. No hardening of the textured yarns is observed as the loading is increased, because of the fact that none of the individual filaments are continuously subjected to the applied load during the entire test period. All filaments exhibit loops and entangled sections intermittently along their lengths, but these are separated by straight portions of filaments. At any section of the yarn, at any particular instance, only these straight portions will resist the applied load. However, when the loops associated with these particular filaments have been pulled out under the applied load, their effective lengths are increased; consequently, some other less straight slack filaments in the same region may become subject to the applied load and in turn contribute to carrying it

It is most likely that within the length of yarn tested, a particular section will be more effectively entangled than the rest and will have fewer load carrying filaments; these will be surrounded by filaments which exert lateral forces that increase the interfilament friction at this section. These load-bearing, firmly entangled filaments will rapidly reach the breaking point simultaneously within a short time and consequently an almost instantaneous breakage of the yarn will occur at this section. The load-elongation curves of an air-jet textured yarn and a spun staple cotton yarn are very similar as shown in Figure 4.40, suggesting that other physical characteristics of these yarns may also be similar.

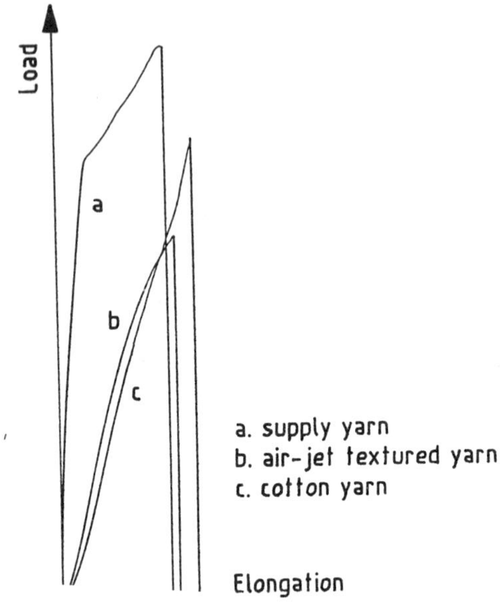

Figure 4.40 Comparison of the load-elongation curves

In general, synthetic untextured filament yarns have higher tenacities than air-jet textured yarns. This is because all the individual filaments share the applied load, and virtually there is no lateral force on the filaments. However, an air-jet textured yarn has a reduced strength because the applied load is borne only by a small fraction of the individual filaments.

Therefore, the effectiveness of the nozzle and thus the texturing process could be observed as a reduction in tenacity (i.e., a better textured yarn –one with many small compact and entangled loops) will exhibit a greater decrease in tenacity when compared with the supply filament yarn from which it is produced.

4.2.11 Relations between Process Variables and Textured Yarn Properties

4.2.11.1 Air pressure

As the air pressure is increased, the flow velocity that is the main driving force that opens up the filaments to enable them to entangle and texture, also increases. The nonuniformity and turbulence of the air flow are also enhanced by increasing air pressures. An improved texturing of the yarn is, therefore, expected at higher pressures. One might also anticipate a more stable entanglement (i.e., less instability for yarns textured at high pressures) but Figure 4.41 depicts a slight increase in the of the jet decreases at higher overfeeds and as seen in the SEM photographs of a 30% instability of air-jet textured yarns as the pressure increases. From a comparison of the SEM photographs of two yarns textured at the two extremes of pressure, it is observed that yarn textured at 5 bar contains larger but fewer loops, whereas yarn textured at 9 bar shows a compact yarn core and a greater number of loops because of the improved texturing conditions. The slightly increased instability at higher pressures probably arises from the greater number of loops, which increase the likelihood of loop removal, but in no way can this increased instability be interpreted as a degradation of the texturing quality. The ensuing considerations of other properties such as the linear density increase and the decrease in tenacity substantiate this argument.

Because the amount of overfeed is kept constant at 20%, any variation in the linear density of the yarn must be caused by better entanglement and loop formation at higher air pressures. Figure 4.41 indicates that the linear-density increase of the textured yarn increases from 12% at 5 bar to 14.5% at 7 bar, with a small increase when the pressure is increased beyond this value. These results are typical for the supply yarn and texturing nozzle used (viz. the Heberlein T100 HemaJet) and cannot be generalized for other nozzles. For example, the T341 HemaJet and Taslan XIV nozzles produce textured yarns with linear densities almost linearly increasing with increasing air pressures.

Figure 4.41 also shows that the yarns textured at 5 bar exhibit approximately 60% of the supply yarn tenacity. When the air pressure is increased further up to 9 bar, the reduction in tenacity would reach approximately 50%. This reduction in yarn strength follows from the increased numbers of loops, as depicted by the SEM photographs, which, in turn, reduces the number of load-bearing filaments, thereby causing a reduced strength.

4.2.11.2 Overfeed

When the overfeed is as low as 10%, the excess lengths of filament available to form loops and arcs are, indeed, small. Consequently, the texturing is poor with very few loops on the surface and a slight reorientation in the core of this yarn

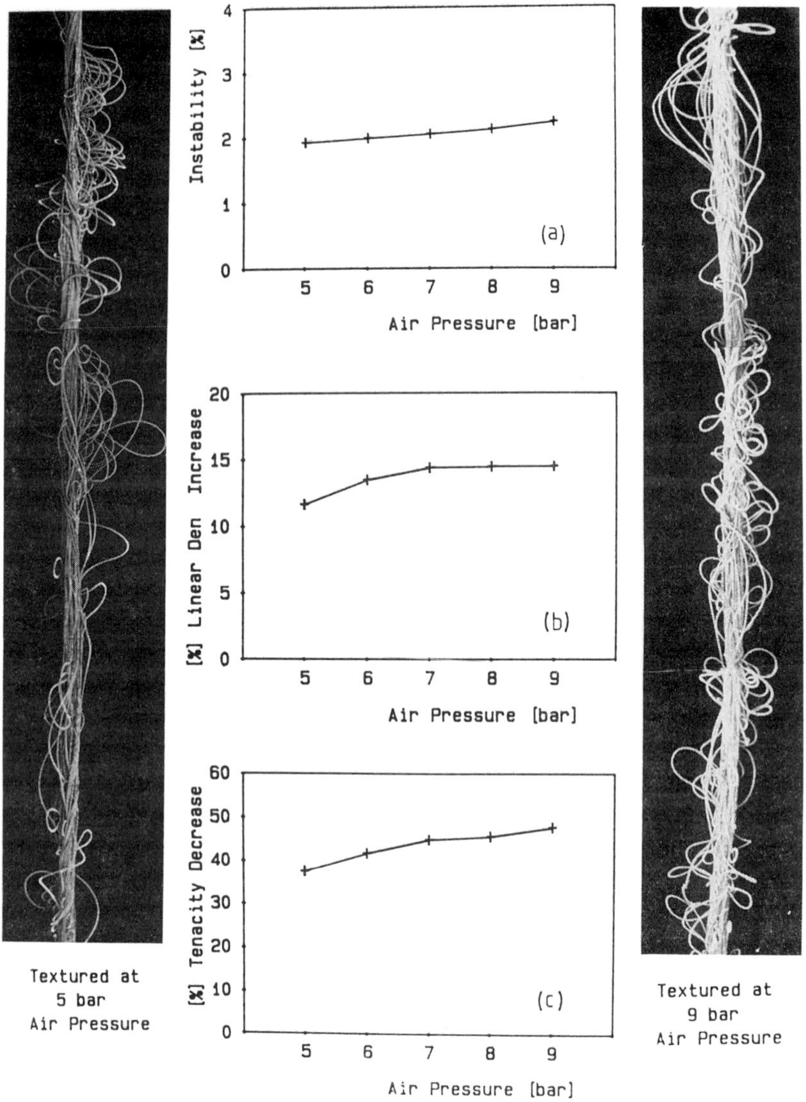

Figure 4.41 Effects of air pressure on properties of air-jet textured yarns

as seen in the SEM photograph in Figure 4.42. It exhibits a low level of instability, due to the presence of more straight unlooped load-bearing filament portions but it would be unacceptable for many end uses.

As the overfeed is increased more excess lengths of filament are available to form loops that tend to cover the yarn surface. This increases the instability of the yarn without discrediting its acceptability. High-speed shadow graphs of the emerging air jet during texturing process have revealed that the flow becomes more disturbed as the overfeed is increased. Consequently, the texturing power of the jet decreases at higher overfeeds and as seen in the SEM photographs of a 30% overfeed textured yarn (Figure 4.42), its surface is covered with many

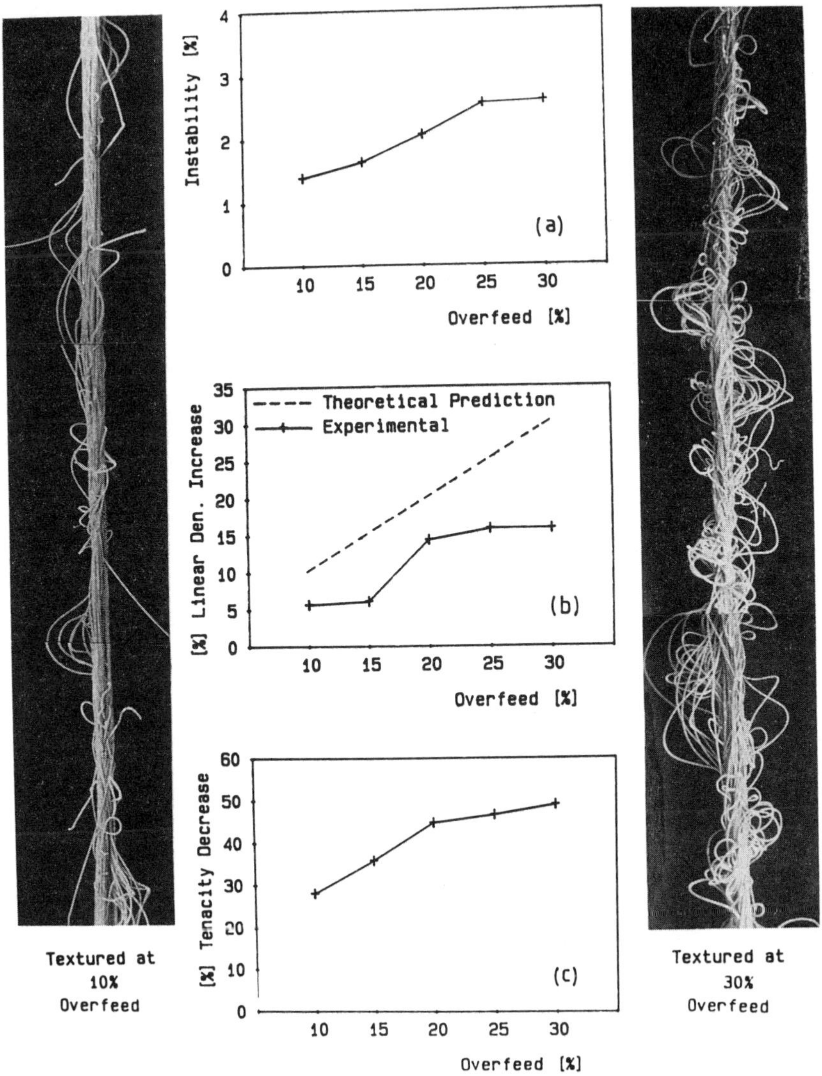

Figure 4.42 Effects of overfeed on properties of air-jet textured yarns

slack and large loops. For the given supply yarn and reference process conditions, the overfeed that yields suitable texturing from the loop appearance viewpoint varies between 15% and 25%. However, higher overfeeds may as well be deployed for special purposes such as core-and-effect, fancy, and slub yarns.

At low overfeeds, the excess lengths of filaments are insufficiently long to form stable loops and arcs. Conversely, as the overfeed is increased beyond the capability of a particular nozzle, effective entanglement of the filaments is unattainable. Therefore, for any particular nozzle, there is a range of overfeed over which effective texturing can be achieved. Figure 4.42 shows that the overfeed has paramount influence on the linear density of the textured yarn. Provided that the other supply yarn and process parameters are acceptable, a desired linear density can be obtained by varying the amount of overfed filaments introduced to the nozzle.

Figure 4.42 shows that, at 10% overfeed, only a 29% decrease in tenacity occurs due to the poor entanglement and loop formation. In the resultant yarn, most of the filaments are approximately straight and hence can carry more load. As the overfeed is increased up to 20% the effectiveness of texturing is rapidly enhanced and because of better filament entanglement the tenacity of the yarn is correspondingly decreased by 45%. Above 20% overfeed, the rate of tenacity decrease is not as high because of the less effective texturing referred to earlier.

4.2.11.3 Texturing speed

Figure 4.43 reveals only a slight increase in yarn instability with increasing texturing speeds. Because the resultant forces and torques on the filaments are mainly generated by the relative velocity between the filaments and the surrounding air flow,[71] higher forces and torques are exerted on the individual filaments at lower-texturing speeds. These greater fluid forces cause a better entanglement and a more firmly fixed loop formation. As the texturing speed is increased, the filament speed is also increased, hence the resultant forces decrease. Consequently the texturing becomes less effective, thereby resulting in large and unstable loop formation as depicted by the SEM photograph of yarn textured at 600 m/min.

For the particular nozzle used, the texturing at 400 m/min is almost as effective as at 200 m/min (See Figure 4.43), although at speeds greater than 400 m/min the texturing quality is reduced because of large and unstable loop formation. Only approximately one-third of the overfed lengths of filaments supplied could be retained in the yarn at 600 m/min texturing speed, whereas at 200 m/min texturing speed, approximately three-quarters of overfed lengths appear to be fixed in the yarn, which exhibits stable loops and a highly entangled core structure.

4.2.11.4 Stabilizing ratio

The purpose of the stabilizing extension, which is performed "on-line" on the machine subsequent to the texturing zone, is to pull out those loops are not

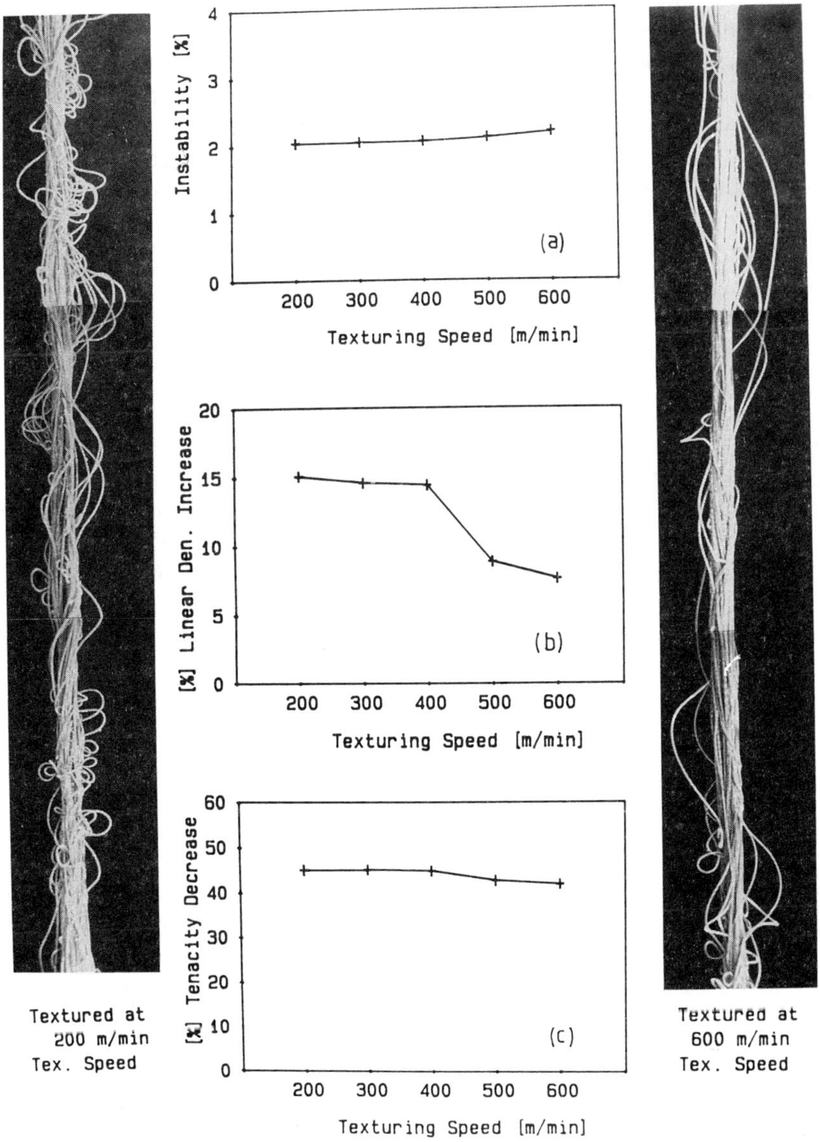

Figure 4.43 Effects of texturing speed on properties of air-jet textured yarns

firmly fixed to the core of the yarn. The level of this extension usually varies from 2% to 6%, because at higher levels permanent yarn damage or even breakage could occur. For the purpose of investigation, however, the stabilizing extension is here varied from 0% to 10%.

Figure 4.44 illustrates a steady decrease in instability with increasing stabilizing extension. This is due to the removal of large and loose loops on the yarn before the application of increasing stabilizing extensions. At 10% stabilizing extension, few loose loops stay intact to the core of the yarn as illustrated on the SEM photograph of this yarn (Figure 4.44) where it can be seen that some of the loops were broken up to form free fiber ends because of the high tension build-up in the yarn. Further evidence for this was the accumulation of fluff around the take-up rollers. It is, therefore, recommended that the stabilizing extension should not be so high as to cause broken filament ends, and a range of about 2% to 4% seemed reasonable for the particular supply yarn and process conditions used.

Figure 4.44 shows the experimental linear density increase with increasing stabilizing extensions together with the theoretically predicted linear density increase, which is the overfeed (20% in this case) minus the amount of stabilizing extension. Because all linear density measurements have inevitably been carried out under a slight tension, some of the loops and entanglement of air-jet textured yarns may have been removed by the effect of this tension particularly for low-stabilizing extensions (less than 2.5% for the conditions considered). At higher stabilizing extensions, the loops and filament entanglement resist the applied tension, and the yarn extends partially elastically so that when the load is removed it recovers some of the extension and, therefore, possesses a higher linear density than theoreticaly predicted.

Figure 4.44 indicates that the amount of stabilizing extension applied during the process has little effect on the resultant yarn tenacity.

4.2.11.5 Wet and dry processing, and use of an impact element

Water application to the filaments before their entry to the nozzle is now a well recognized requirement for better texturing. However, the use of impact elements in the form of a cylindrical bar, spherical element or flat plate, placed at the exit of the nozzle, are only recommended for particular applications, such as low linear density yarns or high speed operations. Figure 4.45 shows that the instability increases both with wet processing and with the use of an impact element of spherical shape. Such increases, provided that they are within acceptable limits, are indicative of improved yarn quality. The SEM photographs also support this argument in that dry textured yarns exhibit larger and fewer loops than wet textured yarns, as do wet textured yarns with respect to yarns textured with the impact element.

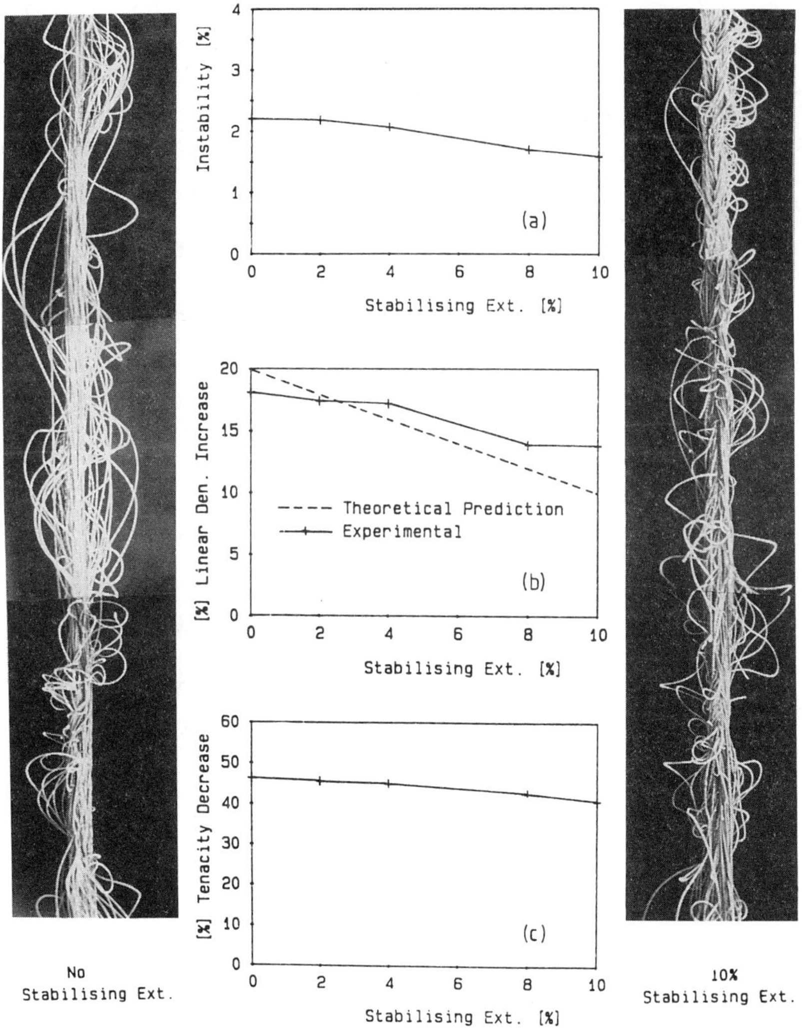

Figure 4.44 Effects of stabilizing ratio on properties of air-jet textured yarns

Although the use of an impact element is not essential, wetting of the supply yarn before the nozzle is a well-accepted requirement of the air-jet texturing process.

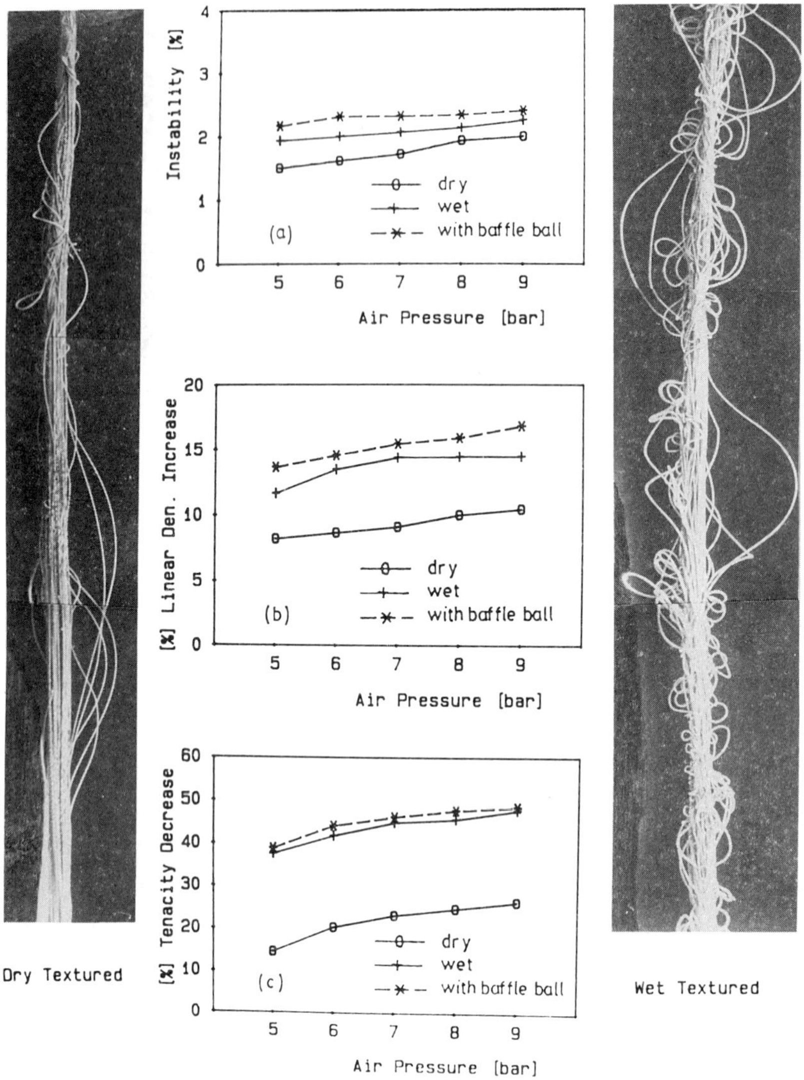

Figure 4.45 Effects of wet and dry processing, and use of impact element

Texturing is enhanced by the water application, because the overfed filaments are better entangled and surface loops are firmly fixed to the compact core of the yarn. This better entanglement causes only a small fraction of the linear density increase in the yarn to be lost under both the stabilizing and winding-up tensions (Figure

4.45). Dry texturing, however, yields poor texturing in that most of the loops and entanglements are removed under such tensions.

As seen in Figure 4.45, when the yarn is dry textured at 9 bar air pressure, the texturing is so poor that only about 26% decrease in tenacity is observed, thus indicating little change from the basic straight filament supply yarn. However, when it is wet textured at the same pressure, the decrease in tenacity, indicative of better texturing, goes up to about 48%. The same trends can be observed at other air pressures, and this improved effectiveness of texturing is due to reduced interfilament friction.

The deployment of an impact element does not seem to make any further significant improvement of wet textured yarns. When the linear densities of such textured yarns are compared, it is concluded that the impact element contributes little to the process, and the lack of a further significant reduction in the tenacity because of the use of impact element adds further evidence to support this conclusion.

4.2.11.6 Nozzle type

As far as the instabilities of the textured yarns are concerned, the T100 and T341 HemaJets produce similar yarns. The Standard-Core HemaJet, produces yarns with reduced instability whereas the yarns textured by the Taslan Type XIV nozzle possess the highest instability (Figure 4.46). Nevertheless, the instabilities of the yarns produced by the four nozzles considered are in the same close range, particularly at high pressures. Figure 4.46 also includes the typical SEM photographs of two yarns textured by T341 HemaJet and Taslan Type XIV nozzles. The distinctive nature of the yarn textured by the Taslan XIV is that the large loops and arcs are dominant on the surface of the yarn, whereas the yarns textured by the HemaJets appear to possess greater numbers of smaller sized loops.

Figure 4.46, depicting similar instabilities for the yarns textured by the T100 and T341 HemaJets, suggests that these two yarns are of the similar quality. However, when their linear densities and tenacities are compared, it will be appreciated that the two yarns produced by two different nozzles have different characteristics.

Figure 4.46 illustrates that the yarns textured by the T341 HemaJet attain the highest linear density corresponding to an increase of about 17%. This clearly indicates that the overfed filaments are so effectively entangled that even a stabilizing extension of 4% had little adverse effect, and this satisfactory texturing could be attributed to the higher air velocities achievable by this nozzle at the cost of a greater compressed air consumption than the other nozzles considered. This argument is also supported by the result for the Taslan

XIV nozzle, which possesses the second highest flow velocities. The strength test results indicate that the T341 HemaJet, which produces the highest air velocity, displays the greatest decrease in tenacity, approximately 50% as seen in Figure 4.46. From Figure 4.46, it will also be noticed that the nozzles

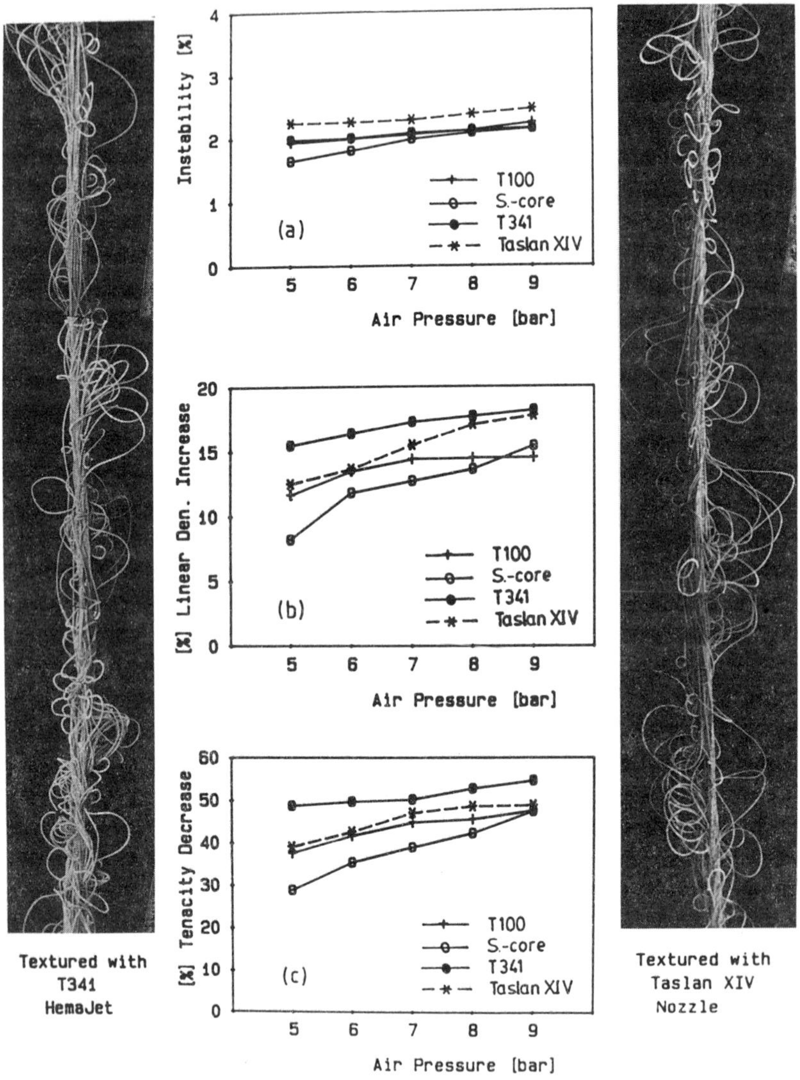

Figure 4.46 Effects of nozzle type on properties of air-jet textured yarns

considered show similar trends as far as the linear densities and strengths of the yarns are concerned.

4.2.12 Relations between Supply Yarn and Textured Yarn Properties

The effects of process parameters and supply yarn properties on the properties of air-jet textured yarns (which include the final linear density or linear density increase; strength-related properties, such as breaking elongation, load-elongation characteristics, and tenacity; instability; loop size, and frequency; bulk; boiling-water shrinkage; unwinding performance from a package; and cover features in a woven or knitted fabric) are probably the most investigated aspect of the air-jet texturing process.

4.2.12.1 Supply yarn material

Two different polymer yarns (i.e., polyester and polyamide) were used. The filament linear density and the number of filaments, and, hence, the total linear density of these yarns were kept as close as possible, although there was no control over their cross-sectional shapes and the applied spin finishes. They were textured under reference conditions but at varying air pressures.

Figure 4.47 indicates that the polyamide yarn exhibits lower instability, slightly lower linear density increase, and lower percentage tenacity decrease indicating that the polyester filament yarn is more suitable for air-jet texturing, although both can be textured effectively. This is also manifested in the presence of more undesired parallel filaments in the core of the polyamide yarn as seen from the SEM photographs. However, it could be argued that the difference in texturing quality is caused by differences in spin finish and filament cross-section.

4.2.12.2 Filament linear density

Supply yarns with approximately same number of filaments but with varying filament linear densities were tested to analyze how this affected the textured yarn properties.

Acar et al[71] have shown that the total drag force on the filaments is dependent on the surface and projected areas of the filaments in the air flow, and greater drag force acts on filaments with increased areas. It is also known that bending and torsional stiffnesses are directly proportional to the second moment of area about a diameter and to the polar second moment of area, respectively. Therefore, it was concluded that the smaller the second moments of areas, the smaller the forces and torques required to bend and twist the filament, respectively.

The findings and conclusions of Acar et al showed that, for an equal total yarn linear density, yarns consisting of finer filaments require smaller fluid forces to displace and entangle them than those consisting of coarser filaments. This is due to the reduced bending and twisting rigidities, and increased surface and

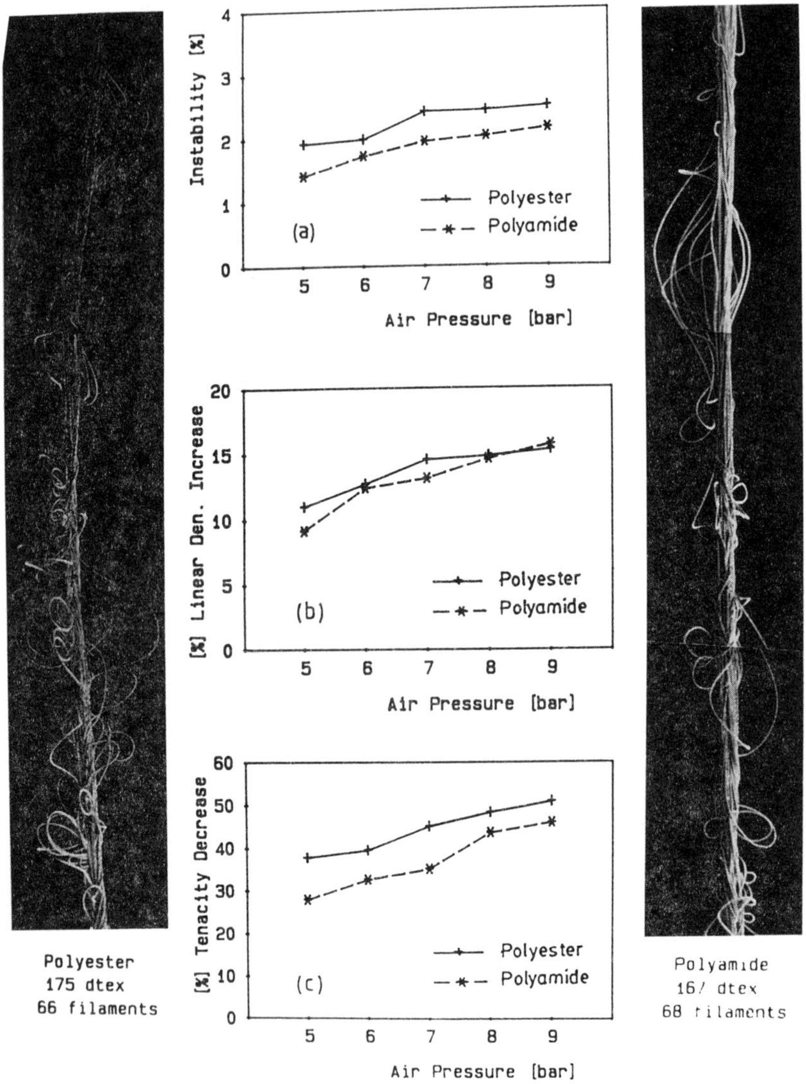

Figure 4.47 Effects of yarn material on properties of air-jet textured yarns

projected areas of finer filaments. Therefore, it could be anticipated that finer filaments will be better textured. This point is verified by Figure 4.48, which shows reductions in textured yarn linear density and in tenacity decrease with

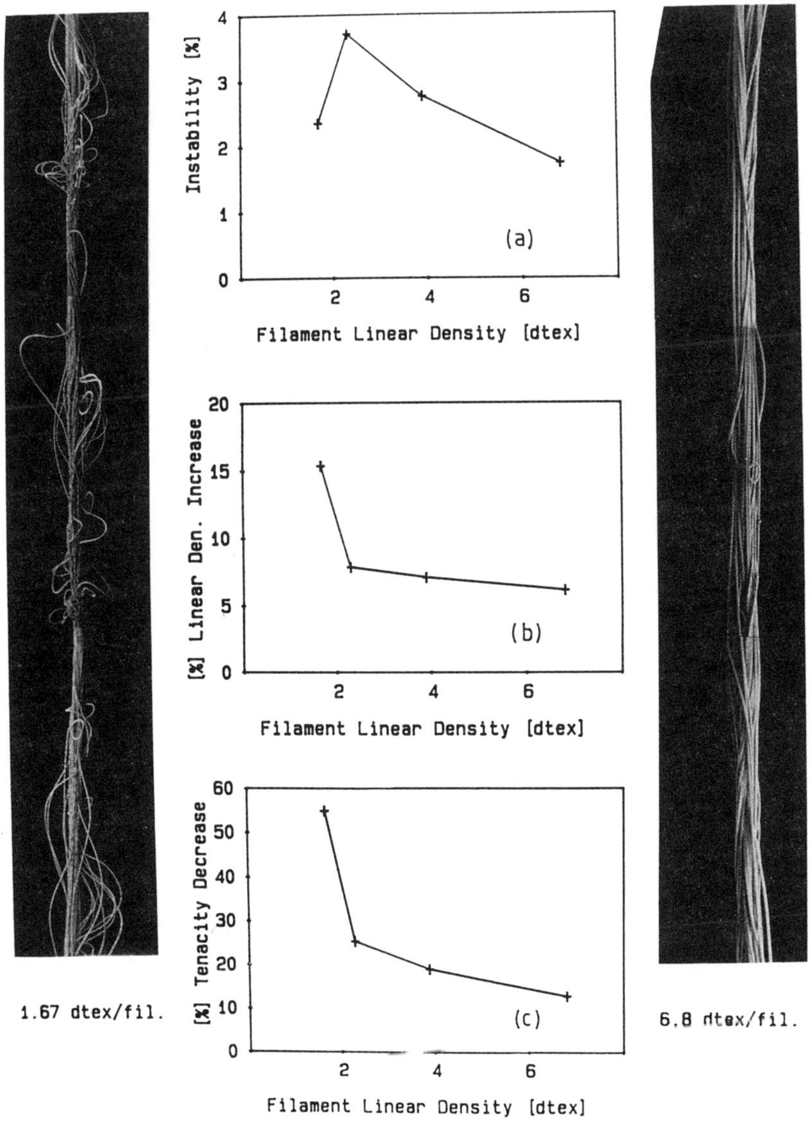

Figure 4.48 Effects of filament linear density on properties of air-jet textured yarns

increasing filament linear density, thereby indicating that finer filaments produce better textured yarns.

Conversely, the instability curve (Figure 4.48) indicates an increase with the increasing linear density up to approximately 2 dtex and then displays a reduction. The low instability at the fine filament end of the range may be because of the enhanced texturing effect, which may give rise to the interfilament friction that holds the entangled filaments and loops together under the applied loads. As the filaments get coarser the entanglement and loop formation deteriorate producing yarn with fewer loops and a poorly entangled core, resulting in a reduction in the yarn instability. This indicates that improved yarn stability can be obtained together with improved loop and entanglement formation, if the individual filament linear density is kept below a certain value. For the particular conditions considered, the most suitable supply yarns for air-jet texturing should comprise filaments finer than 2 dtex linear density, but this could, of course, be different with other process and supply yarn parameters and other texturing nozzles.

The better texturing obtained with finer filaments is also evidenced by the SEM photographs (Figure 4.48), where 1.67 dtex per filament yarn displays better entangled core and surface loops, whereas 6.8 dtex per filament yarn has only mingled and disorientated filaments with virtually no loops present.

4.2.12.3 Total yarn linear density

As the number of filaments increases, an enhancement in yarn quality is rightly expected because potential for filament entanglement is increased. However, the experiments with different types of nozzles have revealed that the receding statement holds only for the number of filaments which a particular design of nozzle can texture effectively at given process conditions. When this optimum number of filaments is exceeded a detorioration in yarn quality is observed for two reasons: (1) the air flow is progressively more disturbed by the presence of increasing numbers of filaments, as is also observed in the case of high overfeeds; and (2) the increased number of loops, arising from the presence of more filaments, increases the likelihood of the loops to be pulled out under applied tension.

The number of filaments of the yarn to be textured should always be remembered, together with other process and supply yarn parameters, before a suitable nozzle is chosen from the range available.

4.2.12.4 Fluid forces and effects of filament cross-section

In the air-jet texturing process the forces acting on the filaments are the primary flow forces, F_p, the secondary flow forces, F_s, and the frictional forces, F, as shown in Figure 4.49. Fluid forces resulting from the primary and secondary air flows in a texturing nozzle, as well as the frictional forces, all affect the resultant force acting on the filaments. In particular, frictional forces play an important role in the texturing process. In the following analysis of the fluid forces, only primary flow effects will be considered, noting that the approach also applies to the secondary flow.

4.2.12.4.1 Fluid forces acting on the filaments

In order to analyse the fluid forces acting on the filaments, a single fiber in an air flow is considered as shown in Figure 4.49.[48] This fiber is assumed to be rigid, its motion is two-dimensional (i.e., parallel to the *xy* plane). The resultant forces and moments on the fiber are caused by the relative velocity between the fiber and the surrounding air flow. Velocity components of its center of mass, G, are u_G and v_G, and V_n and V_t are the relative normal and tangential velocity components, respectively, at a point on the fiber, where *u* and *v* are the velocity components of the air flow at the point on the fiber, and *r* is the distance from point G to the point on the fiber. Hence, with reference to Figure 4.50, V_n and V_t can be calculated as follows:

$$V_n = (u - u_G)\sin\phi + (v - v_G)\cos\phi + r\phi'$$
$$V_t = (u - u_G)\cos\phi + (v - v_G)\sin\phi$$

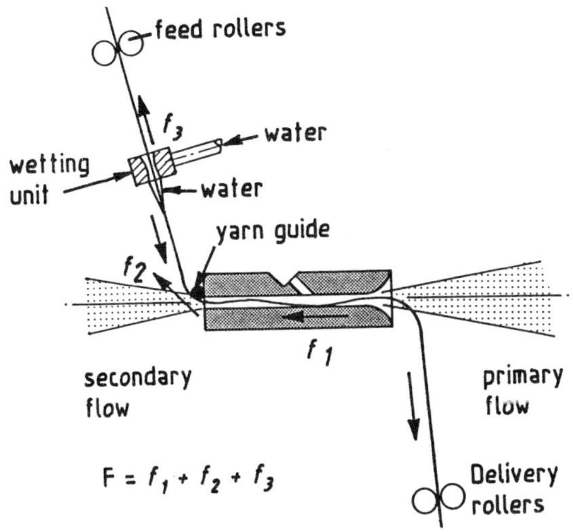

Figure 4.49 Fluid forces resulting from primary and secondary air flows, and the frictional forces acting on filaments

Note that V_n and V_t vary from point to point along the fiber because of the spatial variation of u and v, and because of r.

Fluid forces acting on the filaments can be expressed in terms of the pressure and friction drag forces. The pressure drag, D_p, is caused by the pressure difference between the two sides of the filament in a cross flow (V_n).

Hence, the fluid forces in the x and y directions are

$$D_x = D_p \sin\phi + D_f \cos\phi$$
$$D_y = D_p \cos\phi + D_f \sin\phi$$

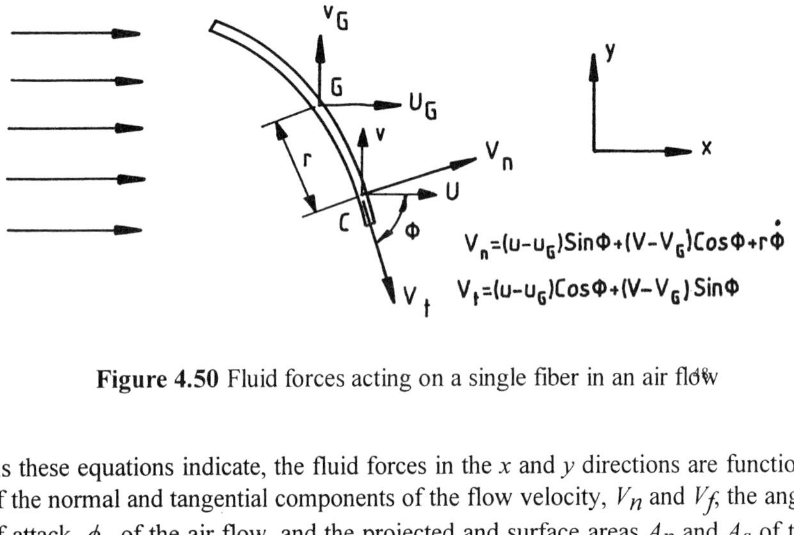

Figure 4.50 Fluid forces acting on a single fiber in an air flow

As these equations indicate, the fluid forces in the x and y directions are functions of the normal and tangential components of the flow velocity, V_n and V_f, the angle of attack, ϕ , of the air flow, and the projected and surface areas A_p and A_s of the fiber within the flow.

The resultant force acting on a filament in an air-jet texturing nozzle is determined by the fluid forces acting on the segment of the filament within the nozzle and by the frictional forces due to contacts it makes with the nozzle internal surfaces, f_1 , with the yarns guides, f_2, and with the wetting unit, f_3 (See Figure 4.49).

Because the primary flow is stronger than the secondary flow because of the greater momentum transferred to it from the incoming jets because of the geometry of their inlet bores, the resultant fluid force acting on the filaments is in the direction of the primary flow, i.e., $F_p > F_s$. Therefore, the resultant force acting on a filament is

$$F_r = F_p - F_s - F$$

4.2.12.4.2 Factors effecting fluid forces

The drag force acting on a filament in a given flow conditions is a function of the local air velocity, and of the filament surface and the projected areas. The air velocity is mainly determined by the air supply pressure, whereas the velocity gradient is determined by the design of the nozzle. The surface and projected areas are determined by the diameter (for a circular filament), length of the filament exposed to the air flow, and by the position of the filament across the nozzle.

Filament position. The surface area of a filament is determined by the length of the filament exposed to the flow, and the projected area depends on where the filament is in the nozzle at any instant. Although the surface area of a filament subject to the flow in the nozzle does not vary significantly with its position, its projected area does vary considerably. A filament within the upper part of the nozzle has a greater projected area than a filament within the lower part and, therefore, is subject to higher fluid forces. Thus the position of the filament in the nozzle has a significant effect on the fluid forces acting on the filaments.

Therefore, it can be argued that forces acting on the filaments vary considerably with the varying position of them in the nozzle owing to the variations in the local flow velocity and to the varying projected areas subject to pressure drag. Consequently, these varying fluid forces acting on the filaments at different positions in the nozzle at any instant cause them to travel at different speeds and hence to be displaced longitudinally relative to each other.

Filament diameter. Filaments with larger diameters will cause the drag forces to rise because of the increase in the projected and the surface areas of the filaments, i.e., F_r is proportional to d_f. Conversely, the mass flow of the filament ($\dot{m}_f$), which the fluid forces have to transport, will increase as a function of the filament cross-sectional area. Therefore, the inertial resistance of the filaments to the fluid forces, i.e., the rate of change of momentum required to move these filaments $\dot{m}_f V_f = (\pi/4)\rho_f d_f^2 V_f^2$ is a function of the filament cross-sectional area, i.e., proportional to d_f^2. Therefore an increase in filament diameter causes the fluid forces to increase as a function of filament diameter, and the force required to move this filament will increase as a function of the square of the filament diameter. Consequently, an increase in the filament diameter will cause it to be blown out at a relatively slower speed and hence the capability of the nozzle to conditions will be reduced. Therefore, it can be concluded that supply yarns that are composed of finer filaments texture into better yarns than the coarser filament yarns.

Effects of the filament cross-section. Area-dependent mechanical properties of the filaments vary with the shape of the cross-sectional areas. This would have an effect on the forces required to deform the filaments when they make a right angle turn as they emerge from the nozzle. The fluid forces acting on the filaments would also vary because of the varying surface and the projected areas caused by different cross-sectional shapes. These may have an effect on the loop-formation process.

To analyze the effect of the shape of the filament cross-section on the loop formation, filaments with various cross-sectional shape but with the same cross-sectional area (i.e., same linear density) are considered. Hence, regardless of the cross-sectional shape of the filaments under consideration, the momentum required to move the filaments remains the same.

Effects on the drag forces. The total drag force acting on the filaments is dependent on the surface and projected areas of the filaments in the flow. Therefore, varying shapes of the filament cross-sections will affect the drag force.

Because an elliptic filament bends about its major diameter under a bending moment its projected area is $[n^{1/2}]$ times, and its surface area is $[(n^2 + 1)/2n]^{1/2}$ times larger than those of a circular cross-section of equal area, where n is the ratio of the major to minor diameters of an ellipse, i.e., $n > 1$. Therefore, the drag force acting on an elliptic filament will be larger than that acting on a circular filament of equal fineness. The higher the ratio of major to minor diameter, the greater becomes the drag forces. A similar argument is valid for a hollow cross-section.

Consequently, it can be argued that, from the viewpoint of drag forces, elliptic and hollow circular filaments are better suited for the air-jet texturing process than filaments with a solid circular cross-section.

Effects on the mechanical properties. During the texturing process, the filaments emerging from the nozzle make a right angled turn relative to the nozzle exit. Figure 4.51 shows a model which represents a single filament emerging from the nozzle during the air-jet texturing process which assumes a circular arc outside the nozzle while it makes the right angled turn. The leading end (A) of the filament is instantaneously fixed in the textured yarn while the fluid forces are assumed to be acting at the trailing end (B) to cause the filament to be blown out of the nozzle. This filament may also be subject to a twisting action, as was observed during the process, due to the swirling effect of the flow, and therefore it is subject to a bending moment and a torque. These will cause instantaneous vertical and horizontal deflection of point B as well as a torsional deflection of the filament.

It can, therefore, be concluded that the deflection of a filament with increased surface and projected areas and reduced area moment of inertias, such as an elliptical cross section, requires smaller axial forces, bending moments, and torsional moments (torque) than those required to deflect filaments with a solid circular cross-section of equal fineness. In addition, fluid forces acting on such filaments become greater because of the increased projected and surface areas. Therefore, from a consideration of both fluid forces and mechanical deformation, elliptical filaments should be more suitable for the air jet-texturing process than are solid circular filaments.

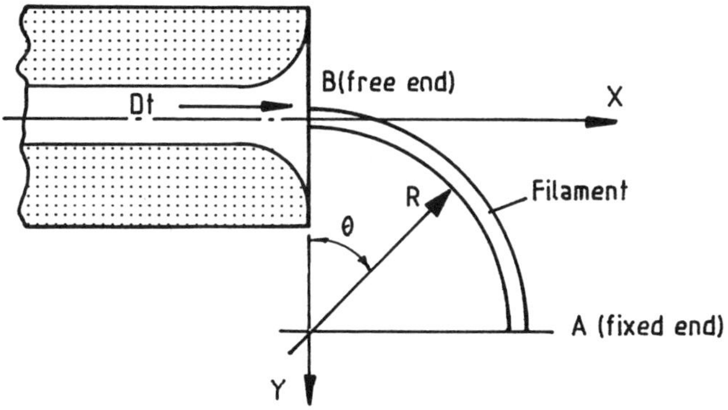

Figure 4.51 Model representing a single filament emerging from he nozzle during air-jet texturing process

In contrast to the elliptical filaments, forces required to deform the hollow circular filaments become higher, whereas the fluid forces acting on them increase when compared with solid circular filaments of equal fineness.Conversely, the hollow circular structure of these filaments might easily collapse under fluid forces when they make the right angled turn as they emerge from the texturing nozzle. This may alter the area-dependent mechanical properties of such filaments and hence reduce their resistance to mechanical deformation. It will also cause a further increase in the fluid forces acting on them due to the increased projected area caused by collapsed structure. Additionally, it can be argued that an advantage of hollow filaments in air-jet texturing could be the increased bulk of the textured yarns due to the hollow structure of the filaments.

4.2.13 Process-Monitoring Techniques

4.2.13.1 Yarn sensors and cutters

To provide production control, the yarn sensors and cutters are essential elements of yarn manufacturing. The yarn sensor is usually positioned closest to the wind-up, whereas the yarn cutter is placed closest to the supply package. If, for any reason, the yarn breaks, the yarn sensor either feels the low tension or the absence of the yarn is sensed. As soon as a yarn breakage is recorded, the yarn sensor activates the yarn cutter and the yarn is cut in the vicinity of the supply package. By this, both the wrapping of the yarn to the rollers, and the yarn waste is eliminated. The yarn sensor also issues a light or sound signal to warn the operator to attend this position immediately.

With the development of electronics, the end-down signals are centrally collected and the production efficiency is, thus, calculated. The central processing of this information also gives a possibility to diagnose any regular breaks; therefore, any faulty machine element as well as the work effectiveness of the operator may be obtained. Yarn sensor and cutters that are used in any yarn manufacturing operation can be also used on the air-jet texturing process too.

4.2.13.2 Tension in the feed zone

Feeding zone is the section that is between the feeding rollers and the nozzle. Apart from the drag force of the air flow on the yarn in the nozzle, no other source of driving force exists in this zone. And the only resistance force is the friction between the yarn and both the yarn guides and the nozzle. The resistance force is reduced by applying water at this zone. Therefore, it could be said that the more powerful the flow is, the higher the tension in this zone becomes. But does this not mean the more effective the nozzle? Because the drag force is a squared function of the air velocity, a powerful air drag means more air consumption, consequently it results in a less effective nozzle or process. Furthermore, the tension in feeding zone is insensitive to other process parameters, such as texturing speed and overfeed.

4.2.13.3 Tension in the texturing zone

Similar to the false twist texturing process, the yarn tension after the jet gives an accurate account of the process effectiveness. Any contamination of the jet, or any change in the qualities of the textured yarn may be identified as a change in the yarn tension in the texturing zone.

Although this fact has been known for many years, until recently a reliable process monitoring technique has been developed. This was mainly due to the lack of appropriate sensor to record the variations in the tension in this zone. Recently, Heberlein Maschinenfabrik A.G. has developed a sensor as well as recording electronics and evaluating software for this purpose. In principle, this system continuously records the tension and calculates the average over a period of time. If and when any long term deviation, or repeated short-term deviation from the average value is recorded, then either the operator is signaled to attend the position, or, if the situation is serious, the yarn cutter is activated and the position is stopped.

4.2.13.4 Tension in the stabilising zone

Because the amount of stabilizing extension is always kept constant, a high average tension in this zone may mean more stable loop formation and, in turn, a better texturing. This result can be acceptable if the necessary extra length of filaments (overfeed) to form into loops exist. Otherwise, however high the tension may be, it does not mean a better texturing. Therefore, observing the average tension in this section at varying overfeed does not provide enough information about the quality of the yarn.Conversely, by keeping the overfeed constant, the effect of the other texturing parameters such as air pressure, texturing speed can easily be observed and the quality of yarn can be assessed.

Monitoring tension in the stabilizing zone is, therefore, a possible means of quality control and it is widely used industrial method. Nevertheless, it cannot be deployed as a universal quality control, criteria due to the adverse results with varying overfeed ratios.

4.2.13.5 Instability tests

Two test methods have apparently been used in industry. They both depend on hanging weights, either at one end of a single yarn or a hank of yarn, but they have basic differences in their method of calculation as a consequence of the different stability concepts involved. Methods based on other techniques for measuring the yarn stability have also been described and used in various research works, these being mainly based on load-elongation curves obtained by tensile tests or on stress-strain curves obtained by a strainometer.

Weight hanging methods

Du Pont method. The Du Pont Company defined a stability test for Taslan yarns[13]. Its purpose is to determine the permanent increase in length of a textured yarn after applying a load and removing it after a certain time is elapsed, this being an indication of the stability of the texturing effect. For this purpose, a simple stability tester is described comprising a vertical board with a toggle clamp at the top on which to hang the textured yarn. At a distance of 100 cm below is a marking notch and beneath this is a centimeter scale. Provision is made for hanging weights on the specimen yarn by means of a weight hanger.

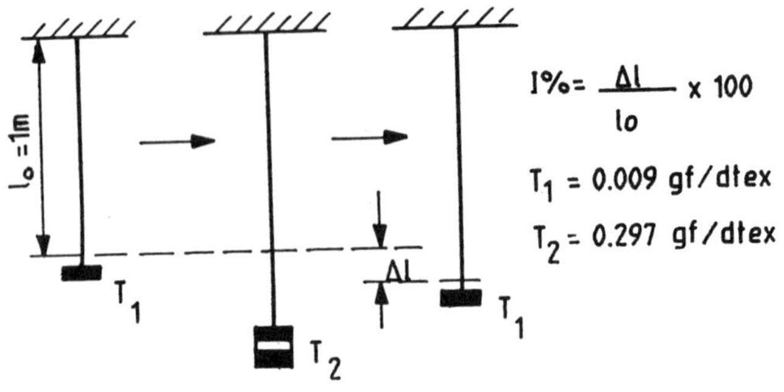

Figure 4.52 Du Pont instability measuring test

The procedure is illustrated in Figure 4.52 and can be described as follows: A weight of approximately 0.01 gf/denier (0.009 gf/dtex), W_1 is hung at the end of the yarn, and this weight is left on the specimen throughout the test. A 100 cm section on thus tensioned specimen is marked. The specimen is then subjected to a further load of 0.33 gf/denier (0.297 gf/dtex), W_2 (based on the textured yarn linear density), for 30 sec. Using the 100 cm mark as datum, the permanent elongation in the length of the specimen is measured 30 sec after the load W_2 has been removed. This percentage elongation is taken as the direct reading of the "stability" Du Pont suggest that a satisfactory textured yarn must have a stability of less than 5%.[13]

As can be seen from the test procedure, a 100 cm long single yarn specimen is used to measure the permanent elongation in the yarn. The applied load, W_2, is 0.297 gf/dtex and it is applied for 30 sec. Although it is not explained why a load of 0.297 gf/dtex is applied and why it is applied for 30 sec it appears reasonable to use a 100 cm specimen length to facilitate a direct percentage reading.

Heberlein method. In another weight-hanging test method,[70] suggested by the Heberlein Company, a hank of textured yarn is used instead of a single-yarn specimen. The yarn is wrapped on a reel of a 100 cm circumference to form a small hank of approximately 2500 dtex, the number of wraps being

$$\text{Number of Wraps} = \frac{2500\ (\text{dtex})}{2 \times \text{supply yarn linear density (dtex)}}$$

(to the nearest whole number)

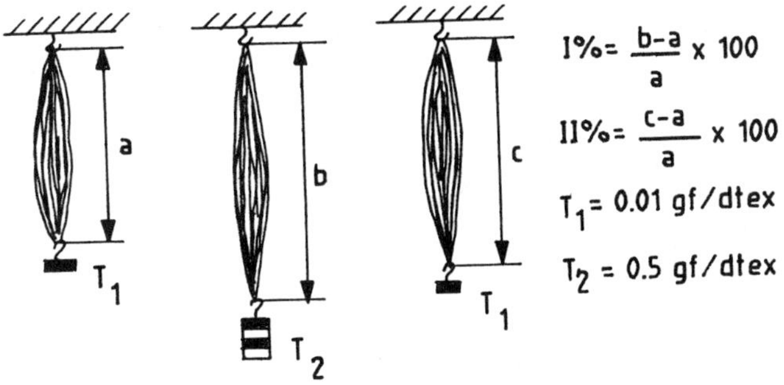

Figure 4.53 Heberlein instability measurement test

As indicated in Figure 4.53, the hank is first tensioned for 60 sec with approximately 25 cN load (corresponding to 0.01 cN/dtex (= 0.01 gf/dtex)), based on the linear density of the untextured yarn, and length *a* is measured. A load of 1250 cN (corresponding to 0.5 cN/dtex (= 0.5 gf/dtex)) is then substituted for this load and applied for 60 sec and the length *b* is recorded. 60 sec after removing this load, the 25 cN (25.48 gf) load is reapplied to the hank and, after a further 60 sec, the length *c* is measured. Two values are then calculated for instability:

$$\text{Insta bility I } (\%) = \frac{b-a}{a} X100$$

$$\text{Insta bility II } (\%) = \frac{c-a}{a} X100$$

Instability I measures the percentage elongation of the yarn while under a certain applied load, which is a different concept to the Du Pont measurement of permanent elongation after the load has been removed. It appears that Heberlein was somewhat unsure about the concept of instability and therefore suggested a second way to measure it. Hence, instability II measures the permanent elongation in the yarn after the applied load has been removed, similar to the Du Pont method.

It should be noted that, in contrast to the Du Pont method, the term instability is used by Heberlein, instead of stability following Wray's earlier suggestion. Wray[14] had stated in 1969 that: "The use of the word 'stability' for the values obtained by the Du Pont test is unfortunate since the lower the percentage value the more stable is the yarn. A more correct term for this measurement would have been 'instability' " (Wray,[14] p. 117)

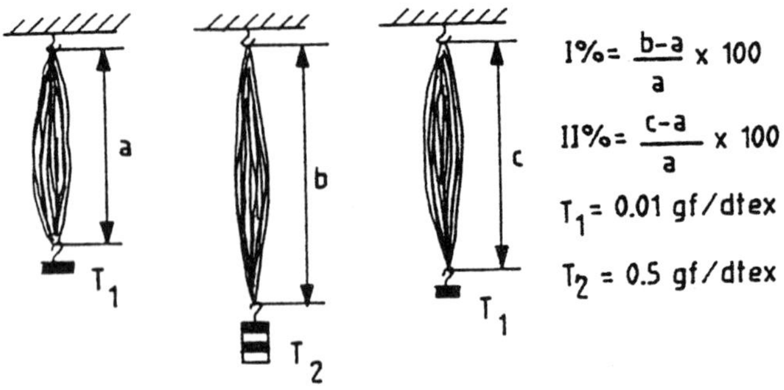

Figure 4.53 Heberlein instability measurement test

The term instability will, therefore, also be used throughout the rest of this study.

Other instability methods

Wray's methods. In 1965, during the early days of air-jet texturing when pretwisted yarns were used as supply material, Wray[14] observed that the textured yarn untwisted under the applied loads. He, therefore, claimed that the value of instability measured by any method that did not prevent untwisting included additional elongation due to removal of twist. Hence he devised an alternative method in which he made use of an Instron constant rate-of-elongation tensile tester, which did not suffer from losing yarn twist during testing. From the load elongation curve, as illustrated in Figure 4.54, he defined the instability of the yarn as follows:

$$\text{Instability (\%)} = e_t - e_p$$

where e_p and e_t are the percentage elongation of the supply and textured yarns, respectively, at a constant load of W = 0.297 gf/dtex (Figure 4.54), with this corresponding to the load used in the Du Pont tests.

This method of instability measurement was an attempt to eliminate the effect of the elastic deformation of straight, load carrying filaments of the textured yarn. Wray claimed that the percentage permanent elongation in the textured yarn was obtained by this method, by subtracting the percentage elongation of the supply yarn (e_p) from the percentage elongation of textured yarn (e_t). It was implied that the elongation of textured yarn under an applied load also included the elastic deformation of the constituent filaments, i.e., elongations of the individual filaments also contributed to the instability of the textured yarn.

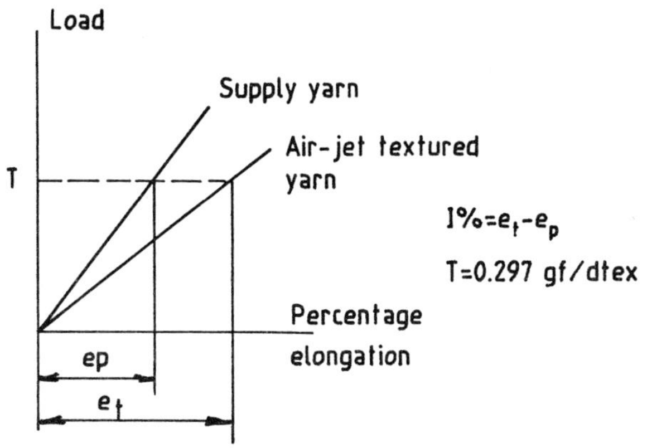

Figure 4.54 Wray's instability concept

Although it is true that the load bearing filaments will elongate under the applied loads (whereas loop forming filaments will not), it is not known what amount of elongation is caused by the pulling out of loops and what amount is caused by the elastic deformation of the individual filaments. Therefore, it is debatable whether the measurements obtained by this method reflect the permanent elongation of the textured yarn.

Wray[14] also devised a quicker method of measuring yarn instability based on principles similar to the Instron tensile tester method. This technique incorporated the use of a strainometer, which imposed a 5% constant strain on the yarn as it passed continuously over rollers. A third roller between these two rollers acted as a sensor to measure the tensions in the yarn, which enabled a calculation of the yarn instability to be made. Unfortunately, this method suffers from the same shortcomings as those described earlier.

Acar's method. Another suggestion for instability measurements of air-jet textured yarns was made by Acar.[60] This was also based on the load-elongation curves from a tensile testing machine of the Instron type. He measured the percentage elongation corresponding to loads of 0.01 gf/dtex and 0.5 gf/dtex, corresponding to the loads used in the Heberlein tests, as a measure of the yarn instability, see Figure 4.55 Hence,

$$\text{Instability (\%)} = \Delta l \,/\, \text{specimen length}$$

Acar's method suggests that the elongation of textured yarn under applied loads should be taken as the measure of yarn instability rather than the difference of elongations of the textured and supply yarns, as suggested by Wray, because the contribution of the extension of load-bearing parallel filaments of the textured yarn to the overall elongation is difficult to account for.

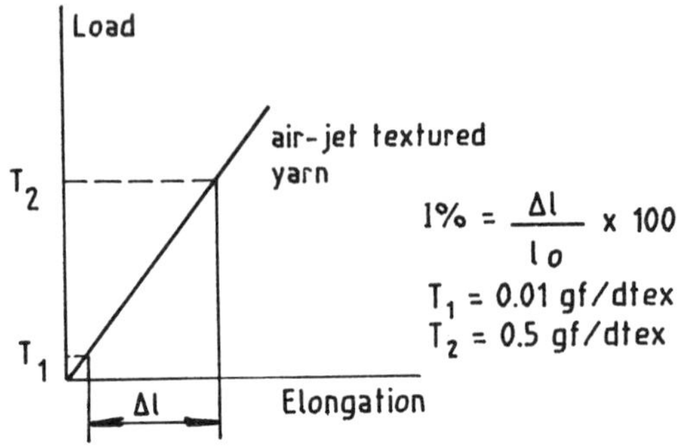

Figure 4.55 Acar's instability concept

Two contrasting yarn instability concepts. As seen in the preceding section, the instability methods used both in industry and by research workers vary considerably not only regarding detail but also regarding basic agreement as to what actually needs to be measured. Nevertheless, it is suggested that they can be grouped in the following two contrasting categories from the viewpoint of the concept of instability:

1. The concept of instability as a measure of the tendency of the textured yarn to elongate under maintained applied loads, thereby corresponding to the loads applied during later processes, such as weaving, knitting, and so on. Therefore, no attempt is made to account for the separate contributions of loop removal and elastic (recoverable) deformation to yarn elongation, because the percentage elongation under a maintained applied load is taken as a measure of instability, e.g., the Heberlein Instability Method I and Acar's method which use an Instron tensile testing machine.
2. The concept of instability as a measure of the lack of permanence of loops formed by the texturing process whereby the elastic (recoverable) deformation of the yarn under the applied load does not contribute to the yarn instability because the load is removed to allow the yarn to recover from such elongations before measurement occurs. Therefore, the percentage permanent elongation is taken as the measure of instability, e.g., Du Pont Method and the Heberlein Instability II method.

It is clear that no consensus of agreement exists regarding the basis for a standard method for instability measurement and inevitably the results reported in various research works do not agree with each other. Therefore, there is a need for a standard instability test for air-jet textured yarns. In the following sections the known instability methods will be critically analyzed, and to initiate discussion to the subject, a method based on one of the preceding two contrasting concepts will be suggested as a standard method.

Effects of test load and duration of applications. The different loads used for the Heberlein and Du Pont stability tests cannot both be representative of the applied tensions during further processing into fabrics. Therefore, a representative standard load should be decided on based on practical fabric forming process conditions and only this load should be used in the instability tests. This will require further investigation but the loads suggested by Heberlein have been used in the tests reported in this work.

Is the duration of application of load important? To answer this question an experiment was designed and carried out by using an Instron tensile tester which can provide very accurate readings. A constant test load of 110 cN (112 gf), equivalent to loads used in Heberlein instability tests (i.e., 0.5 gf/dtex), was applied to a specimen of the "standard" yarn for 15 minutes.

Figure 4.56 is a reproduction of the chart recording from such a test, illustrating that the application of a constantly maintained load for a certain period can have a slight effect on the extension of the yarn, especially during the early stages of the load application period. Nevertheless, the overall effect within the 15 min test duration was insignificant. The initial elongation of the yarn under the applied test load was 1.868%. The additional effects of applying a constantly maintained load for 30 and 60 sec durations were only to increase the elongations to 1.881 and 1.885% respectively and indeed total elongation for a 15 min application of this load was only 1.897%. It can be concluded that the constantly maintained application of the test load has a negligible effect on the

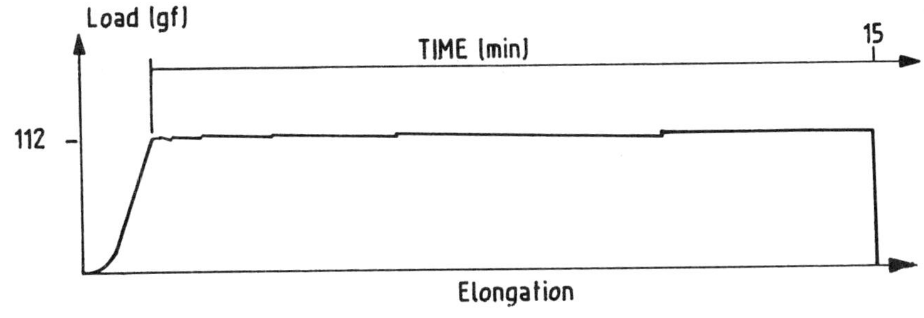

Figure 4.56 Load-elongation curve recorded for certain period under constant load

elongation of the yarn and hence on the instability test results. Consequently, the application of constant loads for any specified time periods can be eliminated.

Effects of the specimen type and length. Heberlein suggests the use of a hank of yarn in the tests, whereas Du Pont uses a 100 cm long single yarn. Does using the hank instead of a single yarn affect the test results? Does the specimen length have a significant effect? Some answers to these questions will be sought in this section.

First the effect of using a hank and a bundle of yarn instead of a single yarn is examined. By using the Instron machine and Acar's test method the following tests were performed:

1. With a single end of yarn fixed in the pneumatic jaws (Acar method).
2. With a hank of yarn hung over hooks fixed in the pneumatic jaws. Because the hank opened out under the applied load, it was not possible to carry out the test, when the two ends of the yarn in the hank were left free. Therefore, although this adverse effect is not made clear by Heberlein, two ends of the yarn had to be tied to make a closed loop hank for the tests to be undertaken and, thus, cause an equal distribution of the load to each constituent loop in the hank to be achieved. The test results show that this closed loop hank method (Heberlein I on Instron) gives slightly higher instability readings than those obtained by using a single yarn specimen (method 1).
3. With a bundle of yarns fixed in the pneumatic jaws. This creates a positive grip of each yarn within the bundle. Tests showed slightly higher instability than the other two methods, which can be attributed to the nonuniform distribution of load to every yarn in the bundle. Therefore, it was concluded that this is not a suitable method for yarn instability tests.

It can be argued that a single yarn specimen, such as method (1), is the most suitable basis for yarn instability tests because a single yarn is easier to handle and the question of nonuniformly distributed load through the bundle does not arise. The only alternative to this could be a hank of yarn with a closed loop (ends tied) hung on low friction hooks as in method (2) and it might be argued that the longer specimen length might give a better overall representation of the yarn characteristics.

Recommended test method. In the light of the work described and conclusions derived from it, the following basis for an improved test method is recommended.

1. A tensile tester of the Instron type should be used to measure the yarn elongation.

2. Single yarn specimens are adequate.
3. The specimen length should be sufficiently small to permit ease of handling: a 30 cm length is convenient for most machines and is, therefore, recommended.
4. Until an agreement is reached regarding the loads that should be applied, the Heberlein suggestion of a 0.5 gf/dtex test load should be used and a load of 0.01 gf/dtex should also be adequate to pretension the yarn.
5. The elongation under this test load should be taken as a measure of the instability (concept (1)) rather than the permanent elongation measured after the load is removed (concept (2)) because it represents the actual case that loads are constantly applied during other textile processes.

The procedure for this improved test would read as follows:

1. Prepare the tensile testing machine for the test. Recommended test conditions are as follows:
   a. Distance between pneumatic jaws, i.e., specimen length, 30 cm
   b. Cross-head speed, 2 cm/min
   c. Chart speed, 10 cm/min
2. Take approximately 50 cm length of textured yarn from a representative package. Use care not to damage the yarn during removal from the package.
3. Clamp both ends of the yarn in the pneumatic jaws. Use care not to overtension the specimen during clamping.
4. Operate the machine and record the elongation up to a value slightly greater than the load corresponding to 0.5 gf/dtex, based on the untextured yarn linear density.
5. From the recorded load-elongation chart calculate the percentage elongation between the loads 0.01 gf/dtex and 0.5 gf/dtex.

This percentage elongation gives the instability of the yarn. This procedure should be repeated for a reasonable number of randomly selected specimens, preferable from different packages of yarn and from different places of one package, to obtain reliable test results.

Test results using this specified method are shown in Figure 4.57, and they are compared with the results of all the other methods described earlier. Although they each yield different results, because of differences in the concepts and assumed parameters on which they are based, they all show the same basic trend, i.e., an increase of instability with increasing values of the air pressure used for texturing.

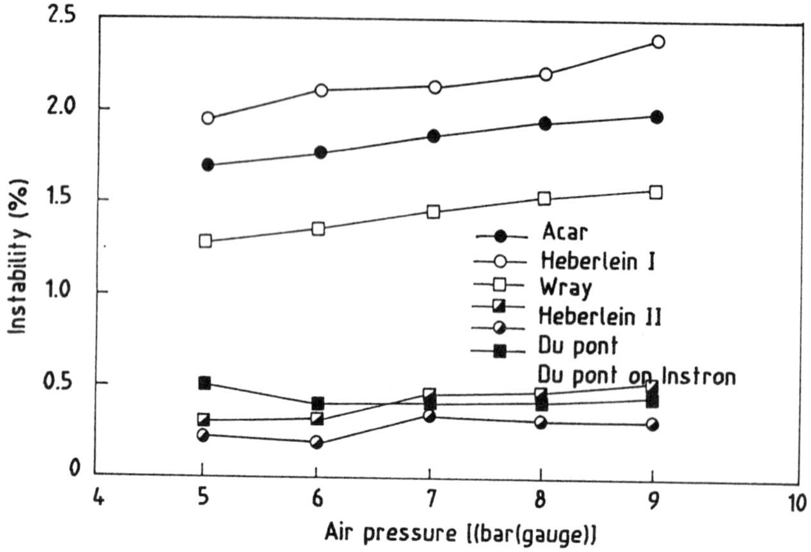

Figure 4.57 Comparison of results of all instability test methods

4.2.13.6 On-line yarn instability measurement

Provided that certain precautions are taken into consideration, the instability of air-jet textured yarns yields valuable information. But these "off-line" techniques are either too primitive, thereby giving unreliable, inconsistent results, or they are too sophisticated, thereby requiring expensive equipment, such as an Instron Tensile Testing machine. They are also tedious and time-consuming.

Using the fundamental principles of the suggested improved instability measurement technique, based on the use of an Instron machine, simple, and easy-to-manufacture "off-line" test device has been developed by Demir.

Like all other off-line instability measuring techniques the improved test is carried out after the texturing is completed, no direct feedback to the texturiser being provided to modify the processing conditions to give any desired texturing effect. Therefore, same instability measuring concept has been used to develop an on-line instability measuring system which offers the potential of providing the necessary feedback to the texturing machine.

Off-line instability measuring device. The fundamental concept of the suggested instability measuring technique consists of applying a certain load to the yarn and measuring the elongation of the yarn under this load. This basic requirement can simply be achieved by stretching the yarn between two clamps up to the specified tension and measuring the elongation of the yarn. To provide for this, a simple device was designed and built. This device consists of two movable clamps with a tension sensing head situated between them. It is referred as the *off-line* instability measuring device.

Figures 4.58a to c show the test results obtained from three different sets of yarns textured with varying amounts of air pressure, overfeed, and texturing speed. All the yarns were tested using both the Instron tensile testing machine and the off-line device, and the overall comparison of these two measurements shows a close matching.

It is suggested that this simple device is suitable for those texturing mills that do not possess a tensile testing machine and for those that require an easy and quick assessment of the yarn quality to provide feedback for the texturing machine operator to make any necessary process corrections.

Microcomputer controlled off-line instability tester. The inherent drawbacks of all conventional instability measuring methods, such as those using weight hanging techniques, also partly exist in the previously described off-line instability tester. These are mainly reading and handling errors, which can be reduced through a microcomputer-controlled system. Therefore, Demir designed and built a device with two yarn clamps, one of which is mounted on a slide, the other attached to a load cell.

Because both the elongation and tension are measured by the computer, errors are expected to be negligible, although the human yarn-handling errors can still exist. The system is less tedious than the previous hand-driven device; hence, more tests could be carried out in any given time.

The results of instability measurements carried out by this microcomputer-controlled instability tester are also included in Figure 4.58, which shows good agreement with the results of the other two test methods, i.e., the Instron method and hand-controlled off-line device.

On-line instability measuring system. It was considered that the instability measuring method would perhaps be implemented in a system, which would provide for instability measurements on running textured yarns. This on-line instability measuring technique was, in fact, achieved by stretching the yarn between two rotating rollers and recording the speeds of these rollers. Similar to the off-line instability measuring device, a tension sensor is required to monitor the tension in the yarn. Because the required tension in the yarn T_2 can only be created by increasing the speed of second roller with respect to the first, the roller speeds need to be adjustable. Because of variations in the yarn quality, the system requires continuous monitoring and feedback to the roller speed control, which will maintain a constant tension of T_2. This can only be achieved by a dedicated microcomputer, or a microprocessor control, together with the necessary hardware and software, as shown in Figure 4.59.

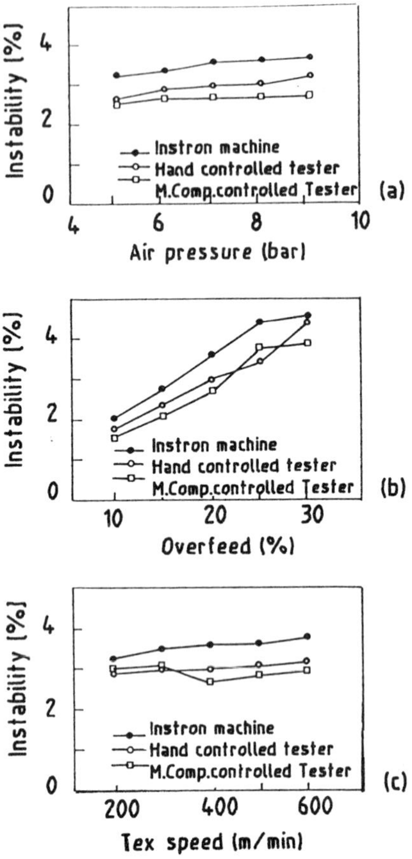

Figure 4.58 Comparison of the instability measurement results obtained with Instron machine, and manual and microprocessor-controlled off-line instability measuring devices

The results obtained from the on-line instability measuring system are compared with the results obtained from the Instron tensile testing machine. A good qualitative agreement with other methods has been observed showing the viability of the system.

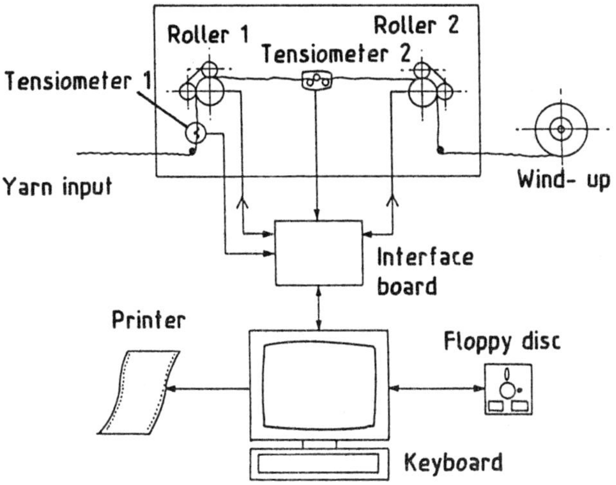

Figure 4.59 On-line instability measuring system

4.2.14 Modifications to the basic process

Three main types of operations are established to produce a range of yarns. These are

1. single-end texturing
2. parallel texturing
3. core-and-effect texturing

4.2.14.1 Parallel texturing

Single-end texturing denotes the situation whereby a single end is overfed to the texturing nozzle, and the resultant textured yarn is withdrawn. In the case of parallel texturing, two or more ends are fed at the same overfeed into the nozzle so as to facilitate a blend of different types of supply material, or of the same material but with different filament linear densities or number of filaments, or different cross-sectional shapes. Yarns produced by the parallel texturing have a reasonably uniform bulk as the overfeeds of all the component yarns are equal.

4.2.14.2 Core-effect texturing

The versatility of the air-jet texturing process is best observed by the core-and-effect operation in which one or more ends of yarns are fed into the nozzle at a low overfeed to form the core, whereas a second group of yarns is simultaneously fed to the same nozzle at a higher overfeed to create the desired bulk. By changing the overfeeds, a wide range of such yarns can be produced. Any of these three main yarn types can be modified by intermittently varying the overfeed conditions so as to cause periodic longitudinal variations in the bulking effect to create fancy effect yarns.

4.2.14.3 Free-fiber ends

Barmag offers its "Texspun" unit in which the yarn wraps round several rolls in succession so that projecting loops of the incoming yarn are sequeezed briefly by the outgoing yarn and thereby broken up (Figure 4.60). The resulting free fiber-ends give the yarn a still more spun-like appearance.

4.2.14.4 Slubbing techniques

Despite the extremely wide variety of these claimed applications, more and more extensions and modifications to the air-jet texturing process have been innovated. Sen and Wray[72,73] endeavored to create yarns similar to the air-jet textured yarns, but without using compressed air. Kollu[70] claims that if the supply yarn is directly fed from the package to the texturing nozzle without any mechanical drive, texturing speeds could be improved. With the alteration of core or effect yarn overfeed by means of devices called Zutex[74] and Fatex[75] varieties of other yarns, such as slub, nub, fancy, and flake yarns, can be produced. Air-jet textured yarns comprising a well defined core and surrounding loops and arcs already closely resemble spun staple yarns, however, by breaking some of the surface loops, this resemblance can be improved. A device to do this known as Texspun is reported[70,76] to be available for most existing machines. If false-twisted yarns are sequentially air textured, a worsted look and handle is said to be achieved. This process is patented under the trade name of Tusma[77] and Milpa[78] by the Teijin Co. of Japan.

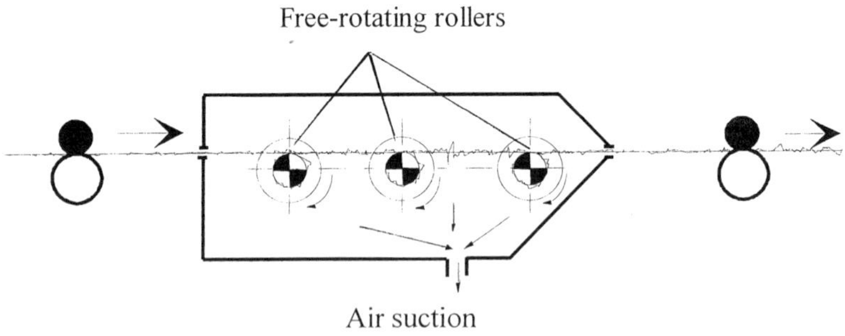

Figure 4.60 Texspun unit to create free-fiber ends

Heberlein has also developed a slub making device, which operates with a higy response servomotor controlled by a computer (Figure 4.61).

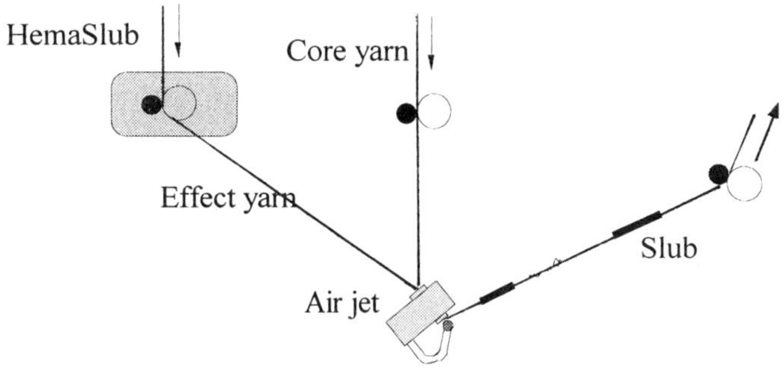

Figure 4.61 HemaSlub: Slub-creating device

4.2.14.5 Post texturing setting

Eltex/FR of Germany has developed a storage heater. Like a weft accumulator, it coils up the yarn and then takes it through a setting chamber. Dwell times up to 2 min are obtained.

4.2.15 Most Recent Advances and the Future of the Process

World textile fiber production is growing, mainly because of rising demands due to increasing population. Production trends show the market share of the natural fibers is decreasingand that of synthetic fibers is increasing,[16-18] and it is predicted to continue along these lines.[19] With the ability to simulate more closely the properties of the natural fiber yarns, the air-jet texturing process might be expected to encroach further on the natural fiber market, and also, depending on its economic competitiveness, to capture some portions of the market currently enjoyed by the false-twist and similar techniques. Therefore success of the air-jet texturing process in the textile industry will rely on reductions in conversion costs which could make it more economical when compared not only with spun yarns, but also with the yarns textured by other methods.

Air-jet textured yarns have a great potential for industrial exploitation on account of their unique characteristics including improved comfort, warmth, and dimensional stability in addition to their better wear resistance and easy-to-care properties. Research leading to a better understanding of the air texturing phenomenon could be worthwhile if it results in further technical improvements and reductions in conversion costs.

The predictions by Wilson[4,5,77] that the air-jet textured yarns will replace 20% of the present spun yarn market by the year 2000 and, furthermore, that they have the potential to replace another 20% of polyester filament yarns, which are textured today by the false-twist method, are considered by Wray and Acar[79] to be overly optimistic. Nevertheless, they also admit that the air-jet texturing is steadily increasing its market share owing to

1. Developments in nozzle design
2. Provision of wider ranges of suitable supply yarns
3. Increased awareness of its unique capabilities to simulate adequately spun yarns cheaply and to blend different types of yarns during processing

4.3 INTERMINGLING/INTERLACING

Interfiber friction is the main cohesive force that hold a bundle of staple fiber to form a yarn. Random distribution of various length of fibers and a small amount of twist applied to this fiber bundle increase the cohesive forces that cause the yarn to carry tension during both yarn and fabric-forming processes as well as during use be it wearing or an industrial application.

Multifilament continuous yarns, however, do not possess this essential cohesive force because of the purely parallel arrangement of the filaments. Although most texturing processes, the air-jet texturing process excepted, impart some sort of irregularity to the parallel arrangement of filaments to enhance hand and optical properties, the resultant textured yarn still lacks of interfilament cohesion. Consequently several difficulties are observed during winding and fabric-forming processes (i.e., knitting, weaving and tufting). Tape-like nature of both untextured and textured filament yarns cause filament snatch by the overlapping layers and creates tension irregularity during the unwinding process. The tension increase may even be so high as to cause filament breakages. The filaments of not very compact textured yarns are also reported to be trapped by the neighboring needles of fine-gauge knitting machines. Similar filament breakages could also occur in this case, if not complete halt of the knitting operation. In a similar way, filaments could be caught by the warp yarns while an uncompact textured weft yarn is traveling through the narrow warp shed, and create irregularities on a woven fabric.

Since the advent of synthetic filament yarns, the preceding difficulties have been faced by the textile industry, and naturally several solutions have also been applied to overcome. True twisting of the filament yarn is one of them, the others being sizing and intermingling. Although twisting and sizing have been employed for many years, the requirements of modern industry seem to favor intermingling solution because of the several disadvantages of twisting and sizing which is a treatment of yarns by a gelatinous film-forming substance to give compactness. If the yarn is twisted, an independent twisting machine is required. This means extra investment and operation costs. If the yarn is sized, on the other hand, again an additional yarn treatment is required and also the applied sizing material needs to be washed away from the final product. This also calls for an increase in the yarn cost. Cost comparison of intermingling process with sizing and twisting is given in Figure 4.62.

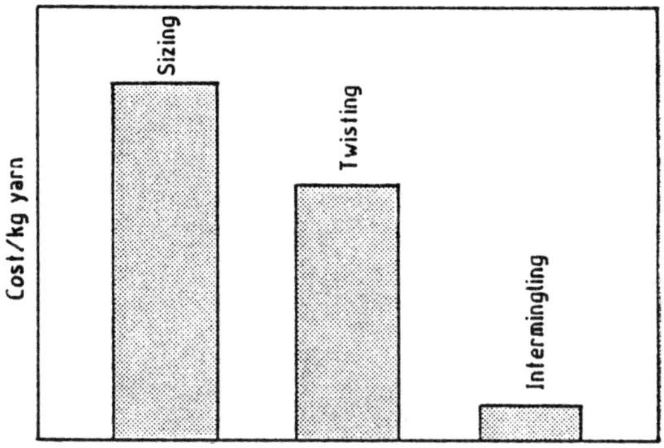

Figure 4.62 Cost-comparison of intermingling process with sizing and twisting

Cost-conscious modern industry needs a compacting process that will be carried out during the yarn production stage whether it be spinning, drawing, texturing or winding. The use of a cold air-jet appears to be an ideal solution, and the intermingling is expected to replace all other compacting or cohesion-imparting techniques in near future.

So what is intermingling? When a loose bundle of either flat or textured filaments are instantaneously subjected to the effect of a perpendicular turbulent cold air jet, the air flow will open up the filaments while at the immediate vicinity of the opened up section, the filaments will be intertwined and mingled with each other to form a compact section. If the yarn travels in the direction of its long axis while the perpendicular air jet is stationary, the opened-up and mingled sections follow each other because the jet fails to open up a section that has already been mingled. Thus a cohesion is imported on an originally lose yarn. No change is created in the basic filament structure both physically and chemically, though the filaments spatial position has been altered. Therefore the overall yarn properties do not change appreciably (i.e., a flat yarn is still flat and a stretch texture yarn is still stretchable). This process of imparting intermittent mingled and opened sections along the yarn by means of a cold air stream is called an intermingling process (Figure 4.63).

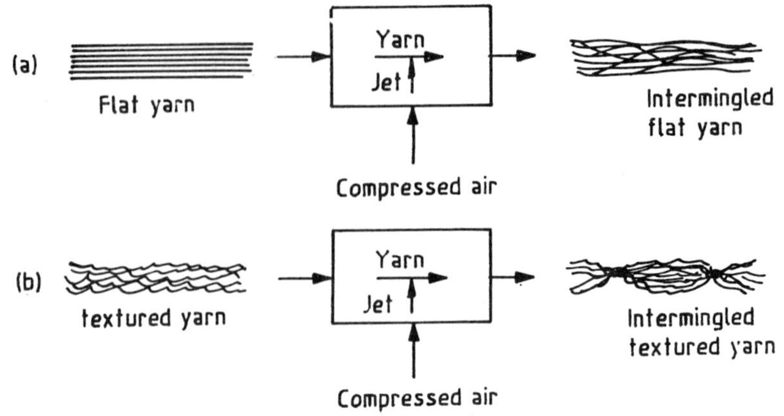

Figure 4.63 Principles of intermingling process

Synonyms of the term *intermingling* are tanglelacing, tangling, entanglement, interlacing, comingling. The latter, however, incorporates intermingling of two end yarn together to form a single intermingled yarn. Each mingled section created by the intermingling process is likewise referred to as "tanglelacing place, tanglelacing point, tanglelacing know, entanglement point, fixed point, interlace, interlacing point and hip."

4.3.1 Historical Development of Intermingling

The earliest traceable patent[80] on intermingling process is filed in 1961, about six years later than the first air-jet texturing patent,[81] again by Du Pont of the United States. Even this indicates that the intermingling process is somewhat a by-product of the air-jet texturing process, both use a cold air-jet to modify the physical yarn structure. In nearly three decades, the intermingling process has been developed by both industry and the research institutions irrespective of their involvement with the air-jet texturing process. Intermingling nozzles, being much simpler than the air-jet texturing nozzles, were even manufactured in house by spinning and false-twist texturing companies to suit their own purposes. Therefore, the intermingling process today enjoys even a wider application than the air-jet texturing process.

4.3.2 Application Fields

In most cases, the intermingling is an intermediate treatment of a continuous filament yarn. A temporary cohesion is imparted to the yarn and is expected to be removed by the applied tension in the final fabric form. This mode of application field is common to both flat and textured yarns. After spinning and drawing, a slight intermingling is given to the flat yarn to facilitate the easy winding up and unwinding.

Flat continuous filament yarns are intermingled in such a way that no visible nips are imparted and the amount of nips per meter yarn is low. This so called "continuous intermingling" is achieved by comparatively simple nozzle designs at low air pressures. It is mostly applied to the yarn during either spinning, drawing, or warping. Figure 4.64 illustrates these application fields.

Textured yarns (air-jet textured yarns excluded), however, are intermingled to form visible nips and open places. This is due to the stretch characteristics of the yarns. When the tension is low, the filaments tend to obtain the minimum energy level that is crimped form. Therefore the open sections become bulkier than the mingled sections (i.e., nips). By the application of a slight tension, the intermingled yarn is stretched, and the nips and open filaments become hardly visible. The mingling of the individual filaments greatly improve the winding-up and unwinding because the bundle of filaments are now cylindrical rather than flat.

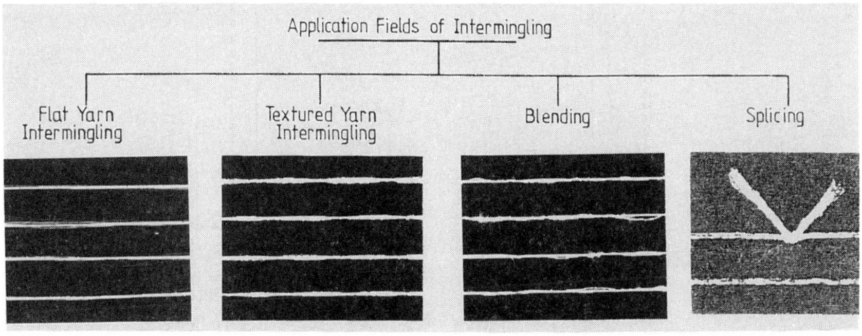

Figure 4.64 Application fields of intermingling

This new configuration of the filament bundle assists further textile processes (i.e., knitting or weaving), yet in the fabric form does not pose any complication.

On a false-twist texturing machine, the intermingling process is carried out by simply positioning an intermingling nozzle at the positions shown in Figure 4.65. Position 1 is a somehow special location and not widely applied. At this location, the intermingling is applied to the flat yarn to facilitate both heat transfer in the heater and twisting at the twisting unit. The position 2 is where the intermingling nozzles are mostly located. In the case of single-heater machines this is the only position (i.e., after intermediate rollers) which defines the texturing speed and before the take-up. However the twin heater machines have two options to position the intermingling nozzle (i.e., before or after the secondary heater). Both positions are widely used depending on the original machine design. In these positions, the intermingling nozzle operates to impart regular nips for the purposes already mentioned.

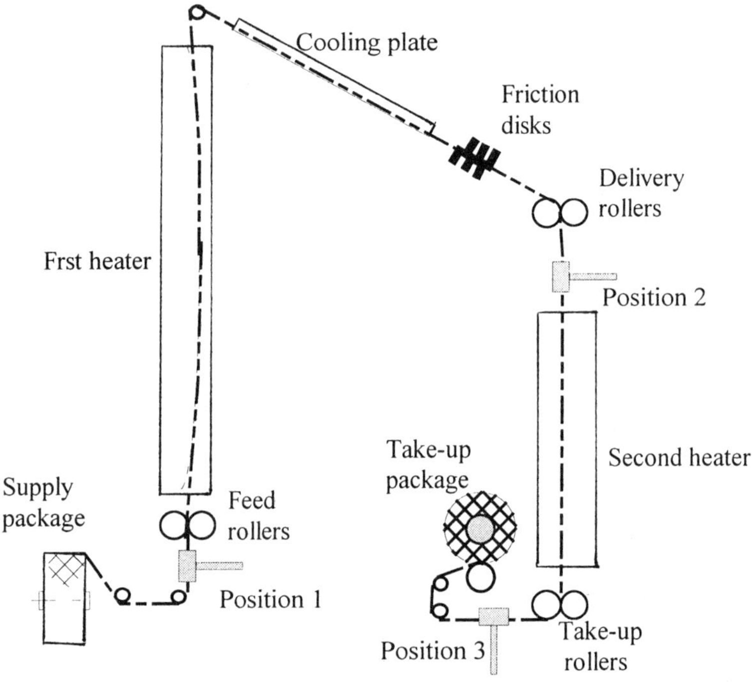

Figure 4.65 Positioning of intermingling nozzle on false-twist texturing machine

Two other application fields of the intermingling process are rather different from applications referred to so far. These are blending of continuous filament yarns with staple yarns and splicing.

The blending process has recently been developed by both intermingling nozzle manufacturing and texturing machine builders. The blending process enables natural fiber-spun yarns to be combined with synthetic filament yarns by an intermingling nozzle. All known materials, such as wool, cotton, viscose, polyester, nylon or silk, are claimed to be used as supply yarns. The blending process also offers the possibility of yarn combinations with elastomeric yarns more economically as a result of the high intermingling speed; 300 to 800 m/min production speeds can easily be achieved depending on the type of supply yarn used.

Another application field for intermingling process is the replacement technique for classical knotting. This alternative technique is known as "splicing." The disadvantages of knots, though indispensable, have been observed throughout the textile history. Knots during weaving exert a considerable strain on the neighboring ends. They also greatly affect the appearance of final fabric. Air splicing (i.e., joining two ends together by means of a nozzle similar to intermingling nozzles) appears to be a favorable alternative to know jointing. Although a knot is much thicker than the yarn, the splice is hardly distinguishable from the yarn itself. Therefore, it is obvious that such a piecing could be expected to produce fewer disturbances during processing. The visual appearance of the fabric is also less affected.

The only difference between the ordinary intermingling and the splicing is the the latter corresponds to a few (mostly one) nips of the former. That is to say the splicing is a discontinuous operation joining two ends together, whereas the intermingling process is a continuous compacting operation even in the case of comingling or blending.

Illustrations of these applications are given in Figure 4.66.

4.3.3 Intermingling Nozzles

In simplest terms, an intermingling nozzle consists of a yarn channel and a jet of air hitting the yarn that travels through the yarn channel. Therefore, a length of a cylindrical tube with a side air entry conforms with these requirements and indeed may perform the task in a limited sense. Because the first intermingling nozzle patent in early 1960's, the textile industry has been trying a variety of configurations to improve the effectiveness of the nozzles, even though very little fundamental scientific research has been carried out.

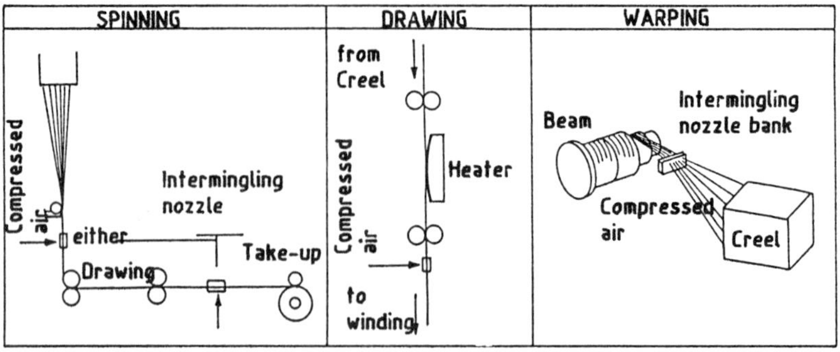

Figure 4.66 Applications of the intermingling process

The classification of this variety of nozzles could be made in several ways. Figure 4.67 makes the widest possible classification and illustrates the samples of each group. A certain nozzle, having various features, can easily be grouped under different titles.

Because the intermingling process is an on-line operation with other textile processes, such as texturing, winding, and so on, the threading of the yarn to the nozzle poses a great importance. For example, while the texturing is running at 600 m/min production rate, it is not wise nor economic to stop the operation, thread the yarn through the intermingling nozzle and restart again. Therefore an easy, slot in, threading facility is mostly required from an intermingling nozzle. So-called open nozzles with a slot-in threading facility, hence, seems to be preferred by most of the textile industry. Closed nozzles, which can only be threaded by means of a needle or by a yarn end, could also find application fields as in draw warping and detorqueing. The design of slot-in threading facility could vary from a thin slot in the wall of the yarn channel to spring loaded flaps closing the main channel.

Intermittent nature of intermingling imparted to the yarn rightly reminds of a pulsating air supply. But it appears that no successful nozzle has so far been developed to achieve this, though certain proposals was made. Consequently, it is obvious that the current technology only deploys a continuous compressed air to the nozzle and the yarn is subjected to continuous air-jet within the yarn channel.

To control the reflection of the incoming jet so that it may create an effective operation, several different cross-sectional shapes have been given to the main yarn channel. Although the choice of some of the shapes have some research backing, many others appear to be proposed conceptually. The cross-sectional

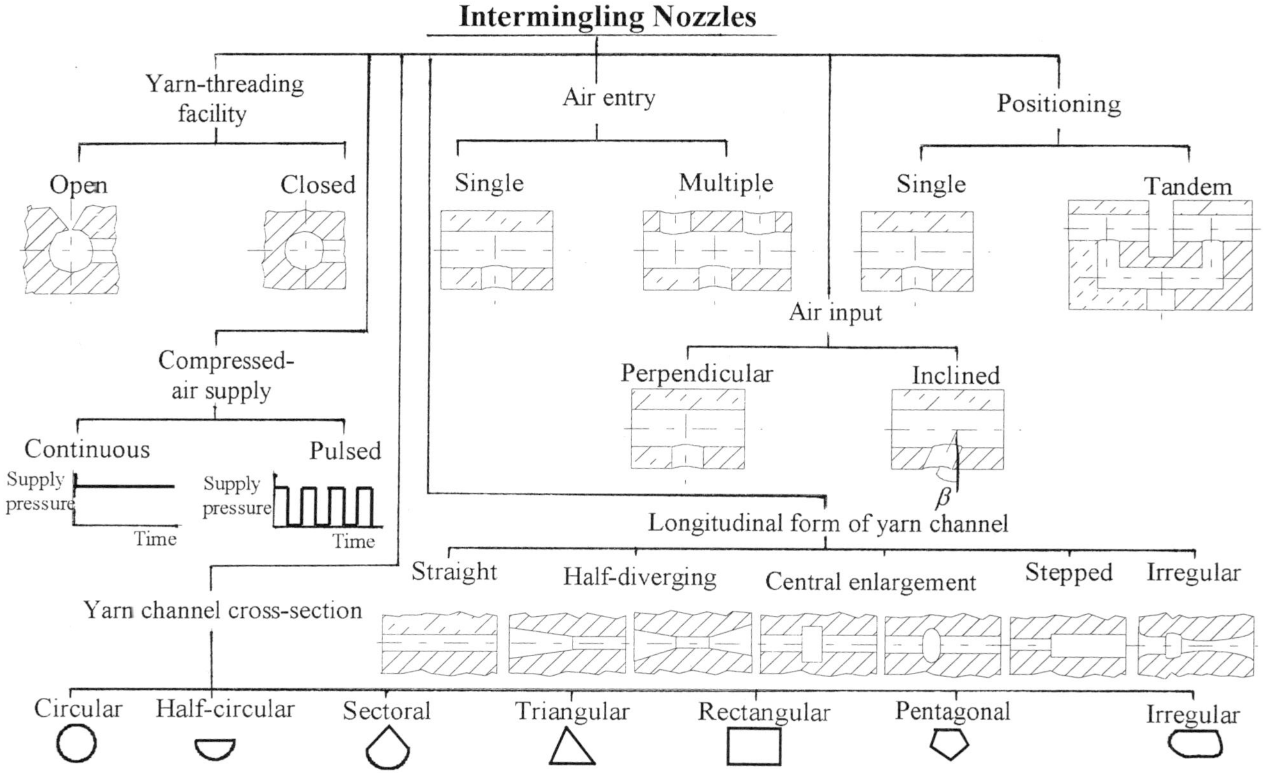

Figure 4.67 Classification of the intermingling nozzles

shapes vary from regular circular, semi-circular, quarter-circular, triangular, rectangular, pentagonal to irregular undefined shapes as seen in Figure 4.67. Conversely, the longitudinal shape of the yarn channel is not always straight. Most of the time the modifications in the channel is made to exhaust the air soon after it hits the filaments though no explanation for this has been offered. The yarn trajectory in these channels is not always straight either. Curved yarn path seems to be preferred by most of the nozzle manufacturers. This is achieved by either a curved channel or the positioning of yarn guides before and after the nozzle with respect to the nozzle itself.

Although most of the intermingling nozzle possess only one air inlet hole in the middle of the yarn channel, nozzles with multiple inlet holes either on the same circumferential line or on staggered positions. Because more holes mean increased air consumption, the tendency seems to be towards single inlet holes. The angle between the axis of the main channel and the axis of the inlet hole is almost always 90° because of the nonforwarding requirements of the intermingling nozzles. However, slight inclination toward the yarn entry side of the main channel is claimed to increase the nozzle effectiveness. This inclination angle could rarely be as high as 20°. The nonforwarding requirement of the intermingling nozzles stems from the add-on or on-line nature of the intermingling process. Usually a textile process such as texturing, winding operates on its own, and a certain yarn tension is required for it. When an intermingling nozzle is added on this line the tension should not be affected considerably for it affects the main process. Therefore, neither an increase nor decrease in yarn tension because of the intermingling nozzle is desired.

Above, all the working pressure range for the intermingling nozzles is usually 1 to 6 bar (gauge). The yarn speed these nozzles can process is up to 800 m/min, but the intermingling density and frequency decreases with increasing yarn speed. Although it has been reported that many nozzles require 1% to 3 % yarn overfeed to the nozzle, it is a means of tension control and by no means a necessity.

4.3.4 Nozzle Configuration

The basic nozzle configuration for the intermingling process is a length of cylindrical tube with a perpendicular air inlet hole in the middle. Most researchers used this configuration as a starting point and developed improved designs. The main concern of the researchers is the number of nips inserted in a meter of yarn and its uniformity. Therefore, many researchers evaluated the effects any nozzle parameters changed by simply the number of nips inserted.

The research carried out on the intermingling nozzles demonstrated that the following parameters influence the mingling process:

1. Area of the inlet hole
2. Area of the yarn channel
3. Cross-sectional shape of the channel
4. Angle between the inlet hole and the channel
5. Length of the yarn channel
6. Surface finish of the yarn channel
7. Longitudinal shape of the yarn channel
8. Positioning of the yarn guides before and after the nozzle.

4.3.5 Filament Behavior and Nip-Formation Mechanism

An understanding of nip-formation mechanism is the major prerequsite for the development of improved intermingling nozzles. This fact has been appreciated by many researchers and, thus, attracted great attention.

To establish a nip-formation mechanism, the visualisation of the filament behavior within the yarn channel and thus constructing an explanation for air flow-filament interaction is essential. Transparent nozzles either perspex or glass, or replacement of flat surfaces with transparent walls are the experimental objects mostly used. Because the yarn speed is usually as high as 600 m/min, the frequency of the nip formation is in the range of 800 to 1,200 Hz. This high frequency, by no means, can be observed by naked human eye. Therefore, high-speed viewing techniques are required. Single- shot high-speed photographic techniques as well as high speed cine filming techniques are necessarily used.

The observations and conclusions suggest that the filament are opened up by the incoming jet and due to the different levels and direction of air flow within the channel, a kind of plaiting or false-twisting action is occurring thus, the nips are formed at both sides of the jet. Because the yarn as a whole is moving perpendicular to the incoming jet the nip sections are also passing through the jet, but they cannot be opened up due to the mingling of the filaments. As soon as the mingled section passes a new process of opening the filaments and mingling thus starts.

4.3.6 Process Parameters and Their Effect on Mingling

Air pressure, yarn speed, and yarn tension are the process parameters for the intermingling operation, whereas the total yarn linear density (number of filaments), linear density per filament, cross-sectional shape of the filaments, yarn material and yarn surface properties are the yarn parameters. The air pressure range for the intermingling nozzles is usually lower than that of the air-jet texturing process, i.e., the working pressures for the intermingling process ranges from 0.5 to 5 bar, whereas the minimum pressure for the air-jet texturing process is 5 bar. Most of the researchers agree that up to about 3 bar pressure, the intermingling density increases with increasing air pressure. As the air pressure increased further, however, the nozzle performance as illustrated by the nip frequency either saturates or even deteriorates. As an explanation to this, Lünenschloss and Zilg[82] suggest the change in flow conditions from subsonic to supersonic. But further explanation given appears to be inaccurate and insufficient. The researchers claim that in the supersonic case a break-up in the jet occurs and therefore thus released kinetic energy is unusable for the intermingling. The break-up of a supersonic jet in the form of explosion as stated by the researchers imply a kind of pulsation that will never occur in a continuous flowing jet.

To conclude, it could be said that the results obtained using different nozzles are specific to the nozzles used and an extra attention is required to generalize the results to the rest of the nozzles.

When the wide ranging applications of the intermingling process is considered, the yarn speed through the nozzle varies from 0 (e.g., splicing) to 4,500 m/min (e.g., POY spinning). Therefore it is hardly possible to define a working yarn speed. Research reports illustrate some sort of contradiction between each other. Although Sparkes[83] and Lünenschloss and Zilg,[82] working in similar speed range (i.e., up to 800 m/min) show a deterioration in nip frequency or difficulty in mingling, Weinsdorfer,[84] working with some now higher speeds (i.e., up to 1,500 m/min) indicates an increase in nip frequency with increasing yarn speed. Weinsdorfer obtains this result from a round channelled nozzle, the results with a second nozzle with a semicircle cross-sectional shape, however, slightly differs (i.e., a decrease in nip frequency is observed with the speeds exceeding 1000 m/min). This contradiction could be arise from the speed ranges employed by different reserachers. However, it is more likely that different nozzle configurations and yarn properties result in different nip frequencies.

The nip formation by the influence of an air-jet is in fact due to the random lateral vibration of the individual filaments and hence randcm mingling of them. Therefore, the yarn tension i.e., filament tension as a special role in the process. When the tension is too high to let the filaments vibrate freely, the mingling fails to occur. Conversely, if the tension is too low, then the whole bundle of filament could easily slip out of the jets potential area causing again a failure of intermingling. This naturally calls for an optimum tension. But the research results reported by Weinsdorfer[84] and Lünenschloss and Zilg[82] unfortunately fail to provide a common optimum value. It can, therefore, be deduced that the optimum tension value is a function of the nozzle, yarn, and process conditions.

4.3.7 Yarn Test Methods and Instruments

Because the intermingling process only inserts nips and open places to a yarn, the number, uniformity and strength of the nips (i.e., mingled sections are the only criteria for quality assessment). Of these three criteria, the number of nips per meter yarn (i.e., the nip frequency is the most used one) and many ways of determining this quality have been developed. The other two criteria, however, are somehow less used and the evaluations are almost always done by manual techniques, being far away from any standard.

The nip counting (i.e., the determination of nip frequency) is accomplished by four different methods.

1. Visual counting
2. Needle insertion
3. Thickness measurement
4. Optical visualization

4.3.7.1 Visual nip counting

In this simplest technique, the intermingled yarn is placed on a dark background and the number of nips within a certain length are counted. This way of evaluation is subjective, tedious, and laborious. To facilitate constant tensioning, simple tensioning devices could be used, but the choir of counting still exists. Although it suffers from many human errors, it is perhaps the most widely used evaluating method in the industry. A quick glance of the yarn also gives a subjective impression of intermingling uniformity.

In the case of a continuously intermingled flat yarn; however, the nips are hardly discernible. If a free length of such a yarn is placed on the surface of a stagnant water, the filaments will tend to spread out, and, hence, the nips become visible.

4.3.7.2 Needle insertion method

At the open sections of an intermingled yarn, the filaments are parallel. This property of the yarn is used to determine the length of an open section; thus, the nip frequency is determined. There are two different ways of accomplishing such measurement.

1. Sampling technique
2. Continuous needle insertion.

Figure 4.68 illustrates both the three steps of sampling technique and the continuous technique. As seen, a certain length of yarn is taken and hang over a pulley and tensioned with the weights of *W1*. At the start of the measurement procedure, a needle is inserted and secured so that the needle does not move. In the second step of the test procedure a weight of *W2* applied to the end to which the need has been inserted. The amount of turn of the pulley indicates the length of one side opening. The weight on the right-hand side is then removed and applied to the left hand side of the yarn. The amount of pulley rotation thus indicates the full opening length of the yarn. This procedure is repeated many times and an average of open section length is calculated. The nip frequency is thus obtained. As will be anticipated, this technique is also quite laborious for manual operation. But the same technique is also accomplished by an equipment developed by Toray Co of Japan.[85] However, the technique still suffers from random sampling errors, needle insertion failures and slow measuring speed (200 measurement per hour).

The principles of continuous needle insertion technique is also seen in Figure 4.68. In this technique, the yarn moves at a certain speed and a needle inserted into the yarn. At an open section, the yarn runs smoothly, as the open section ends with a nip, a force is applied to the needle. When the applied force exceeds a predetermined level, the counter increases one, and the needle swings

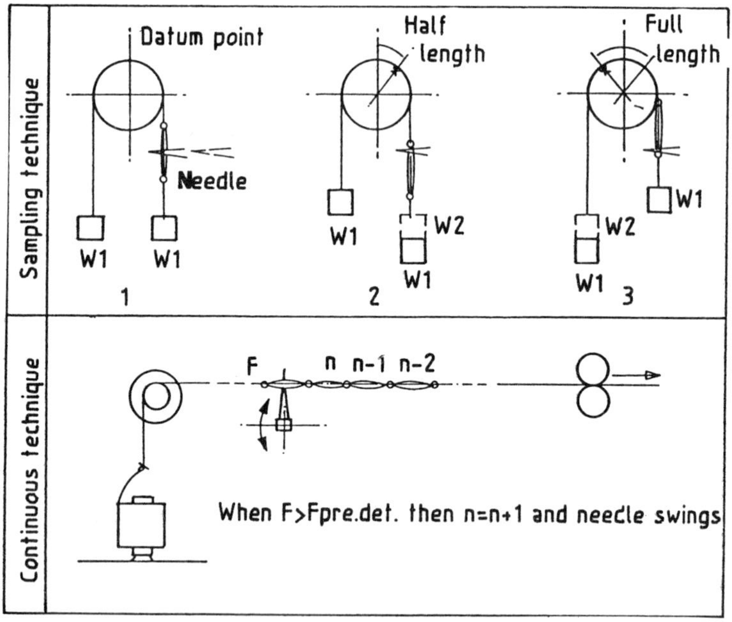

Figure 4.68 Nip-counting techniques depending on pin insertion

backward and repenetrates into the yarn as it swings forward. At low and medium level of intermingling frequencies, the technique works reasonably. However as the nip frequency increases, the needle is liable to omit a small open section just after the one detected. Also the failure of needle penetration, especially with low linear density yarns, is not unusual.

Although this technique also has advantages and disadvantages over the other types of nip counting methods, it is one of the mostly used techniques in industry.

4.3.7.3 Thickness measurement technique

When an intermingled yarn is pressed against a flat surface with a fingertip, it will be sensed that the open sections of the yarn is flat and the nips are round which is thicker than the open section. Weinsdorfer and Teichmann[86] have developed an instrument to count the nips using this principle. The Reutlingen Interlace Counter[87] (Figure 4.69), as its called, scans the yarn mechanically for thickness under spring pressure by a measuring head. The head consists of a fixed metal block which also acts as a yarn guide and a spring loaded, movable tongue. The parallel filaments are pressed flat in strips by the yarn guide selected and the spring pressure. Since the nips can not be squeezed, they are registered as thick places when a certain adjustable threshold value is exceeded.

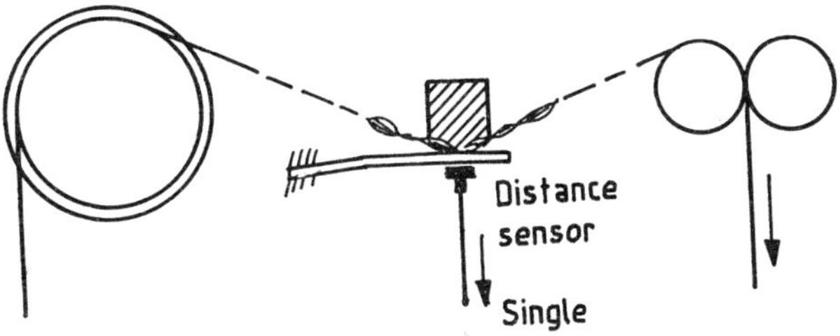

Figure 4.69 Measuring principle of the Reutlingen Interlace Counter

Provided that a suitable measuring head and correct threshold value are selected for the yarn to be tested, the Reuthingen Interlace Counter is reported to give accurate, reproducible and reliable results for nip counting.

4.3.7.4 Optical visualization technique

During both needle insertion and thickness measurement techniques, a contact between the measuring head and the yarn occurs. This contact inevitably disturbs or destroys the yarn structure. Therefore, the measurement results may not always reflect the real yarn structure (i.e., number of nips per meter).

To eliminate the drawbacks of measuring head contact, two different optical techniques have been developed. These are

1. Light transmittance technique as applied on Enka's Obestat[88] tester
2. Charge coupled device (CCD) technique as applied on Loughborough[19] and Microdynamics[89] yarn scanners.

The Obestat instrument is based on the principle of electrostatic charges (Figure 4.70). The yarn specimen is first loaded with a high voltage DC of about 50 kV. To facilitate good conductivity, the yarn is also humidified. Such loaded filaments spread out as soon as they go out of the electrostatic field. Thus, the structure of the intermingled yarn becomes visible.

The distance of this opening is determined by optical scanning in its spread condition by means of a light barrier. The differences in shadow between the open sections of the yarn and the nips are registered in the photodetector and transformed into impulses which can be tuned up and counted electronically.

The principle of CCD-based yarn scanners is seen in Figure 4.70. Acar et al[19] in Loughborough and Microdynamics Co. of the United States developed similar devices using this principle to analyse almost all types of yarn structures. Intermingled yarns having open sections and nips i.e., bead-like structure can conveniently be scanned and nip frequency, nip uniformity could be obtained. On these instruments, the yarn is moved at a steady speed by a yarn handling device. The CCD sensor is positioned at right angles to the axis of the yarn. The shadow of the yarn is cast on the sensor by a light source.

As the yarn moves relatively to the sensor, the latter progressively scans the shadow of the yarn so as to build up data providing an image of a representative length of the yarn. This was only possible with the help of advanced microcomputer technology because of the huge demand of memory to form the image of a reasonable length of yarn. Then the microcomputer calculates the nip frequency, as well as information about the nip uniformity.

Although the CCD based yarn scanners are well suited for the quality assessment of the intermingled yarns, the mere cost of the equipment because of sophistication is hindering its wide use in the industry and research.

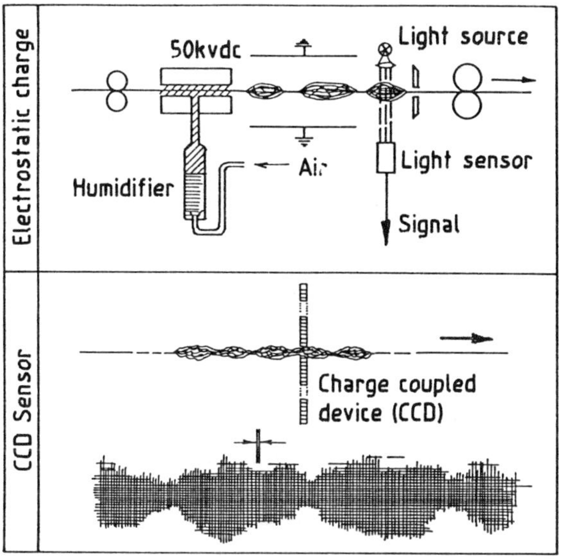

Figure 4.70 The principles of Obestat instrument and CCD based yarn scanner

The strength or stability of nips is mostly determined manually by extending the yarns a few times and comparing the number of nips still intact in the yarn. This is very subjective assessment and handling errors are gross. The only mechanical stability tester is defined by Weinsdorfer.[84] The principle of the tester is shown in Figure 4.71. To determine the nip stability, Weinsdorfer subjects the yarn to strain on this stability tester. The yarn is wound round itself under tension and is also drawn over a sharp edge. The stability value results from the ratio of the number of nips measured after this straining to the number measured before.

Overall, it could be concluded that the intermingling industry lacks of a standard, effective and economic quality assessment technique. It may, perhaps, be hoped that the advances in microcomputer technology bring down the prices of optical visualization method and further research into this method could lead to a standard, universally accepted technique.

4.3.8 Future of the Intermingling/Interlacing

Increasing economic constraints on the textile industry brought forward alternative and less expensive methods to conventional techniques. Intermingling as an alternative to sizing, twisting in texturing, drawing, spinning, and knotting in splicing, and also blending as a new process. Therefore, the intermingling process, provided that the interest problems are overcome, appears to be promising for the future of textile industry.

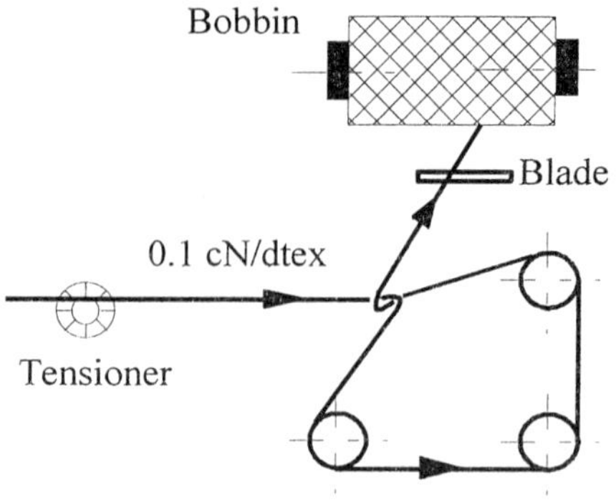

Figure 4.71 Principle of mechanical stability tester

REFERENCES

1. Brehm G., "Texturing of Polyester and Other Filament Yarns," *Chemiefasern/Textilindustrie*, August 1979, Vol. 28/80, 706-708.
2. Fischer K. E., and Wilson D. K., "Air-Jet Texturing: An Alternative to Spun Yarn Production," in *Textile Machinery: Investing for the Future*, The Textile Institute, Manchester, 1982.
3. Glen J.M., "Texturing: The Future Is Now," *Fiber Producer*, 1979, Vol. 7 (4/2), 45, 46, 67.
4. Wilson D. K., "Apparel Fabrics: What Chances Have Air-Jet Textured Yarns?," *Textile Institute and Industry*, 1979, Vol. 17 (5), 170-174.
5. Wilson D. K., "Lighter Fabrics from Air-Textured Yarns," *Textile Institute and Industry*, 1979, Vol. 17 (10), 360-362.
6. Piller B., and Lesykova E., "Structure and Economical Aspects of Air-Jet Textured Filament Yarns," *Melliand Textilberichte* (Eng. Edition), 1982, Vol. 63/11 (7), 485-493.
7. Fischer K., and Kessler J., "Air Texturing: Spunlike Yarns Made of Filament," *Melliand Textilberichte* (Eng. Edition), June 1986, Vol. 12 (6), 383-387.
8. Wilson D. K., "Economic Aspects of Air-Jet Texturing," *Air-Jet Texturing Process: One Day Symposium*, Loughborough University of Technology, 11 July 1984.
9. Longenecker F. R., *Air Jet Texturing Conference*, Clemson University, Greenville, South Carolina, September 1984.
10. Wilson D. K., "Air Jet Texturing: The Latest Technological Developments," *Air Jet Texturing Conference*, Clemson University, Greenville, South Carolina, September 1984.

11. Kollu T., UMIST, Personal Communication, May 1985.
12. Wray G. R., "*The Construction and Resultant Properties of Air- textured Filament Yarns*," Ph.D. Thesis, University of Manchester, 1965.
13. Du Pont, Technical Information Bulletin, X-154, "Characteristics of Taslan Textured Yarns," October 1961.
14. Wray G. R., "The Properties of Air-Textured Continues Filament Yarns," *Journal of Textile Institute*, 1969, Vol. 60 (3), 102-126.
15. Wray G. R., "A Rapid Method of Measuring Yarn Diameters," *Journal of Textile Institute*, 1970, Vol. 57, T134-T136.
16. Fischer K., "Aerodynamic Processes in Filament Yarn Production," *International Textile Bulletin*, 1979, (1), 17-38.
17. Bock G., "Texturing Filament Yarns in an Air Flow Tangling Mechanism," *International Textile Bulletin/Spinning*, 1981, (4), 359-389.
18. Kollu T., "*Structure and Properties of Air-Jet Textured Yarns*," MSc. Thesis, UMIST, Manchester, 1982.
19. Acar M., King T. G., and Wray G. R., "Textured Yarn Quality," *Textile Asia*, 62-66, November 1986.
20. Þen H., and Wray G. R., "A Water-Absorption Test for Assessing the Physical Bulk of Air-Jet-type Bulked Yarns," *Journal of Textile Institute*, Manchester 1970, Vol. 61, 237-240.
21. Rozmarynowska K., and Godek J., "Effects of Processing on the Properties of Air Bulked Yarns," *Bulk, Stretch and Texture*, The Textile Institute, 1966, 29-41.
22. Heberlein Yarn Technical Centre, "Description of Test Methods for Air-Jet Textured Filament Yarns," August 1980, 6110, 13/8/80.
23. Wray G. R., "The Construction of Air-Textured Filament Yarns," in *Bulk, Stretch, and Texture*, The Textile Institute, Manchester, 1966, 18-28.
24. Acar M., Alexander A. J., Turton R. K., and Wray G. R., "Loop Formation Mechanism in the Air-Jet Texturing Process," *International Textile Bulletin*, (3), 1983.
25. Artunç H., "The Air Texturing of Drawn and of High-speed Polyester Yarns," *Chemiefasern/Textilindustrie*, 1981, Vol. 31/83, 289-297, E29.
26. Piller B., "Development Trends in the Production of Air-Jet Texturised Filament Yarn with Spun yarn Characteristics," *Melliand Textilberichte* (Eng. Edition), 1979, Vol. 60/8 (2), 128-136.
27. Kollu T., "Mechanics of Air-Jet Textured Yarns," *4th International Textile*
28. Holden G. A., "Air Jet Textured Yarns for Woven Apparel Fabrics," in *Alternatives to Spun Yarns in Apparel Fabrics Conference*, Shirley Institute Publication, S34, Manchester, 15 February 1979.
29. Wilson D. K., "On-Line Control is Coming," *Textile Horizons*, 1981, (9), 26, 27.
30. Price S. T., "Bulked Yarns Produced by Air Texturing," in *Modern Yarn Production*, Ed. G. R. Wray, Columbia Press, Manchester, 1960, 139-151.
31. Wilkinson G. D., "Textured Yarn Technology: Production, Properties and Processing," Monsanto Textile Division, Vol. 1, 1967.

32. Phillips M., "Air Texturing: Where Are the Fabrics?," *America's Textiles*, 1983, Vol. 12 (5), 17-20.
33. Hoffsommer K. P., "Air Texturing for the 1980's: A Look Backward and Forward," *Fiber Producer*, 1980, Vol. 8 (6), 22-77.
34. 732 929. Du Pont de Nemours, Dec. 15, 1952.
35. Price S. T., "Air Texturing," *Modern Textiles*, 1976, Vol. 57 (7), 28-31.
36. Isaacs III A., "Air Texturing Prepares for Lift-off Again," *Textile World*, 1984, (10), 48-55.
37. Piller B., "*Bulked Yarn*," The Textile Trade Press, Manchester, 1973.
38. 762 630. Du Pont de Nemours, 19 July, 1954.
39. 893 020. Du Pont de Nemours, 20 May, 1960.
40. Wray G. R., and Entwistle J. H., "The Modification of a Taslan Jet to Operate Low Air Pressures," *Journal of Textile Institute*, 1969, Vol. 60 (10), 411-419.
41. 1 282 148. Du Pont de Nemours, 27 June, 1969.
42. 1 279 908. Du Pont de Nemours, 9 November, 1970.
43. 1 448 100. Du Pont de Nemours, 14 December, 1973.
44. Taslan Jet Information Sheet Type XIV and Type XV.
45. 4 574 436. Du Pont de Nemours, 29 April, 1985.
46. 4 097 975. Heberlein Maschinenfabrik, 19 July, 1977.
47. Acar M., "*Analysis of the Air-Jet Texturing Process and Development of Improved Nozzles,*" Ph.D. Thesis, Loughborough University of Technology, 1984.
48. Demir A., "*The Air-Jet Texturing Process with Particular Reference to the Nozzle Design and Improved Test Methods,*" Ph.D. Thesis, Loughborough University of Technology, 1987.
49. Wray G. R., and Entwistle J. H., "An Investigation of the Air-Jet Bulking Process," *Journal of Textile Institute*, 1968, Vol. 59 (11), 122-136.
50. Þen H., "*A Study of the Air-Jet Type Bulked Filament Yarn Process,*" Ph.D. Thesis, Loughborough University of Technology, 1970.
51. Sivakumar V. R., "*The Mechanism of Bulking of Air Textured Yarns,*" Ph.D. Thesis, UMIST, Manchester, 1975.
52. Bock G., and Lünenschloss J., "Air Flow and Loop Formation during Aerodynamic Texturing," *Chemiefasern/Textilindustrie*, 1981, Vol. 31/83 (3), 202-204 and (5), 380-384, E41.
53. Bock G., and Lünenschloss J., "An Analysis of the Mechanism of Air-Jet Texturing," *Textile Machinery: Investing for the Future*, The Textile Institute, Manchester, 1982.
54. Bock G., and Lünenschloss J., "Der Verwirbelungsprozess beim Luftblastexturieren," *Textile Praxis International*, 1984, Vol. 39 (6), 551-558.
55. Bock G., and Lünenschloss J., "Einfluss von Druck und Dusenenstellung beim Luftblastexturieren," *Chemiefasern/ Textilindustrie*, 1986, Vol. 34/86 (7-8), 492-496.

56. Acar M., Turton R. K., and Wray G. R., "An Analysis of the Air-Jet Yarn Texturing Process, Part II: Experimental Investigation of the Air-flow," *Journal of Textile Institute*, 1986, Vol. 77 (1), 28-43.
57. Acar M., Turton R. K., and Wray G. R., "Air Flow in Yarn Texturing Nozzles," *Trans. of ASME: Journal of Engineering for Industry*, August 1987, Vol. 109, 197-202.
58. Demir A., "A Study of Air-Jet Texturing Nozzles: The Effects of Nozzle Configuration on the Air Flow," *Transactions of ASME: Journal of Engineering for Industry*, February 1990, Vol. 112, 97-104.
59. Versteeg H. K., Bilgin Þ., and Acar M., "Effects of Geometry on the Flow Characteristics and Texturing Performance of Air-Jet Texturing Nozzles," *Textile Research Journal*, 1994, Vol. 64/4, 240-246.
60. Acar M., Alexander A. J., Turton R. K., and Wray G. R., "The Mechanism of the Air-Jet Texturing Process," in *Texturing Today*, Shirley Institute Publications S.46, Manchester, 1983, 207-230.
61. Zucrow J. M., and Hoffman J. D., "*Gas Dynamics*," John Wiley & Sons, New York, 1976.
62. Shapiro B., "*The Dynamics and Thermodynamics of Compressible Fluid Flow*," The Ronald Press Company, New York, 1963, 1185.
63. Anderson J. D., "*Modern Compressible Flow, with Historical Perspective*," McGraw-Hill, New York, 1982, 466.
64. Krenzer, E., "Air-Jet Texturing and Its Products Today," Second International Conference on *Air-Jet Texturing and Intermingling,* Loughborough, 27-29 September 1989.
65. Whitehead B., Milliken Ltd., Manchester, Personal communication, November 1984.
66. Acar M., Turton R. K., and Wray G. R., "An Analysis of the Air-Jet Yarn Texturing Process, Part VI: The Mechanism of Loop Formation," *Journal of Textile Institute*, 1986, Vol. 77 (6), 371-376.
67. Olsen J. S., "Frictional Behaviour of Textile Yarns," *Textile Research Journal*, 1969, Vol. 39/1, 31-37.
68. Schick M. J., "Friction and Lubrication of Synthetic Fibers," in *Surface Characteristics of Fibers and Textiles*, Ed. M.J. Schick, Marcel Dekker, 1977, 1-65.
69. Schultz K., "New Machinery for Air-Jet Texturing," *Chemiefasern/ Textilindustrie*, October 1975, Vol. 25/77 (10), 929-932.
70. Kollu T., "*Air-Jet Textured Yarns*," Ph.D. Thesis, UMIST, Manchester, 1985.
71. Acar M., Turton R. K., and Wray G. R., "An Analysis of the Air-Jet Yarn Texturing Process, Part IV: Fluid Forces Acting on the Filaments and the Effects of Filament Cross-sectional Area and Shape," *Journal of Textile Institute*, 1986, Vol. 77 (4), 247-254.
72. Þen H., and Wray G. R., "Apparatus for Manufacturing Yarns of the Air-Jet-bulked Type without the Use of Air," *Journal of Textile Institute*, 1970, Vol. 61, 77-87.

73. Þen H., and Wray G. R., "The Properties of Yarns of the Air-Jet-Bulked Type Produced without Use of Air," *Journal of Textile Institute*, 1971, Vol. 62, 335-349.
74. Enterprise Machine and Development Corp., Sidewinder Air-Texturing Machine Catalog.
75. Eltex Reutlingen, Air Texturing Machine Catalog.
76. Acar M., "Current State of Technology in Air-Jet Texturing," *4th International Textile Symposium*, Izmir, Turkey, 15-17 October 1986.
77. Wilson D. K., "Breakthrough of Air Textured Yarns," *Fiber Producer*, August 1982, Vol. 10 (4), 86-88.
78. Tani M., and Sasaki Y., "Spunlike Texturing as a Substitution of Spinning of Staple Fibers," *24th International Man-made Fibers Congress*, Dornbirn, Austria, 25-27 September 1985.
79. Wray G. R., and Acar M., "Past, Present and Future of the Air-Jet Texturing," *Air Jet Texturing Conference*, Clemson University, Greenville, South Carolina, 18-19 September 1985.
80. Weinsdorfer H., "Production and Testing of Tanglelaced Yarns," *Chemiefasern/Textilindustrie*, Vol. 36/88, 1986, 289-295, E41-E44.
81. USP 2985995, Du Pont de Nemours, 1961.
82. Lünenschloss J., and Zilg J.P., "Intermingling of Filament Yarns in an Air Stream," *Chemiefasern/Textilindustrie*, Vol. 30/82, October 1980, 809-820 E91.
83. Sparkes A. R., "Intermingling/Co-mingling Jets," *Internal Research Report*, Rieter-Scragg Ltd, Langley, Macclesfield, 1980.
84. Weinsdorfer H., "Mechanism of the Intermingling of Filament Yarns," *Chemiefasern/Textilindustrie*, Vol. 31/83, March 1981, 198-202, E21-E22.
85. Toray Engineering Co. Ltd, Osaka, Japan, "ET-500 Entanglement Tester."
86. Weinsdorfer H., and Teichmann K.H., "Rational Production Control of Tanglelaced Filament Yarns with the Reutlingen Interlace Counter," *Chemiefasern/Textilindustrie*, Vol. 29/81, 1979, 862-864, E120.
87. Institute fur Textil-und Verfahrenstechnik, Denkendorf, W Germany, "Reutlingen Interlace Counter, RIC II."
88. Kloss R., Latzke P. M., Schwab M., and Welfers J., "Obestat Testing Apparatus for the Tangle Configuration and Tangle Distance of Filament Yarns," *Chemiefasern/Textilindustrie*, Vol. 28/80, 1978, 788-793, E136.
89. Microdynamics Corporation, Atlanta Georgia, USA, "Digital Interlace Analyser DIA-1000."
90. Demir A., Acar M., and Wray G. R., "Instability Tests for Air-Jet Textured Yarns," *Textile Research Journal,* 1989, Vol. 56/3, 191-202

CHAPTER 5

OTHER TEXTURING TECHNIQUES

5.1 INTRODUCTION

In late 1960s, the texturing route of self bulking, based on heterofil spinning, was a major route. This, in practice, has made an impact only on hosiery, and it does not have a major part to play in the technology of texturing.

5.2 BICOMPONENT TEXTURING

Bicomponent fibers consist of two polymers of the same generic family, e.g., nylon 6 and nylon 6.10. Biconstituent fibers however are made from components that are not of the same generic family, e.g., nylon 6.6 and polyester.[1]

5.2.1 Bicomponent Filament Production

Bicomponent fibers were invented by the sheep! The self-crimping property of wool results from its unusual structure. Photomicrographs of cross-sections of wool fibers show that, unlike any other natural fiber, wool fibers are composed of two halves, each roughly semicircular in cross-section. These two components (the orthocortex and paracortex) adhere strongly to each other and rotate spirally around each other as they run the length of the fiber. The orthocortex and paracortex differ in fine structure, i.e., in the nature and arrangement of the structural units (fibrils, crystalline regions); it is this difference that causes the paracortex to shrink slightly less than the orthocortex when the fiber is dried. The difference in shrinkage leads to the helically crimped configuration of the wool fiber.[2]

Although the manufacture of synthetic bicomponent materials is not strictly a texturing process, it is a means of introducing crimp into the continuous filament yarns, giving properties associated with textured yarns.[3] Bicomponent spinning involves the extrusion of two different polymer materials through the same spinneret holes. Three different classes of fiber structure are produced by using three different methods for presenting the two polymer components to the holes of the spinneret during extrusion. Figure 5.1 illustrates these methods in principle and the fiber structure produced by them. In principle, any of the basic spinning processes can be adopted for bicomponent spinning, but in practice, the melt spinning of thermoplastic polymers is by far the most used.[4]

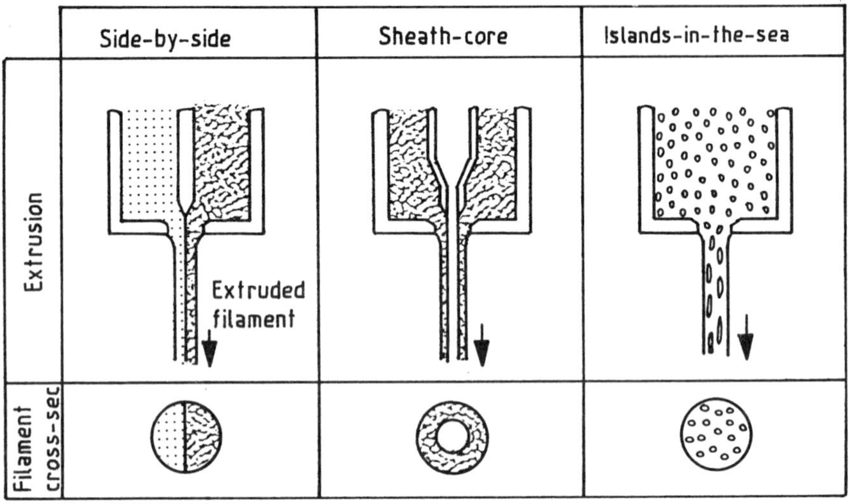

Figure 5.1 Principles of bicomponent spinning

The constituents are combined before extrusion by feeding each to a defined point in the cross-section in such a way that each remains continuous along the length of the filament. There are many variations in the distribution of constituent materials in the cross-section making bicomponent fibers. However, the two simplest are side-side and sheath-core (Figure 5.2)

The ordered arrangement of the components is controlled by the distribution of the polymer melt immediately behind the spinneret. Separate melting of the polymers, separate pumping and separate filtration are necessary, and the two constituents are normally combined just before the orifice and are extruded as a single filament.

The complexity of the above system is obvious, especially when the extrusion of multifilament yarns are concerned. It is therefore more usual to form the polymer streams into layers some distance from the spinneret. The spinneret holes are then arranged so that they intersect the interfaces between the molten polymer layers. Side-by-side bicomponents are then produced, although in general the distribution of components will not be the same in every fiber, for which reason the crimping properties will therefore vary from filament to filament. Alternatively, the two molten polymer components are formulated into a multilayered or laminated structure and no attempt is made to arrange the spinneret orifices in any systematic way. This technique results in the production of a mixture of filaments, some containing 100% of one component, some 100% of the other, and some with varying proportions of both[5] (Figure 5.3).

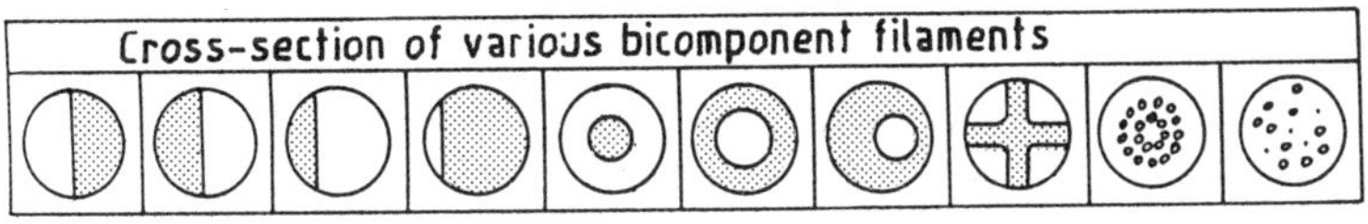

Figure 5.2 Cross-sectional shapes of bicomponent fibers

Figure 5.3 Mixed bicomponent fibers

The main commercial outlet for the bicomponent fibers is in the production of self-crimping yarn. The two components are present in the fibers in a side-by-side or eccentric sheath-core arrangement: the self-crimping action depends upon the differential shrinkage of the two components when the fiber is relaxed from a stretched, oriented condition.[2]

5.2.2 Bicomponent Filament Texturing

The self crimp develops in the wool fiber presumably because of the greater tendency of the paracortex to shrink with respect to the orthocortex. Further, differential shrinkage develops in the wool through commercial textile processes (wetting and drying). In the case of synthetic fibers, the bicomponent fiber is such that the material on one side tends to shrink preferentially during either its manufacturing history or on "activation" during texturing or some equivalent process. The essential feature of both classes of fibers is that they contain at least two parallel elements[6].

Side-by-side bicomponent filaments are most commonly used for self crimping; two polymers differing in chemical structure, melting point, comonomer content, molecular weight, or some other property produce an oriented filament in which the two oriented components undergo differential shrinkage when subjected to some thermal or other relaxing treatment. After this treatment, both strips will shrink to a compromise shrinkage and, hence, will exert a force on each other because one has shrunk less than it could have and the other has shrunk more than it should. This situation is highly unstable, as both sides have free surfaces that can assume shrinkages more freely than the surfaces which are fused together. Thus, there is a redistribution of the strains. that are fused together have the same strains; free surfaces try to reduce the strain. The free surface of the more shrinkable component will try to reduce the compressive strain caused by the other component and will form the inner side of the final, bent configuration. Similarly, the free surface of the less shrinkable component tries to reduce the tensile strain caused by other component and takes the other side of the final, bent configuration.[7] The product, hence, develops a helical crimp (Figure 5.4).

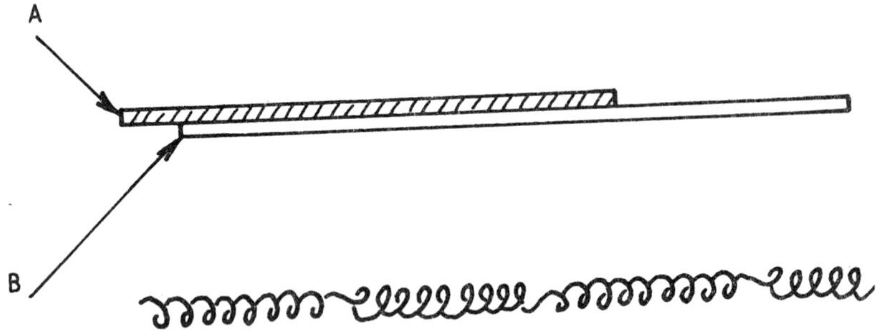

Figure 5.4 Development of helical crimp

There are two methods of texturing bicomponent filaments.

1. This is by far the most important type. The crimp is helical in character and is produced by a difference in the amount of shrinkage of the two components of the side-by-side (or eccentric sheath-core) bicomponent filament, these two components adhering together throughout the whole process. Nearly all commercially available textured bicomponent filaments are of this type.[2]

2. In this type, one of the two side-by-side components is a hard polymer (e.g., a polyester, a polyamide or an acrylic) and the other component is an elastomer. After the spinning and drawing processes the two components separate, either partly or completely, because of their poor adhesion; the elastomer contracts a great deal, and the hard component is pulled into a bulked, loopy structure.

High-linear-density bicomponent yarns for carpets are being produced in significant quantities together with small amounts of monofilament and multifilament for ladies' stockings.[5]

5.2.2.1 Theoretical aspects of self-crimping behavior

When a bicomponent filament is crimped, several factors might be expected to affect the curvature developed in the filament.

1. Difference in the shrinkage between the components
2. Difference in modulus between the components
3. Overall cross-sectional shape of the filament and the individual cross-sectional shapes of the two components
4. Thickness of the filament

Most analyses of self-crimping behavior adopt an approach basically similar to the studies on the curving behavior of bimetallic strips. Brown and Onions[8] investigated the crimping behavior of wool fibers. In their analysis the fibers were considered to possess an idealized elliptical cross-section; the changes in crimp curvature on wetting and drying were calculated. They showed that the ratio of the moduli of the two components plays an insignificant part in determining the fiber curvature. It was shown that the change in curvature on wetting the fiber is (1) proportional to the differential change in component lengths, (2) inversely proportional to the thickness of the bicomponent fiber, (3) dependent on the relative thickness of the two components[2].

Brand and Backer[9] and Fitzgerald and Knudsen[10] have developed more generalized theories. Brand and Backer also concluded that the values of the moduli of the two components are not important in determining crimp, and their equation again leads to conclusion of Brown and Onions that the crimp curvature is proportional to the differential length change and inversely proportional to the filament thickness.[2] All these effects may be formulated a follows:

$$\frac{1}{R} = \frac{K}{h} C_p \frac{\Delta l}{L}$$

where K is a constant (slightly dependent on the ratio of the moduli of the components and the overall shape of the filament), h is the thickness of the filament, C_p is a crimp potential function dependent on the relative amounts of the two components and their distribution, and $\Delta L/L$ is the differential change in length of the two components on wetting or drying. The application of this theory leads to several general conclusions applicable to all cross-sections. These conclusions are

1. The maximum crimp potential is found in filaments with only one interface (Figure 5.5a).
2. The crimp potential is a maximum when the interface passes through the center of area of the fiber cross-section (Figure 5.5b).
3. The crimp potential is zero for any distribution with a center of symmetry (Figure 5.5c).

5.2.2.2 Mechanical properties of crimped bicomponent filaments

5.2.2.2.1 Extension-recovery behavior

A bicomponent filament developing crimp under practical conditions (i.e., in the presence of some longitudinal restrains and with the fiber ends usually prevented from rotating) will be divided into sections of alternate clockwise and anticlockwise helix rotation (See Figure 5.4). The points of reversal rotate about the filament axis as the filament is extended or relaxed, thus effectively increasing or decreasing the number of helical turns. The crimp frequency of a filament in a yarn, therefore, depends on three factors.

1. Potential crimpability of the individual filaments
2. Conditions of restraint during the development of the crimp
3. Extension subsequently imposed on the yarn

In view of these complexities, no simple theory of spring-like behavior can be employed to explain the extension-relaxation properties of bicomponent filaments in yarn form. The best that can be expected is a qualitative explanation of behavior.[2]

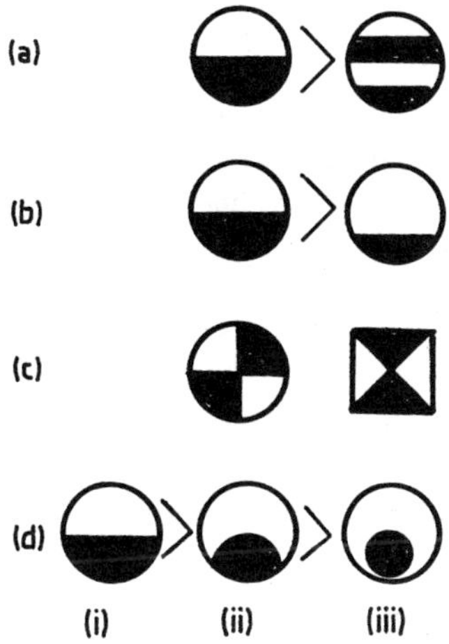

Figure 5.5 Crimp potential as function of shape and number of interface

A further complication in any theoretical treatment of the mechanical properties of crimped bicomponent filaments is the nonelastic behavior of the material involved, and in some cases a lack of detailed knowledge of the tensile properties of the materials.[2]

5.2.2.2.2 Bulkiness

In a yarn of helically crimped fibers, the bulkiness will clearly tend to be reduced by any interpenetration and interlocking of adjacent helical coils. Interpenetration largely depends on two factors, the crimp frequency of the filaments in the yarn, and the number of helix reversals occurring in the filament. It is easy for two helices of the same pitch and same direction of rotation to interpenetrate, and is almost as easy for two filaments of the same pitch but opposite direction. Reversal points, however, will obviously tend to reduce interpenetration. It is obviously an advantage, if a high bulkiness is required, to intermingle filaments with a wide range of crimp frequencies, since helices of varying pitch will obviously not readily interpenetrate.[2]

5.2.2.3 Comparison of crimped bicomponent yarns with other textured yarns

False-twist textured yarns and crimped bicomponent yarns both have a basically helical structure. However, the equilibrium state of bicomponents is a helix of fixed radius and the individual filaments are torque-free, whereas false-twist textured filaments have an equilibrium state which is an untwisted helix varying in radius, and the filaments are torque-lively. As a result of this, the retractive power of false-twist textured yarn would, in general, be greater than crimped bicomponent yarn of the same yarn and filament linear density. The frequent reversals and variations in helix radius in false-twist textured yarn tend to discourage interpenetration of the helical coils, and so little yarn bulkiness is lost on this account.[2]

The zig-zag crimp of the stuffer box texturing gives a product of relatively low contractive power and, therefore, a yarn of a soft bulky character without a great deal of stretch; it is quite different from crimped bicomponent yarns.

The characteristics of the crinkle (gear-crimped or knit-deknit textured) yarns are the "in-phase" and two-dimensional crimp of the filaments. The crimp structure is different from that of crimped bicomponent yarns.[2]

Edge crimped yarns are closely comparable with the crimped bicomponent yarns because their crimp is due to a structural asymmetry (the filaments are in fact a type of bicomponent filament) and is helical in character.

5.3 DIFFERENTIAL SHRINKAGE TEXTURING

Although shrinkage is often a nuisance in textiles, it can be used to advantage in some applications. In the production of high-bulk staple-spun yarns, for example, substantial shrinkage is encouraged. Two types of fibers, one with a high potential shrinkage and one with little potential shrinkage, are blended together before the spinning process. When the resulting yarn is heated under relaxed conditions the differential shrinkage causes the low shrinkage components (up to 20% longer than the other component) to buckle into crimps and loops in the yarn, causing it to become bulky. Acrylic fibers have the capacity to accept very high extensions and then to be able to shrink on heating, and are used in the production of high-bulk yarns in this way. The resultant yarns, which are wool-like in appearances, are used in knitting.[11]

5.4 CHEMICAL TEXTURING

All texturing processes excluding air-jet texturing involve mechanical deformation of a filament or a bundle of filaments and dissipation of internal strains induced by mechanical stressing to obtain a permanent or quasi-permanent, stress-free, distorted configuration. The dissipation of the elastic strain imposed on the matrix by the mechanical field is brought about by providing a thermal environment in the case of texturing of thermoplastic yarns.

The events that occur in a thermosetting process may be summarised in the following three stages:

1. Weakening the intermolecular links in the fiber and increasing the mobility of its structural elements by raising its temperature

2. Formation of a more even structure as a result of thermal relaxation and reduction of internal stresses
3. Stabilization of the even structure by reformation of bonds in an energetically favorable configuration; usually this is attained by cooling the yarn sharply[12]

The rate and the completion of the thermorelaxation process are determined mainly by five factors.

1. Chemical structure of the polymer chain
2. Polymer dispersion
3. Temperature
4. Medium in which the process takes place
5. Mechanical tension applied to the filament[12]

The thermomechanical texturing process involves a high-thermal input to raise the temperature of a yarn of low thermal conductivity relatively close to its melting point. The electrical power requirements is high and constitutes a major cost in the operation. Moreover, the heat is dissipated in air during cooling.[12]

The strain-dissipative environment need not, necessarily, be thermal in nature. It may be a suitable solvent, the suitability being dependent on the solubility parameters of the polymer and the solvent.[13] In this case, since there is no cooling, the heat losses during this operation would be avoided, making the process more economical.

Successful texturing of polyester, nylon 66, and acetates with the aid of suitable solvents has been reported.[14] Texturing with solvents becomes all the more appealing when one realises that setting through a rearrangement of morphological order need no longer be confined to thermoplastic fibers. Successful texturing of nonthermoplastic fibers, such as jute by chemomechanical texturing process has also been reported.[15,17]

The process. Sengupta[14] describes the chemomechanical texturing process as follows; the yarn is passed through a solvent bath, for example, 3% concentration (v/v) of phenol in water for 116 dtex, 24 filament nylon 6 yarn, for approximately 2 min immersion time and twisted by means of conventional ring frames up to the desired twist rate. The twisted yarn is then transferred onto a perforated cone, washed with a 5% solution of sodium carbonate, followed by two more washes with plain water. After drying in air at room temperature for 24 hours, the yarn is detwisted on the ring frame in the reverse direction at a slightly higher twist rate than first twist so that a small amount of twist keeps the textured filaments together.

The crimp rigidity values of conventional and solvent textured yarns are found to be comparable by Sengupta.[14] Although the breaking elongation values of the two textured yarns are similar, there appears to be a significant difference in their tenacity values. In the conventional process, there is a loss in strength of about 8%, whereas with solvent texturing the loss is more than 20%. This considerable loss in strength suggests a possibility of chemical degradation during the solvent texturing process. However, it is found that the dye-uptake of solvent textured yarns is phenomenally higher than the conventionally textured yarns. It is possible that leaching-out of oligomers causing voids and selective surface etching due to phenol may be responsible for both the poor strength and high dye-uptake of solvent textured yarns.[14]

The chemical texturing process can also be applied to staple yarns, for example texturing of jute staple fiber, otherwise called woolenisation, is an established process. The properties of these textured jute fibers are markedly different from those of raw jute fibers. The blends of jute with other synthetic materials, such as polypropylene can also be chemically textured whereby the polypropylene fibers would remain unaffected by the alkali treatment, whereas the jute fibers would experience shrinkage that results in bulking and texturing.[16]

All jute and jute-polypropylene blend staple yarns may be textured by treating the yarns with 18% (w/w) NaOH solution at room temperature. The material to liquor ratio may be chosen as 1:20, and the residence time is 45 to 60 minutes. The material should be removed from the caustic solution, washed with running water thoroughly, and neutralized with 1% to 1.5% acetic acid solution. The yarn should once again be washed thoroughly with running water and dried. No tension is applied during texturing.[16,17]

Gupta et al[16] observed that the previously described chemical treatment brings about weight loss in both all-jute and jute-polypropylene blended yarns. This weight loss is due to the removal of batching oil added to the jute fibers before processing, and also to partial removal of the hemicellulose and lignin component of the jute fiber.

The chemical-texturing process also brings about shrinkage and increase in yarn diameter in both all-jute and jute-polypropylene yarns. The extent of shrinkage for all-jute yarn was more than that for blended yarn, but increase in diameter for blended yarn was more than that for all-jute yarn. As a result, the available air space of textured blended yarn was higher than that of all-jute yarn. In addition, as these blended yarns were treated with 18% (w/w) sodium hidroxide solution, the high-shrinkage jute fiber migrated to the core, while non-shrinking polypropylene fibers migrated to the surface of the yarn and formed loops and kinks there.[16] These crimps are three-dimensional in contrast to two-dimensional crimps of stuffer-box and gear-mesh textured yarns.[17]

REFERENCES

1. Saunders J. H., "Fibers, Identification," in *Encyclopedia of Polymer Science and Technology,* 2nd Edition, Vol. 11, 410-445.
2. Jeffries R., "*Bicomponent Fibers*," Merrow Publishing Company, Watford, 1971.
3. Denton M. J., "Texturing of Continuous Filament Synthetic Yarns," *The Journal* (Bradford Textile Society), 1987, 51-56.
4. Cumberbirch R. J. E., "How Man-Made Fibers Are Produced," *Textiles*, 1972, Vol. 1, No. 1, 25-28.
5. Denton M. J., "The Future of Textured Yarns: Draw-Texturing and Bicomponents," *Textile Institute and Industry*, January 1974, 5-9.
6. Gupta B. S., and George W., "A Theory of Self-Crimping Bicomponent Filaments," *Textile Research Journal*, April 1975, Vol. 45, No. 4, 338-349.
7. El-Shiekh A., Bogdan J. H., and Gupta R. K., "The Mechanics of Bicomponent Fibers," *Textile Research Journal*, Part I: Theoretical Analysis, April 1971, Vol. 41, No. 4, 281-297. Part II: Experimental Investigation, November 1971, Vol. 41, No. 11, 916-922.
8. Brown T. D., and Onions W. J., *Journal of Textile Institute*, 1961, Vol. 52, T101.
9. Brand R. H., and Backer S., *Textile Research Journal*, 1962, Vol. 32, 39.
10. Fitzgerald W. E., and Knudsen J. P., in *Bulk, Stretch, and Texture*, The Textile Institute, Manchester, 1966, 134.
11. Morris W. J., "The Use of Heat in the Production of Synthetic Fibers," *Textiles*, 1981, Vol. 10, No. 1, 3-9.
12. Sengupta A. K., "Studies on Texturing with the Aid of Solvents," *Indian Journal of Textile Research*, 1976, Vol. 1, 53-59.
13. Sengupta A. K., Gianchandani J. K., and Jindal S. K., "A Comparative Study of Thermomechanical and Chemomechanical Texturing of Nylon 6 Multifilament Yarns," *Textile Research Journal*, June 1979, 360-364.

14. Sengupta A. K., "Studies on Texturing with the Aid of Solvents," *Indian Journal of Textil Research,* 1976, Vol. 1, 53-59.
15. Sengupta A. K., "A New Method for Producing All-Wool Stretch Yarns," *Indian Journal of Textil Research,* 1976, Vol. 1, 81-82.
16. Gupta N. P., Majumdar A, Bhattacharya G. K., Sur D., and Roy D, "Chemically Texturizing Jute and Jute Polypropylene Blended Yarns," *Textile Research Journal*, November 1982, 694-702.
17. Sinha A. K., Mathew M. D., and Roy D., "Bulk Characteristics of Jute/Polypropylene yarns Textured by Sodium Hydroxide Solution," *Textile Research Journal*, September 1990, 507-509.

BIBLIOGRAPY

1. McIntyre J. E., "Man-Made Fibres: Manufacture," *Encylopaedia of Polymer Science and Technology*, John Wiley & Sons, 1971.

2. Hetherington C., "Higher Productivity in Nylon and Polyester Filament Yarn Manufacture," *International Textile Machinery*, 1979, 58-61

3. Enka A.G., "Man-Made Fibres in 1986-World Picture," *Textile Month*, April 1987, 39-42.

4. Denton M. J., "Synthetic Fibre and Yarn Production: Part I." *Textile Month*, December 1987, 33-37:; Part II: January 1988, 30-33, ITMA 87 Report.

5. Denton M. J., "Air Texturing and Intermingling," *Textile Month*, January 1988, 34-37.

6. Denton M. J., "False-Twist Draw Mexturing Machines," *Textile Month*, February 1988, 37-42, ITMA-87 Report.

7. "ITMA-Technology," *Textile Month*, February 1988, 43-45.

8. Slack I. D., "High-Speed Spinning of Fine Deniers," *Textile Month*, October 1988, 21-22.

9. Kumar S., "Compressive Strength of High Performance Fibres: Part I," *International Fibre Journal*, February 1989, 4-6.

10. Bennett D., "Coluration of Polypropylene for Textile Applications," *International Fibre Journal*, February 1989, 8-16.

11. Morris W. J., "Nylon," *Textiles*, 30-34.

12. Morris W. J., "Polyester Fibres," *Textiles*, 30-35.

13. Morris W. J., "Polyester Fibre," *Textiles*, 1986, Vol. 15 No. 3, 59-66.

14. Anderson D. F., "Polypropylene Fibres," *Textiles*, 1980, Vol. 9, No. 3, 8-11.

15. Goswami J., Martindale J. G., and Scardino. F., "*Textile Yarns, Technology, Structure and Applications*," John Wiley, New York, 1977.

16. Wray G. R., "Textured Yarns," in *Encylopaedia of Polymer Science and Technology*, 1st Edition, Vol. 13, 747-763.

17. "Spinner Unit Makes Impact on Texturing Trade," *Textile Month*, May 1970, 67-84.

18. Akyüz H., "Tekstürize Ýplik," *Mühendis ve Makina*, 1. Bölüm, Eylül 1971, Cilt 15, No. 173, 61-67. 2. Bölüm, Ekim 1971, Cilt 15, No. 175, 139-145.

19. Dyson E., Iredale, J. A., and Parkin, W., "Yarn Production and Properties," *Textile Progress*, 1974, Vol. 6, No. 1.

20. "Versatility Is Still UK Textured Yarn Trade's Strongest Weapon," *Textile Month*, January 1974, 41-49.

21. "Specialty Processing Is Still the Bastion of the Independents," *Textile Month*, January 1976, 35-37.

22. Bowen D. A., "Texturing on the Road to Recovery," *Textile Industries*, July 1980, Vol. 144, 30-42.

23. Harding W. A., "Future Developments in Textured Polyester," *Fibre Producer*, August 1980, Vol. 8, 38-66.

24. Koslowski H. J., "Structural Change in the Texturing Industry: A Comparison of PES Texturised Yarns in Western Europe and the USA," *Chemiefasern/Textilindustrie*, January 1981, Vol. 31/83, 25-26, E7.

25. Hearle J. W. S., "Tomorrow's Yarns," *Tomorrow's Yarns: Manufacture and Markets*," Symposium 26-28 June 1984, Department of Textiles, UMIST, Manchester.

26. Kollu T., and Wilson, D. K., "Texturing with Reference to Air-jet Texturing," *International Symposium on Textiles*, 2-4 November 1981, Izmir, Turkey.

27. Quynn R. G., "The False-Twist Texturing of Various Fibres," *Journal of Textile Institute*, 1971, Vol. 62, 510-511.

28. "Draw-Texturing a Hot Topic in the U.S.," *Textile Month*, September 1971, 65-67.

29. Batra S. K., "The Normal Force Between Twisted Filaments: Part I: The Fibre-wound-on-cylinder Model - Analytical Treatment," *Journal of Textile Institute*, 1973, Vol. 69, 209-222; Part II: "Experimental Verification," *Journal of Textile Institute*, 1973, Vol. 69, 363-373.

30. Hearle J. W. S., and Yegin A., "The Mechanics of Torque-stretch Yarns: A Correction," *Journal of Textile Institute*, 1973, Vol. 69, 601-606.

31. Miura Y., "The Mechanics of False-Twist," *Textile Research Journal*, January 1974, 21-23.

32. Quynn R. G., "Pre-oriented Polyester as a Feed Yarn for Simultaneous Draw-Texturing," *Textile Research Journal*, February 1974.

33. Sengupta A. K., Singh R. K., and Majumdar A., "Effect of Drawing and Texturizing on the Molecular-Weight Distribution of Nylon 6 Multi-filament Yarn,," *Textile Research Journal*, March 19074, 155-163.

34. "Advances in False-Twist Texturing Process," *Textile Month*, March 1974, 48-52.

35. Denton M. J., "Twisting-rate Variations in the False-Twist Threadline: Part I: Background and the Effect of a Step-Change in the Twisting Rate," *Journal of Textile Institute*, 1975, Vol. 66, No. 8, 282-288; Part II: "The Effects of a 'Rectangular Pulse' Transient Change and a Sinusoidal Variation in the Twisting Rate," *Journal of Textile Institute*, 1975, Vol. 66, No. 8, 289-296.

36. Denton M. J., "Tension Ratio and Slip in Friction-Twisting," *Journal of Textile Institute*, 1975, Vol. 66, No. 8, 303-306.

37. Konopasek M., "A Mechanical Model of Helical Elements in Textured Yarns," *Textile Research Journal*, April 1976, 278-283.

38 Backer S., Brookstein, D. S., and Thwaites, J. J., "Transient Threadline Behaviour in False-Twist Texturing," *Journal of Textile Institute*, 1976, Vol. 67, No. 6, 224-226.

39. Lünenschloss J., Fischer K., and Osswald W., *Chemiefasern/Textilindustrie*, August 1976, 664-668.

40. Greenwood K., Tension Ratio and Slip in Friction-Twisting," *Journal of Textile Institute*, 1976, Vol. 67, No. 10, 362-366.

41. Thwaites J. J., "Linear Density Variations in Filaments from Draw-Textured Partially Oriented Yarns," *Journal of Textile Institute*, 1976, Vol. 67, No. 10, 341-347.

42. Hooper C. W., and Thwaites J. J., "Measuring Linear-density Variations in Fine Filaments," *Journal of Textile Institute*, 1976, Vol. 67, No. 10, 348-354.

43. Tikker P., Johnson D., Dobb M., and Sikorski J., "Properties and Structures of Polyethylene Terephaltalate Yarns in Conventional and Simultaneous Draw-Texturing," *Textile Research Journal*, January 1977, 29-38.

44. Vaidya A. A., and Nigam J. K., "Minimising Barriness in Textured Nylons," *Textile Institute and Industry*, March 1977, 96-101.

45. Gupta V. B., "The Role of Winding Tension in Rendering Crimp Latent in False-Twist Textured Yarn," *Textile Research Journal*, September 1977, 634-635.

46. Familant H. M., "Heat Transfer in a False-Twist Texturing Process," *Journal of Textile Institute*, 1977, Vol. 68, No. 12, 425-477; Jones CR, "Heat Transfer in a False twist-Texturing Process-Reply," *Journal of Textile Institute*, 1977, Vol. 68, No. 12, 427-428.

47. Smith S. C., and Bakcer S., "Pseudo-False-Twisting in the Second Heater During the Texturing of Thermoplastic Yarns," *Journal of Textile Institute*, 1978, Vol. 69, No. 12, 385-390.

48. Mukherjee A. K., Sengupta A. K., and Bhattacharya L., "Effect of Tension Level on Texturising of Nylon 6 Multifilament Yarn," *Textile Research Journal*, January 1979, 40-42.

49. Cannon C. G., "The Packing of Filaments in False-Twist Draw-Texturing Process," *Journal of Textile Institute*, 1979, Vol. 70, No. 3, 105-106.

50. Hearle J. W. S., "Studies in Friction Twisting: Part I: A Simple Treatment of the Slipping Mode," *Journal of Textile Institute*, 1979, Vol. 70, No. 7, 287-297.

51. Hearle J. W. S., and Beech S. R., "The Positive Mode," *Journal of Textile Institute*, 1980, Vol. 71, No. 5, 225-235.

52. Hearle J. W. S., "The Difference Between Pin-Twisting and Friction-Twisting," *Journal of Textile Institute*, 1979, Vol. 70, No. 7, 321.

53. Denton M. J., "Understanding Stretch Fabrics," *Textiles*, 1980, 25-28.

54. Sengupta A. K., Bajaj P., and Sen K., "New Methods of Texturing Polyester/Viscose Blended Yarns by Simultaneous Setting of the Components," *Textile Research Journal*, May 1980, 275-578.

55. Quyn F. G., and Poisson W. H., "Thermal Shrinkage and Mechanical Crimping during False-Twist Texturing," *Textile Research Journal*, May 1980, 283-288.

56. Greenwood K., "The Difference Between Pin-Twisting and Friction Twisting," *Journal of Textile Institute*, 1980, Vol. 71, No. 6, 313-314.

57. Bajaj P., Sengupta A. K., and Sen K., "Texturing of Polyester Viscose Blended Yarn by Batch and Continuous Process," *Textile Research Journal*, October 1980, 610-617.

58. Hou T. H., and Familant H. M., "The Behaviour of Twisted Yarn Passing over a Heater," *Journal of Textile Institute*, 1981, Vol. 72, No. 1, 39-41.

59. Denton M. J., "The Behaviour of Twisted Yarn Passing Over a Heater," *Journal of Textile Institute*, 1982, Vol. 73, No. 1, 42-44.

60. Hou T. H., and Familant H. M., "The Behaviour of Twisted Yarn Passing Over a Heater-Reply," *Journal of Textile Institute*, 1982, Vol. 73, No. 1, 44-46.

61. Bajaj P., Sengupta A. K., and Sen K., "Application of Diisocyante and Plasticizing Solvents for Texturing of Polyester/Viscose Blended Yarn," *Textile Research Journal*, September 1981, 614-615.

62. Gupta D. K., and El-Shiekh A., "Load-Deformation Behaviour of Two-dimensional Crimped Filaments," *Textile Research Journal*, October 1982, 621-632.

63. Beech S. R., "The Behaviour of Twisted Yarn Passing Over a Heater," *Journal of Textile Institute*, 1983, Vol. 74, No. 2, 97-98.

64. Thwaites J. J., "Snow-deposit Patterns on Friction-Twisting Discs," *Journal of Textile Institute*, 1984, Vol. 75, No. 1, 76-79.

65. Naik A., "Crimping Force: A Parameter to Characterise Polymeric Yarns," *Melliand Textilberichle* (Eng. Edition), February 1985, 127-131.

66. Greenwood K., and Grigg P. J., "The Development of Twist in a False-Twist Texturing Machine with a Friction Twister," *Journal of Textile Institute*, 1985, Vol. 76, No. 4, 244-263.

67. Buckley C. P., Hearle J. W. S., and Mandal R., "Heat-setting of Drawn Polymeric Fibres: Anomalous Twist Recover," *Journal of Textile Institute*, 1985, Vol. 76, No. 4, 264-274.

68. Kang T. J., "Transient Threadline Behavious in the High Speed Falsle-Twist Texturing," *Journal of the Korean Society of Textile Engineers and Chemists*, October 1985, 48-58 (English).

69. Gacen J., Maillo J., and Baron A. B., "Influence of Texturing Process on the Fine Structure of Heat-set Polyester," *Melliand Textilberichte*, 1/1989, 12-14, E5-E6.

70. Gill G. F., "A Detailed Survey of False-Twist Heater Temperatures and Their Subsequent Effects on Yarn Quality," *HATRA Special Report.*

71. Kothari V. K., Sengupta A. K., Srinavasan J., and Goswami T. J., "Air-jet Texturing of Cotton-filament Composite Yarns for Better Apparel Comfort," *Textile Research Journal*, May 1989, 292-299.

72. Buchanan D. R., and Walters J. P., "Glass Transition Temperatures of Polyamide Textile Fibres," *Textile Research Journal*, June 1977, 398-406.

73. Wray G. R., "An Appraisal of the Platt Kosmatex Draw-Texturizing Machine," *Textile Institute and Industries*, December 1973, 342-344.

74. ARCT, "New Draw Texturing Machine Conception," *International Textile Bulletin/Spinning* 1/79, 74-75.

75. Mutschler G., "Converting Draw-Twisters into Draw-Texturing Machines," *International Textile Bulletin/Spinning*, 1/79, 99-109.

76. Sommerfeld A., "Yarn Tension Measuring System: Stationary Sensor Controls the Texturing Process Continuously," *International Textile Bulletin/Yarn Forming*, 2/87, 51-54.

77. Isaacs III McA., "How Unifi Stays on top in Textured Polyester," *Textile World*, February 1988, 55-57.

78. Artunç H., and Weinsdörfer H., "Experimental Texturing Machine at Denkendorf," *Textile Praxis International*, 1988, Vol. 43, No. 6, 581.

79. Wilshaw R, "A Period of Consolidation," *African Textiles*, August/ September 1987, 10-14.

80. Schellenberg H., "Economical Aspects of BCF Texturing," *Chemifasern/ Textilindustrie*, Vol. 38/90, October 1988, 902-903, E105.

81. Lünenschloss J., and Bruske J., "Increasing the Performance of False-Twist Texturing Machines: Problems and Solutions," *Chemiefasern/ Textilindustrie*, October 1988, Vol. 38/90, 882-883, E102.

82. Wilson D. K., "False-Twist Texturing Machines and their Features," Personal Information.

83. Rieter-Scragg Ltd., "A Brief History," November 1983.

84. Rieter-Scragg Ltd., *Operational Notes*, Scragg EMI Air-Jets.

85. Rieter-Scragg Ltd., *Operational Notes*, R-SL, "Super-Twist Polyurethane Disc Systems."

86. Rieter-Scragg Ltd., *SDS1100*, Texturing Machine Technical Data

87. Singh R. K., and Vohra J. N., "A Study of Process Mechanics and Yarn Characteristics Using Fabricated Stuffer-box Crimper," *Textile Research Journal*, March 1976, 164-170.

88. Backer S., and Batra S., "On the Mechanics of Bi-Component Texturing," *Textile Research Journal*, December 1971, 1008-1012.

89. Abhiraman A. S., "Mechanics of Crimp Formation on a Bicomponent Fibre Model," *Textile Research Journal*, April 1974, Vol. 44, No. 4, 319-323.

90. Batra S. K., "A Generalised Model for Crimp Analysis of Multi-Component Fibres: Part I: Theoretical Development," *Textile Research Journal*, May 1974, Vol. 44, No. 5, 377-385.

91. Denton M. J., "The Crimp Curvature of Bicomponent Fibres," *Journal of Textile Institute*, 1982, Vol. 73, No. 6, 253-263.

92. Textechno Herbert Stein, Textile Testing Instruments, "Crimp Contraction Tester, TEXTURMAT M, Trade Catalog.

93. "Textured Yarn Review," *Modern Textiles*, July 1973, Vol. 54, No. 7.

94. Hearle J. W. S., "Structural Mechanics of Torque-Stretch Yarns: The Mechanism of Snarl Formation," *Journal of Textile Institute*, 1966, Vol. 57, T441-T460.

95. Thwaites J. J., "The Dynamics of the False-Twist Process: Part I: The Process Surveyed," *Journal of Textile Institute*, 1978, Vol. 69, No. 9, 269-275.

96. Thwaites J. J., "The Dynamics of the False-Twist Process: Part II: A Theoretical Model," *Journal of Textile Institute*, 1978, Vol. 69, No. 8, 276-286.

97. Thwaites J. J., "The Dynamics of the False-Twist Process: Part III: Experiments with Fully Drawn Yarns," *Journal of Textile Institute*, 1981, Vol. 72, No. 6, 239-248.

98. Thwaites J. J., and Hooper C. W., "The Dynamics of the False-Twist Process: Part II: Draw-Texturing Experiments," *Journal of Textile Institute*, 1981, Vol. 72, No. 6, 249-257.

99. Thwaites J. J., "The Dynamics of the False-Twist Process: Part V: Response to Variations in Feedstock," *Journal of Textile Institute*, 1984, Vol. 75, No. 2, 112-122.

100. Thwaites J. J., "The Dynamics of the False-Twist Process: Part II: Tension and Torque Generation in the Disc Spindle," *Journal of Textile Institute*, 1985, Vol. 76, No. 3, 157-170.

101. Backer S., and Yang W. L., "The Dynamics of the False-Twist Process: Part II: Experimental Observations of Transient Behaviour," *Textile Research Journal*, October 1976, 724-733.

102. Brookstein D., and Backer S., "The Dynamics of the False-Twist Process": Part III: Experimental Observations of Torsional Behaviour of the Texturing Threadline for Pre-Drawn PET Yarns," *Textile Research Journal*, November 1976, 802-809.

103. Thwaites J. J., "The Dynamics of the False-Twist Process: Part IV: The Origin and Significance of the Torsional Behaviour of the False-Twist Threadline," *Textile Research Journal*, December 1976, 886-892.

104. Brookstein D., and Backer S., "The Dynamics of the False-Twist Process: Part V: Steady-State Mechanics of Draw Texturing," *Textile Research Journal*, April 1977, 256-66.

105. Bell C. E., and Backer S., "The Dynamics of the False-Twist Process: Part VII: Cyclic Behaviour During Pin-Spindle False-Twist Texturing," *Textile Research Journal*, December 1978, 722-732.

106. Arghhyros S., and Backer S., "The Dynamics of the False-Twist Process - Part VII: An Experimental Study of Heat Setting," *Textile Research Journal*, May 1982, 295-312.

107. Cook J. G., *Handbook of Textile Fibres,* Vol. 2: *Man-Made Fibres*, Merrow Publishing, Durham, North Carolina, 1984.

108. Snider D. E., and Richardson R. J., "Polyamide Fibres," *Encylopaedia of Polymer Science and Technology,* 1st Edition., Vol. 10, 347-460.

109. *Alternatives to Spun Yarns in Apparel Fabrics*, Shirley Institute, S.34, Conference Proceedsings, 15 February 1979.

110. Damyanov G., *Texturing*, Moscow, 1980.

111. Landoll L. M., "Olefin Fibres," *Encylopaedia of Polymer Science and Technology,* 2nd Edition, Vol. 10, 373-395.

112. Grobzig, "New Spinnerets From Grobzig," *Man-Made Fibre, Year Book*, 1987, 32.

113. Davies S., "*The Man-Made Fibre Industry in Western Europe*," The Economist Intelligenece Unit, Special Report No. 1107, London, October 1987.

114. Autorenkollektive, "*Technologie der Texturierung,*" VEB Fachbuchverlag, Leipzig, 1987.

115. Galanti A. V., and Mantell C. L., "*Polypropylene Fibres and Films,*" Plenum Press, New York, 1965.

116. Greer R., and Hearle J. W. S., "Snarl Spliting in Polyethylene," *Polymer*, 1970, Vol. 11, 441-442.

117. Morris W. J., and Roberts A. S., "Some Causes of Variation in False-Twist Bulked Yarns," *Journal of Textile Institute*, Vol. 57, No. 5, T217-T221.

118. Denton M. J., "The Twist-liveliness of Twist-Set-Untwist or False-Twist Bulked Yarns," *Journal of Textile Institute*, Vol. 57, No. 8, T372-T374.

119. Krause H. W., "The Twist-liveliness of Twist-Set-Untwist or False-Twist Bulked Yarns-Reply," *Journal of Textile Institute*, Vol. 57, No. 12, T586-T589, Denton M . J. Reply, T589-T590.

120. Lünenschloss J., and Iyer C., "Texturiertafel," *Textile Praxis*, 1968, Vol. 23, 804.

121. Quynn R. G., "Sticking Behaviour of Preoriented PET Yarns," *Journal of Applied Polymer Science*, 1972, Vol. 16, 3393-3394.

122. Denton M. J., The Enthalpy and Specific Heat of Nylon 6.6," *Journal of Textile Institute*, 1966, Vol. 57, T265-T268.

123. Denton M. J., "The Structure of Twist-Set-Untwist Crimped Yarns," *Journal of Textile Institute*, 1966, Vol. 57, T265-T268.

124. Gay T. H., "False-Twist Draw Texturing," *Asia and World Textiles*, Textile Institute, World Conference, 20-25 November 1973, Hong Kong.

125. Rieter, "Neue Rieter-Strecktexturiermaschine für Teppichgarne" *Chemiefasern/Textilindustrie*, 1979, Vol. 27/79, 890-895.

126. UK Patent Application, GB 2 214 937 A, Glass Fibres Ltd, 09.02.1988

127. Clements J. L., "Relationship of Dye Uniformity of False-Twist Polyester Yarns to Fabric Streakiness," *Knitting Times*, 1972, Vol. 41, 72-78.

128. Meredith R., "The Torsional Rigidity of Textile Fibres," *Journal of Textile Institute*, 1954, Vol. 45, T489-T503.

129. Simmens S. C., "Filament Surface Investigatigation by a Simple Interferometric Method with Particular Reference to Nylon Filaments," *Journal of Textile Institute*, 1954, Vol. 45, T569-T574.

130. "*Advances in False-Twist Texturing Process*," Shirley Institute Publication, S12, Shirley Institute, Manchester.

131. Warwicker J. O., “The Structural Causes of the Dying Variations of Nylon 66 Yarn Subjected to the False-Twist Texturing Process,” *Journal of Applied Polymer Sciences*, 1975, Vol. 19, 1147-1161.

132. Thwaites J. J., “Character of Draw-Textured Yarns and Significance of Denier Variation,” *Knitting Times*, 1974, Vol. 43, 69-72.

133. Sebina L., Usenko V., Nefedov B., and Smirnov V., “Changes in the Structure and Properties Produced in Polyester Yarn by the Heat Treatment in the Texturing Process,” *Fibre Chemistry*, 1972, Vol. 4, 179-181.

134. Morton W.E., “The Arrangement of Fibres in Single Yarns,” *Textile Reseacrh Journal*, 1956, Vol. 26, 325-331.

135. Dumbleton J. H., “Spin Orientation Measurement in Polyethylene Terephtalate,” *Textile Research Journal*, 1970, Vol. 40, 1035-1041.

136. Denton M. J., “The Crimp Frequency of False-Twist Textured Yarns,” *Journal of Textile Institute*, 1968, Vol. 59, 550-553.

137. Hall E.E., and Rawlings G. D., “Knits of Textured and Draw-Textured Dacron Polyester Yarns,” *Knitting Times*, 1970, Vol. 43, 73-76.

138. Hensen F., “Neue Technologien und Anlagen zur Herstellung von Spinnfasern und Filamentgarnen aus vorzugsweise Polypropylen,” *Chemiefasern/Textilindustrie*, January 1978, Vol. 28/80, 36-40.

139. Malcomete O., Blascu V., and Ciocan C., “Views on the Rheological Behaviour of Some Textured Yarns by using Modern Technologies,” *Buletinul Institutului Politehnic Din Lasi*, 1989, Vol. XXXIX, No. 1-2, Section: VI-Textile, 1-4.

140. Weinsdorfer H., “*Die Strecktexturierung von Polyester-faden nach dem Simultan- und Sequential-verfahren,*” Dissertation, Universitat Stuttgart, 1974.

141. Brookstein D. S., “*On the Dynamics of the Draw Texturing,*” Ph.D. Thesis, Massachusetts Institute of Technology, 1976.

142. "How to End High-Speed Instability in Texturing,” *Textile World*, April 1983, Vol. 4, 50-51.

143. Sasaki T., Kuroda K., and Suminokura T., "Twist Distribution in False-Twist Texturing Process," *Journal of Textile Machinery Socity of Japan, Transactions*, 1970, Vol. 23, 77-80.

144. Weigert H., "Present State and Future Prospects of Twisting Systems in Torsion Texturizing," *Melliand Textilberichte*, December 1978, 887-891, 945-948.

145. Du G. W., and Hearle J. W. S., "Threadline Instability in the False-Twist Texturing Process: Part III: Stability of the Yarn Path and Surging Mechanism," *Journal of Textile Institute*, 1990, Vol. 81, No. 1, 36-47.

146. Dhingra R. C., and Postle R., "Measurement of Torque in Continuous Filament Yarns: Part II: The Effect of Yarn Tension," *Journal of Textile Institute*, 1974, Vol. 65, 171-181.

147. Hearle J. W. S., Gupta P.K.S., and Mathews A., "Changes in Length and Mechanical Properties Due to Single and Double Heat Treatment of Nylon Yarn Under Tension," *Fibre Science Technology*, 1971, Vol. 3, 167-185.

148. Vilcu M., Radu A.S., and Vasile C., "Study of the Changes in the Morphology and Properties of the Textured Polyester Fibres with the Processing Conditions," *Buletinul Institutului Politehnic Din Lasi*, 1989, Vol. XXXIX, No: 1-2, Section: VI-Textile, 11-16.

149. Morton W. E., and Hearle J. W. S., *Pyhsical Properties of Textile Fibres*, Heinemann, London, 1975.

150. Hearle J. W. S., Wilding MA, Auyeung C., and Ihmayed R, "A Note on the Heat-setting of Nylon 6.6," *Journal of Textile Institute*, 1990, Vol. 81, No:2, 214-217.

151. BP 762 630. DuPont de Nemours, 19 July 1954.

152. Schubert G., "Vortexing of Flat Yarns," *Chemiefasern/Textilindustrie*, October 1980, Vol. 30/82, 820-823, E92-E93.

153. Wood J. I., "The Development of Air-splicing Techniques for the Joining of Woollen-Spun Yarns for Tufted-Carpet Manufacture" *Journal of Textile Institute*, 1982, No. 2, 71-79.

154. Lünenschloss J., and Zilg J.P., "Intermingling of Synthetic Filament Yarns in an Air Flow," *Textile Praxis International*, No. 9, 934-936 + VII-IX, 1982.

155. Weinsdorfer H., and Wolfrum J., "The Interlacing of Textured Yarns -Mechanics and Yarn Characterisation," *Texturing Today*, Shirley Institute Publication S.46, Manchester, 1983, 253-275.

156. Jian Z., Baishen J., and Weilian P., "Investigation of the Movement of the Interlacing Yarn in the Yarn Duct," *Journal of the East China Institute of Textile Science and Technology*, No. 1, 67-73, 1985.

157. Lazauskas V., Lukosaitis A., and Matukonis A.., "Effect of the Design Parameters of the Aerodynamic Device on the Tanglelacing Intensity of Filament Yarns," *Tekhnologiya Tekstil'noi Promyshlennosti*, No. 3 (177), 27-30, 1987.

158. Heberlein Maschinenfabrik AG, Switzerland, *Interlacer FP600,* November 1984.

159. Heberlein Maschinenfabrik AG, Switzerland, *FP600, FPF, FPE,* September 1986.

160. Fiberguide Ltd, Macclesfield, UK, *V-Jet for Intermingling and Co-mingling Fibres.*

161. Comoli Ferno and Figli SAS, Italy, *Entanglement Device for continuous Filament Yarn.*

162. Slack and Parr Special Products Inc, N C, USA, "*Innovation, Performance, Quality.*"

163. USP 3389444, Eastman Kodak Company, 22 November 1963.

164. USP 3837155, Celanese Corporation, 2 November 1971.

165. USP 4004329, Burlington Industries Inc, 18 September 1971.

166. BP 1558091, Toray Industries Inc, 16 July 1976.

167. BP 1564885, Snia Viscosa SpA, 13 February 1978.

168. UK Patent Application, GB 2019906A, Snia Viscosa SpA, 24 April 1979.

169. USP 4251904, Toray Industries Inc, 31 October 1979.

170. UK Patent Application, GB 2118581A, Mitsubishi Rayon Co Ltd, 20 September 1982.

171. UK Patent Application, GB2123859A, Murata Kikai Kabushiki, 19 May 1983.

172. UK Patent Application, GB2128644A, Barmag Barmer Maschinenfabrik, 7 October 1983.

173. European Patent Application, 0140526, Fiberguide Ltd, 29 August 1984.

174. UK Patent Application, GB2178072A, Rieter-Scragg Ltd, 21 May 1986.

175. Evans M., Dunlop Aerospace, *Personal communication*, April 1987.

176. Steinman A. J., "Air Jet Texturing," *Air Jet Texturing Conference*, Clemson University, Greenville, South Carolina, 18-19 September 1985.

177. Holden G. A., "Advances in Air Texturing," *Bulked Yarn Conference*, UMIST, Manchester, 1975.

178. Wilson D. K., "The Production of Textured Yarns by Methods other than the False-Twist Technique," *Textile Progress*, 1977, Vol. 9, No. 3, 1-63.

179. Wray G. R., and Acar M., "The Production and the Properties of Air-jet Textured Continuous Filament Yarns," *International Textile Symposium*, Izmir, Turkey, 2-4 November 1981, (Proceedings published by Ege University Press, Izmir, 1984).

180. Gülal B., "*Blending Synthetic Yarns by Air-Texturing*," M.Phil. Thesis, Leeds University, Leeds, 1983.

181. Acar M., and Wray G. R., "An Analysis of the Air-jet Yarn Texturing Process: Part I: A Brief History of Developments in the Process," *Journal of Textile Institute*, 1986, Vol. 77, No. 1, 19-27.

182. Acar M., Turton R. K. and Wray G. R., "An Analysis of the Air-jet Yarn Texturing Process: Part V: The Effects of Wetting the Yarn," *Journal of Textile Institute*, Vol. 77, 1986.

183. Schweizer U., "FK6T Air Texturing Machine," *Textured Yarn Association of America, Inc. Convention*, Myrtle Beach, South Carolina, 3-5 August 1978.

184. Patent Application, G.B. 2 079 189 A Heberlein Maschinenfabrik AG, 25 June 1981.

185. Lünenschloss J., and Helli J. G., "Die Texturierung von Chemiefaden in Luftstrom: Part I-IV," *Textile Praxis*, 1969, Vol. 24, (8-11), 515-725.

186. Datye K. V., and Bose C., "Air-Texturing," *Man-Made Textiles in India*, March 1982, Vol. 25, No. 3, 158-163.

187. Hes L., and Piller B., "Heat Transmission in Air-Jet Textured Filament Yarns," *Melliand Textilberichte* (Eng. Edition), 1982, Vol. 11, No. 10, 678-682.

188. Bose C., and Govindarajulu R., "Effect of Denier on Air Textured Yarn," *The Indian Textile Journal*, 1982, Vol. 93, No. 9, 97,98.

189. Bock G., and Lünenschloss J., "Heat Treatment of Air-jet Textured Polyester Filament Yarns," *Chemiefasern/Textilindustrie*, 1983, Vol. 33/85, No. 2, 103-106, E12.

190. Sengupta A. K., Kothari V.K., and Roy A. K., "Stability of Air Textured Yarns," *Textile Research Journal*, 1984, Vol. 54, No. 2, 125-126.

191. Wilson D. K., "Texturing by Air and other Methods, Process and Products," *Unpublished Paper*, 1986.

192. Pritts M.A., "How to Assure Economical Air-Jet Texturing," *Textile World*, 1983, No. 11, 47-50.

193. Ingersol-Rand Co. "Air Compressors Are Key to Texturing of Synthetic Fibres," *Fiber Producer*, June 1980, Vol. 8, No. 6, 48-51.

194. Rilling R., "Further Developments in the Field of Air Texturing," *Chemiefasern/Textilindustrie*, 1981, Vol. 31/83, No. 4, 298-301, E30-E31.

195. Denton M. J., Eaton D. C., and Baker A. H,, "Noise in False-Twist Texturing Plant," *Texturing Today*, Shirley Institute Publication S.46, Manchester, 1983, 65-107.

196. Holder D.W., and North R.J., "*Schlieren Methods*," Her Majesty's Stationery Office, London, 1963, 1-106.

197. Benedict R.P., Wyler J.S., Dudek J.A., and Gleed A.R,, "Generalised Flow Across an Abrupt Enlargement," *Trans. Of the ASME: Journal of Engineering for Power*, July 1976, 327-334.

198. Che-Haing C., "Axially Symmetric Supersonic Turbulent Jets Discharged from a Nozzle with Underexpansion," *Turbulent Jets of Air, Plasma and Real Gas*, Ed. G. N. Abramovich, Translated from Russian, Consultants Bureau, New York, 1969.

199. Gaunter J.W., Livinggood N.B., and Hrycak P., "Survey of Literature on Flow Characteristics of a Single Turbulent Jet Impinging on a Flat Plate," *NASA Technical Note*, February 1970, NASA TN D-5652.

200. Gürevich M., "*Theory of Jets in Ideal Fluids*," London, Academic Press, 1965, 353-366.

201. Soo S. L., "Gas Dynamic Processes Involving Suspended Solids," *A.I.Ch.E. Journal*, September 1961, Vol. 7/3, 384-391.

202. Abraham O., Binn J. H., Deboer B.G., and Stein G. D., "Gas Dynamics of Very Small Laval Nozzles," *Physics of Fluids*, June 1981, Vol. 24/6, 1017-1031.

203. Marks M. R., ICI Fibers Co., Gloucestershire, UK, *Personal Communication*, 1985.

204. Simmen S. C., Heberlein Maschinenfabrik A.G., Switzerland, *Personal Communication*, March 1986.

205. Herrmann R., "Der Kondensationsstoss in Überschall-Windkanal-düsen," *Info 19 201*, 1942.

206. Oswatitsch K., "Kondensationserscheinungen in Ueberschalldusen," *ZAMM,* 22 H1, 1942.

207. *Air-Jet Texturing*, One Day Symposium, Loughborough University of Technology, 11 July 1984.

208. Wegener P. P., and Pouring A. A., "Experiments on Condansation of Water Vapor by Homogeneous Nucleation in Nozzles," *The Physics of Fluids*, March 1964, Vol. 7/3, 352-361.

209. Pouring A. A., "Thermal Choking and Condensation in Nozles," *The Physics of Fluids*, October 1965, Vol. 8/10, 1802-1810.

210. Kang S. W., "Analysis of Condensation Droplet Growth in Rarefied and Continium Enviroments," *AIAA Journal*, July 1967, Vol. 5/7, 1288-1295.

211. Saltanov G. A., Selezenev L. I., and Tsiklauri G. V., "Generation and Growth of Condensed Phase in High-velocity Flows," *Int. J. Heat and Mass Transfer*, 1973, Vol. 16, 1577-1587.

212. Sislian J. P., and Glass I. I., "Condensation of Water Vapor in Rarefaction Waves: I. Homogeneous Nucleation," *AIAA Journal*, December 1976, Vol. 14/12, 1731-1737.

213. Kotake S., and Glass I. I., "Condensation of Water Vapor in Rarefaction Waves: II. Heteregeneous Nucleation," *AIAA Journal*, February 1977, Vol. 15/2, 215-221.

214. Glass I. I., Kaira S. P., and Sislian J. P., "Condensation of Water Vapor in Rarefaction Waves: Part III. Experimental Results," *AIAA Journal*, May 1977, Vol. 15/5, 686-693.

215. Matsuo K., Kawagoe S., Sonoda K., and Setoguchi T., "Oscillations of Laval Nozzle Flow with Condensation," *Bulletin of JSME*, September 1983, Vol. 26/219, 1556-1562.

216. Matsuo K., Kawagoe S., Sonoda K., and Sakao K., "Studies of Cendensation Shock Waves," *Bulletin of JSME*, July 1985, Vol. 28/241, 1416-1422.

217. Wray G. R. (ed.), "*Modern Yarn Production from Man-Made Fibres and their Conversion into Fabrics*," Columbia Press, Buxton, 1976, 202 pp.

218. Wison D. K., "The Production of Textured Yarns by the False-Twist Technique," *Textile Progress*, 1978, Vol. 10, No. 3, 1-66.

219. VanAken F. J., "Air-Textured Filament Yarns for Outerwear," *Chemiefasern/Textilindustrie*, 1979, Vol. 29/81, No. 2, 108-116.

220. Fischer K. E., "Air-Jet Texturing," *International Textile Bulletin/Spinning*, 1980/3, 287-292.

221. Williams J., "Air-Jet Textured Yarns for Automobile Upholstery," *Texturing Today*, Shirley Institute Publication S.46, Manchester, 1983, 199-206.

222. "World Production of Certain Textile Fibres," *Textile Organon*, 1966, Vol. 37, No. 6, 97.

223. "World Production of Certain Textile Fibres," *Textile Organon*, 1975, Vol. 46, No. 6, 65.

224. "World Production of Certain Textile Fibres," *Textile Organon*, 983, Vol. 54, No. 6, 81.

225. Predicasts Inc., "Worlwide Consumption of Man-Made Fibres to Exeed Natural Fibres by 1995," *Canadian Textile Journal*, 1982, Vol. 99, No. 11, 32, 35.

226. Wray G. R., "Bulked Yarn Manufacture in Czechoslovakia: Part 2," *Textile Recorder*, 1961, Vol. 78, 51-52.

227. Heberlein Maschinefabrik AG, "*HemaJet Air Texturing Jet,*" January 1980.

228. Wray G. R., "Yarn Structures Generated by Fluids in Engineered Channels," *Engineering Foundation Conference on New Directions in Textile Technology*, Ringe, New Hampshire, 22-27 August 1976.

229. Artunç H., Bocht B., and Weinsdörfer H., "Air-Jet Texturing Process with

Integrated Drawing and Shrinking Zones," *Chemiefasern/Textilindustrie*, 1979, Vol. 29/81, No. 10, 857-861, E118-E120.

230. Piller B., "Shrinking Behaviour of Air-Texturised, Modified Filament Yarns," *Melliand Textilberichte* (Eng. Edition), 1980, Vol. 61/9, Nos. 4 and 5, 522-526, and 649-655.

231. Massey B. S., "*Mechanics of Fluids*," Van Nostrand Reinhold Company, London, 1979, 405-407.

232. Schlichting H., "*Boundary Layer Theo*ry," 7th Editon, McGraw-Hill, New York, 1978.

233. Wallis G. B., "*One-Dimensional Two-Phase Flow*," McGraw-Hill Book Company, New York, 1969, 1-43, 175-242.

234. Hansen W. W., and Tabor D., "Hydrodynamic Factors in the Friction of the Fibres and Yarns," *Textile Research Journal*, 1957, Vol. 27, No. 4, 300.

235. Schick M. J., "Friction and Lubrication of Synthetic Fibres: Part 1," *Textile Research Journal*, 1973, Vol. 43, No. 2, 103.

236. Schick M. J., "Friction and Lubrication of Synthetic Fibres: Part 4," *Textile Research Journal*, 1973, Vol. 43, No. 6, 342.

237. Schick M..J., “Friction and Lubrication of Synthetic Fibres: Part 3,” *Textile Research Journal*, 1973, Vol. 43, No. 5, 254.

238. Skelton J., “Interfiber Forces during Wetting and Drying,” *Science*, 1975, Vol. 190, No. 4209, 15-20.

239. Wilson D. .K., and Kollu T., “The Production of Textured Yarns by the False-Twist Technique,” *Textile Progress*, 1991, Vol. 21, No. 3.

INDEX

D

E

F

J

K

L

M

N

O

P

Q

R

S

T

Y

Z